ECONOMICS:

Applications to Agriculture and Agribusiness

Fourth Edition

ECONOMICS:

Applications to

Interstate Publishers, Inc.
Danville, Illinois

Agriculture and Agribusiness

Revised by

Randall D. Little, Ph.D.
Associate Professor
Department of Agricultural Economics
Mississippi State University

With contributions from prior editions by

Ewell P. Roy / Floyd L. Corty / Gene D. Sullivan

ECONOMICS:

Applications to Agriculture and Agribusiness

Fourth Edition

Copyright © 1997 by Interstate Publishers, Inc.

All Rights Reserved / Printed in U.S.A.

Prior editions: 1971, 1975, 1981

Library of Congress Catalog Card No. 96-75234

ISBN 0-8134-2949-8

1 2 3 4 5 6 7 8 9 10   02 01 00 99 98 97 96

Order from

Interstate Publishers, Inc.

510 North Vermilion Street	Phone: (800) 843-4774
P.O. Box 50	Fax: (217) 446-9706
Danville, IL 61834-0050	Email: info-ipp@IPPINC.com

This book is dedicated to:

My wife,	*Dulcy*
My children,	*Jonathon*
	Christi
	Benjamin
	Micah
	Anna Lea
And my dad,	*Howard*

Preface

Why study economics? There must be more to it than just another hurdle to clear on the way to a college degree, right? Well, there is. And the potential benefits of a fundamental understanding of economic theory accrue at different levels.

On a personal level, we can apply much of what we gain from *microeconomic* theory every day. For example, we can use rules based on economic theory to guide the decisions we make about how we spend our time. Does the **utility** (satisfaction) exceed the **opportunity costs** (benefits foregone from activities we chose not to do) when we spend an afternoon playing golf instead of studying for our upcoming agricultural economics test? Economic theory provides a framework, an approach, for making time management decisions.

Decisions concerning resource allocation influence the success of a business. Every firm is interested in finding the best way to utilize its scarce resources. Our primary focus in this text is on applying microeconomic principles to firms involved in production agriculture—farms and ranches—and other firms in the agribusiness sector.

We also need a fundamental understanding of how the economy, as a whole, functions and how it is influenced by developments internationally. Aspects of *macroeconomic* theory help us understand these linkages and interrelationships, both between various sectors of the domestic economy and between our economy and other economies in the world market. Recent political changes around the world have created opportunities that are challenging and difficult, yet exciting. Economic laws and principles will play an important role in the development arising out of those changes.

The agribusiness sector is a mainstay in the U.S. economy. However, the structure of this sector has changed dramatically over time—and at all levels. It is hoped that we will gain a better understanding of the changes that have occurred and the effects of those changes on the key players in the agribusiness sector: input suppliers, producers, processors, wholesalers, retailers, etc.

The fourth edition of *Economics: Applications to Agriculture and Agribusiness* is a much-needed updating by Dr. Randall D. Little of a work nurtured through three prior editions over a period of 25 years by its original authors, the late Dr. Ewell P. Roy and Drs. Floyd L. Corty and Gene D. Sullivan. The goal of this book is to serve as a basic text in a beginning economics course, but it can be used by anyone interested in the application of economic principles to agriculture and agribusiness. The book explains the role of agriculture in the general economy, as well as recognized changes in technology, institutions, and policies that have altered the structures of farming and agribusiness in recent decades.

Economic problems unique to agriculture and agribusiness are discussed in common terms, and economic terminology is introduced to help students and others develop a familiarity with, and an appreciation for, economic theory and economic institutions. The book should help readers, through increased understanding, to make more intelligent economic appraisals, evaluations, and decisions.

Part One introduces the subjects of agricultural economics and agribusiness, provides a brief history of U.S. agriculture, and discusses the several economic systems and types of business organizations. It describes the U.S. monetary system and provides a macro-profile of the nation's economy.

Part Two is oriented toward the human, natural, and capital resources available in agriculture and discusses the principal characteristics associated with U.S. farming operations.

Part Three is devoted to the explanation of economic concepts and principles. Among these are comparative advantage, production functions, diminishing marginal returns, costs, returns, optimum output levels, supply, demand, price elasticity, equilibrium price, cobweb theorem, price cycles, perfect competition, imperfect competition, and monopoly.

Part Four is oriented more toward agribusiness and includes a discussion of the farm supply business, marketing, consumption of food and fiber, and farm policies.

Part Five deals with the history, organization, and functions of the Federal Reserve System; input-output functions; principles of agribusiness management; international trade; and economic development.

Both the original authors and the revising author made every effort to produce a book this is readable and applied, yet thought-provoking and stimulating. Suggestions and comments from users are always welcome.

Acknowledgments

Many people have helped bring this project to fruition. I could not have completed it without their contributions. First, I thank Dr. Jasper S. Lee, Editorial Consultant, Interstate Publishers, Inc., for coming to the Department of Agricultural Economics at Mississippi State University when Interstate was looking for someone to undertake the task of updating and revising this textbook. Next, I must thank Dr. John Lee, Head of the Department of Agricultural Economics, Mississippi State University, for recommending me for the task.

Several people gave their time to review drafts of the manuscript. Their insightful comments and suggestions improved the text considerably (but, of course, I take responsibility for any errors that remain). The reviewers included Albert Allen, Michael Allen, Barry Barnett, Ralph Brown, John Lee, Wayne Malone, Bob Martin, Travis Phillips, Lynn Reinschmiedt, and John Robinson. David Schweikhardt also provided much useful guidance as I launched into the agricultural policy chapters.

Several graduate students "volunteered" to do some of the legwork associated with this revision—collecting data, building the Glossary, etc. Eris Culpepper, Mary Loyce Makamson, Dana Reed, and M. Afzaal Kahn deserve specific mention. My secretary, Barbara Wright, certainly deserves thanks for diligent efforts, especially for typing all those tables.

Ronald L. McDaniel, Vice President–Editorial at Interstate, warrants a paragraph to himself. Thanks for your commitment to quality work, Ron. I know this is a better book because of your great editing touch.

Thanks, too, to Kimberly Romine, who worked with Ron in designing a format that is both attractive and reader-friendly.

Finally, I thank my family for their patience, encouragement, and tolerance (and prayers) while I worked on this update. Dulcy, Jonathon, Christi, Benjamin, Micah, and Anna Lea, you guys are great!

Randall D. Little

Contents

Preface vii
Acknowledgments ix

Part One

1. Introduction 3
2. Brief History of U.S. Agriculture 9
3. Types of Economic Systems 23
4. Types of Business Organization in a Capitalistic Economy 39
5. The U.S. Monetary System and Fiscal Policy 51
6. Macro-profile of the U.S. Economy 67

Part Two

7. Human Resources in Agriculture 85
8. Natural Resources 105
9. Capital Resources 127
10. Selected Characteristics of U.S. Farming Operations 155

Part Three

11. Specialization and Comparative Advantage 171
12. Physical Production Relationships 179
13. Costs and Revenue 191

14	Optimum Levels of Output	205
15	The Supply Concept	219
16	The Demand Concept	231
17	The Interaction of Supply and Demand: Prices	247
18	Market Structures and Competition	263

Part Four

19	Economic Aspects of the Farm Supply Business	285
20	Food and Fiber Marketing Functions	301
21	Marketing Channels for Farm Products	321
22	Costs of Marketing Food and Fiber Products	343
23	Consumption of Agricultural Products	357
24	Economic Setting for U.S. Agricultural Policy	383
25	Achieving the Goals of Agricultural Policy	409

Part Five

26	The Federal Reserve System: History, Organization, and Functions	435
27	Input-Output Functions	449
28	Selected Principles of Agribusiness Management	465
29	International Trade and Agriculture	495
30	Economic Development in the Developing Areas of the World	523

Glossary 541

Index 553

PART ONE

1 Introduction

2 Brief History of U.S. Agriculture

3 Types of Economic Systems

4 Types of Business Organization in a Capitalistic Economy

5 The U.S. Monetary System and Fiscal Policy

6 Macro-profile of the U.S. Economy

PART ONE

Introduction

Economics is derived from the Greek word οικονομικος *(oikonomikos)*, which literally means "skilled in household management." Over time, the meaning of economics has been extended to entities besides households to include the management of all types of resources. Economics is now defined as the science of allocating scarce resources (land, labor, capital, and management) among different and competing choices and utilizing them to best satisfy human wants. Viewed in another way, economics is a study of how to get the most satisfaction for a given amount of money or to spend the least money for a given need or want.

Implicit in these definitions is the problem of resource scarcity, namely, the limited supply of resources relative to people's needs and desires. If resources were abundant in every respect, there would be no need for the science of economics. As a general rule, total human wants are insatiable—they can never be completely satisfied. On the other hand, resources for supplying these wants are limited. It is true of human existence that wants beget other wants. People living in a small apartment likely want nicer housing, perhaps a relatively more spacious duplex. After upgrading to a single-family residence, for example, they may aspire to own a home in an exclusive gated subdivision.

At this point, the question arises whether any economy could supply a home in an exclusive subdivision at moderate cost to every family. The answer is a qualified yes. However, doing so would require giving up the thousands of other items—automobiles, CD players, designer clothing, etc.—everyone wants. Thus, economic science is concerned largely with priorities, identifying the most important needs and wants, then attempting to meet those needs and wants first. In reality, consumers try to allocate their incomes among several priorities, the highest of which are food, clothing, and shelter. Other items are of more moderate priority, such as transportation. Still others may have very low priority, such as vacations, sports, and recreation.

Consumers easily dispense with low-priority items if their incomes are insufficient to provide for them.

We do not wish to leave the impression that all consumers are rational and make the best choices every time. Furthermore, we may not all agree on what constitutes a priority. For example, a young couple may choose to allocate all their available cash to a home entertainment center while eating peanut butter sandwiches for months. They could be accused of behaving irrationally, yet they made their choice independently as consumers in control of their own money, feeling no doubt that their actions were rational. Obviously, consumers do not all have the same wants, needs, and desires. Establishing the priorities of individual wants and allocating available goods and services to address those wants is a critical part of economics.

We recognize, of course, that economics alone does not govern all our actions. What appears as irrational economic behavior might be explained by other factors, such as a desire for self-expression, a need for approval and acceptance by one's contemporaries, a desire for personal power over others, or a deep concern for the welfare of others. It is not unusual to witness the accumulation of great monetary fortunes that are subsequently donated to charity, the arts, and religious endeavors. Also, many people choose low-paying professions to provide the freedom and opportunity to engage in charitable or benevolent endeavors. Even though they are largely nonmonetary, economic principles are applicable to these types of decisions.

So far, we have been stressing the individual's choices. The study of individual economic choices is called **microeconomics.** We now proceed to the sum total of these choices and the ramifications thereof. By what common denominator do we as consumers base or allocate our choices? The answer is prices. Any economic system, whether capitalistic or communistic, must have a system of pricing to establish economic order. Whether prices are set by a government agency or a free market, the role of prices is paramount in allocating resources.

For example, if food prices increase *ceteris paribus* (all else remaining unchanged), consumers will have to spend more for food and less for other goods. Higher food prices signal an increase in consumer demand for food, a reduced supply of food, or a combination of both. It is interesting to note that if consumer budgets are already heavily committed to things other than food, a slight increase in food prices will cause much alarm, possible consumer boycotts, and other forms of protest. Such consternation arises partly because consumers have incurred considerable debt obligations for homes, cars, boats, furniture, etc. In such cases, food may become a residual item in the budget, with much lower priority, it would seem, than consumer debt payments, say, for a new car.

Whether prices do their job well depends, of course, on many things that we will examine later in greater detail. Market competition among sellers and buyers is certainly one factor that determines how well prices do their job. Another is the technical knowledge possessed by consumers about products they wish to buy and the amount of information given them at the time they are considering purchases. We will also examine these and other facets in later chapters.

At this point, we must recognize that prices play a dual role. As costs, prices influence consumers to spend their money a certain way. Similarly, as income, prices influence firms to allocate their productive resources a certain way. For example, higher food prices may signal farmers to expand production, while at the same time they may signal consumers to buy less. Thus, supply and demand in a relatively free market interact to establish prices that act as signals for consumers, producers, and others involved in the production-marketing process.

It is logical to expect that the signals prices send in a marketplace will not be satisfactory to everyone. For example, in response to price dissatisfaction, farmers may shift resources to enterprises that appear more lucrative. Likewise, consumers may substitute one food for another, go without, or spend more money on nonfood items, to an extent.

The price level is dynamic, sometimes changing rapidly, as on stock exchanges, while at other times remaining fixed over a long period, like physician fees. We will discuss in detail the matter of prices, price determinants, and price movements in later chapters.

Our knowledge and application of economics should not be confined to our personal lives but should relate to society's welfare as well. Governments, too, are faced with economic problems and priorities. Through our legislative, executive, and judicial systems, decisions are made relative to these priorities. For example, the defense and security of a nation has first priority, for without defense the very existence of a nation may be endangered. Education, transportation, and communications are next likely priorities. Reaching agreement on these priorities is no easy task for individuals and government leaders alike. **Macroeconomics** involves an aggregate look at the performance of a national or international economy. Macroeconomics is the study of the effects of changes in the production of goods and services and employment and how they interact to influence economic performance.

The science of economics has many branches, of which agricultural economics is but one. **Agricultural economics,** a specialized and applied branch of general economics, deals with the allocation and utilization of scarce resources (land, labor, capital, and management) among different choices of crop, livestock, orchard, timber, and other enterprises to produce goods and services that satisfy human wants.

Agricultural economics is one of a family of economic specialties including labor, industrial, business, and consumer economics. **Agribusiness** is also a subfield of economics that includes not only agricultural economics but also segments of labor, industrial, business, and consumer economics pertaining to agriculturally related industries and businesses. Agribusiness includes all the vital activities performed in any of the three broad categories of the food and fiber system—the agricultural input sector, the production sector, and the processing-manufacturing sector.

Agricultural economics also has many subspecialties, including crop, livestock, poultry, land, resource, forestry, marine, and water economics. Also among the many concerns of agricultural or agribusiness economists are marketing, finance, business organization, policy, statistics, and resource development. Agricultural economists must also consider interactive relationships between the agricultural sector and the general economy because all economic relationships are interdependent. Finally, agricultural economists must take a broad cross-profession view, recognizing interrelationships between economics and other professions, such as law, education, political science, history, psychology, sociology, anthropology, engineering, and other disciplines influencing the economic well-being of the agricultural sector.

Agricultural economics is also related to consumer economics. Because the farm home and the farm family are integral parts of the farm business, agricultural economists cannot ignore the economics of the farm home, since the farm family competes for funds with the farm itself and is an integral component of its processes. In addition, sound consumer economics is related to the general well-being of the farm entrepreneur and family. Poor nutrition and poor health of the family that might arise because funds were diverted from proper diets and medical attention to a new, but not necessary, piece of farm equipment are of concern to both consumer and agricultural economists. There are many other such examples.

Consequently, agricultural economists are involved increasingly as advisers or consultants to other sciences and disciplines because of their knowledge of, and concern for, the efficient use of human, natural, and capital resources. There are wide differences in scientific approaches among agricultural economists, however, by virtue of their background, experience, training, personal philosophies, and loyalty to clients. These differences are wider among institutional economists, or those who study economics in terms of qualitative approaches, than they are among economists specializing in quantitative methods.

The scope of agricultural economics extends far beyond the traditional production issues facing farm firms. Agricultural economists address a broad array of issues affecting the agricultural and agribusiness sectors. Some of these issues include rural development; rural community services; job training and retraining; poverty abatement, including welfare, WIC, and food stamp programs; environmental quality in agriculture; international trade; contract farming and farmer bargaining power; world population and world food needs; and food safety issues relating to food additives, food packaging and sanitation, and organic or natural foods.

Finally, we may note that over time, agricultural economists, in general, have become economists working with agriculture rather than agriculturists working with economics.

Topics for Discussion

1. Define "economics."
2. With what is economics concerned?
3. Name at least five specialties in economic science.
4. Discuss the scientific role of the agricultural economist.
5. Why must agricultural economists take a broad view of economic and social problems of rural people?

Problem Assignment

Arrange an interview with a general economist or agricultural economist employed by the public or in private enterprise. Inquire about the economist's work specialty, his or her experiences, and the economic problems encountered.

Recommended Readings

Beierlein, James G., Kenneth C. Schneeberger, and Donald D. Osburn. *Principles of Agribusiness Management,* 2nd ed. Prospect Heights, IL: Waveland Press, Inc., 1995. Ch. 1.

Beierlein, James G., and Michael W. Woolverton. *Agribusiness Marketing: The Management Perspective.* Englewood Cliffs, NJ: Prentice-Hall, Inc., 1991. Ch. 1 and 2.

Colander, David C. *Economics,* 2nd ed. Chicago: Richard D. Irwin, Inc., 1995. Ch. 1.

Gwartney, James D., and Richard L. Stroup. *Economics: Private and Public Choice,* 7th ed. Orlando, FL: The Dryden Press, 1995. Ch. 1 and 2.

Houck, James P. "The Comparative Advantage of Agricultural Economists," *American Journal of Agricultural Economics,* 74(5):1059-1065 (1992).

Roy, Ewell P. *Exploring Agribusiness,* 3rd ed. Danville, IL: Interstate Publishers, Inc., 1980. Ch. 1.

Brief History of U.S. Agriculture

American agriculture and agribusiness have contributed much to the growth of the national economy and continue to do so. Fewer farmers are needed to provide people with food and fiber at lower cost in relation to total income than ever before.

The achievements of American farmers have been possible largely because of basically favorable soils and climate; a combination of skills with crops and livestock from all parts of the world; a unique system of research and education; human motivation and determination in the face of adversity; and the foresight and leadership of outstanding Americans who have devoted their talents and lives to improving agriculture. But a more fundamental reason for the greatness of American agriculture lies in the capitalistic system itself, which allows men and women the freedom to own; to think; to explore; to cultivate; and to reap the bounty of their own efforts and resources.

The first European settlers in what is now the United States found the country sparsely settled by Native Americans, who hunted, fished, and farmed. The Native Americans had developed several crops that are still of great importance in agriculture, including corn, kidney and lima beans, squashes, pumpkins, tomatoes, tobacco, and short-staple cotton. Potatoes, both sweet and white, were developed by South American Indians but did not reach North America until after the first European settlement.

The first Europeans found American conditions difficult. They faced starvation until they adopted the crops and tillage practices of the Native Americans, which differed from practices in Europe. For example, Native Americans in New England taught the Pilgrims to place fish beneath each hill of corn. While the Indians may not have known that the decaying fish released nitrogen, they knew that it helped the corn grow.

Land was abundant in North America, and a pattern of individual ownership was soon established in the colonies. The opportunity to own land probably attracted more settlers than any other factor. Actually, the early colonists at first experimented a little with communism, in the sense that they practiced common ownership of land, tools, etc., for a time. However, after it became obvious that lazy individuals were attempting to exploit hard-working people by living off their produce, the colonists soon abandoned the communal farming plan.

In 1613, John Rolfe shipped a few barrels of tobacco to England. With that, the Virginia colonists discovered a crop with a largely untapped market. Later, export markets for rice and indigo, as well as tobacco, provided much of the capital necessary for growth and development in American agriculture and industry, including fishing, shipbuilding, woodworking, spinning, and weaving.

The colonists also shipped furs, timbers, and grain to England and southern Europe. Grain, flour, and dried fish went from the middle and New England colonies to the West Indies. Other markets opened as time passed.

While the colonists on the East Coast were tied to British rule, the French, Spanish, and Dutch were also involved in colonizing North America. Spain's influence was primarily in western and southwestern parts of the present-day United States. Spanish influence was also evident in Florida and Louisiana. Citrus culture and cattle and sheep production are legacies of Spanish rule. Spanish missionaries, with their far-reaching religious zeal, did much to colonize large parts of North America. Transplanting serfdom from Europe to America resulted in Spanish rule that tended to assimilate Native American culture into a serf-and-serfdom economy. In contrast, British rule, which promoted individual enterprise among the colonists themselves, tended to isolate Native American culture.

The French were excellent explorers in North America. However, because they emphasized fur trade and not more permanent agricultural development, they were poor colonizers. French influence extended from eastern Canada, west to the Great Lakes, and south to the Gulf of Mexico. The French and Spanish empires met in the area west of Natchitoches, Louisiana. French-Indian relationships were strongest among the major power nations of that era. French colonization would likely have been more successful in North America had the rulers in France been better administrators.

Dutch influence was restricted primarily to New York State, particularly the Hudson and Mohawk valleys. The Dutch system of land grants and serfdom likewise became incompatible with the political and economic interests of colonists. Dutch rule ended in 1664 with its capitulation to the British.

At the time of the American Revolution, Native Americans were dispersed throughout much of the present-day United States. There were also Spanish inhabitants in much of the West, the Southwest, and Florida; French in the

Illinois region and in Louisiana; and British in colonies along the eastern seaboard, as far south as Georgia.

Attempts by Britain to control agricultural trade in its colonies and to restrict westward movement were among the causes of the American Revolution. The Revolution was fought mainly by farmers, who comprised over 90 percent of the population. At the time of the Revolution, the average farmer used many of the same implements and followed many of the same practices in use in biblical times. The Revolution, however, with its political, social, and economic changes, marked the beginning of a period of gradual progress and improvement in farming methods. As colonists severed their ties with Britain, farmers gained freedom and independence to trade freely among themselves and with other nations.

George Washington and Thomas Jefferson both emphasized the importance of agriculture to the new nation and urged farmers to improve their practices. Jefferson himself experimented with seeds and livestock, invented improved farm implements, and was active in establishing a local agricultural society.

Jefferson had opposition, however. For example, Alexander Hamilton, who favored industry over agriculture, propounded the "infant industry" theory designed to erect import barriers that would insulate American industries from competition from more highly developed nations of the time.

The Land Ordinance of 1785 was an important step in the development of what was then the western frontier. Townships, consisting of blocks of 36 square miles, were each subdivided into square-mile sections in the lands between the Appalachian Mountains and the Mississippi River. There was also a provision in the act stipulating that as each territory attained a certain population level, it was to be admitted into the Union as a state. Minimum acreage to be sold at auction was 640 acres at a minimum sale price of $1 per acre. Terms of sale were strictly cash until 1787, when one-third could be paid in cash, with the balance due in three months. There were few land sales under this ordinance, even after the change in terms of sale, because not many early settlers had sufficient capital.

The most spectacular early invention was the cotton gin, developed by Eli Whitney in 1793. Whitney's cotton gin boosted cotton cleaning from a few pounds to as much as 50 pounds per person per day. This increased efficiency established short-staple cotton as the major crop of the South and as a primary earner of foreign exchange. Income from cotton exports flowed from the South to the Northwest in exchange for foodstuffs, and to the Northeast for manufactured goods, transportation fees, and mercantile services.

Not only did cotton exports stimulate income growth and initiate expansion of markets, but they also accelerated the western migration of labor and capital. Unfortunately, the cotton economy also led to further development of slave trade, which had begun previously with tobacco, rice, and indigo production. The economy of the South, unlike that of the East, developed more on the basis of land grants, slave labor, and export markets. The rapid growth in cotton production tended to maintain and expand an agricultural economy tied to slave labor.

Jefferson was not the only inventor interested in improving farm implements. In 1819, Jethro Wood patented a cast-iron plow with interchangeable parts. However, the iron plow would not turn the sticky soils of the western prairies. This problem was solved in the 1830s when two Illinois blacksmiths, John Lane and John Deere, made one-piece wrought-iron plows with steel-faced cutting edges on the shares. By the mid-1850s, the John Deere firm was producing about 10,000 plows per year.

The mechanical reaper was probably the most significant agricultural invention introduced between 1800 and 1860. It replaced much human power at that crucial point in grain production. Reapers were patented by Obed Hussey in 1833 and Cyrus H. McCormick in 1834. In the 1850s, McCormick adapted the best features of his and Hussey's reapers into a reasonably dependable and efficient machine. McCormick astutely moved his factory to Chicago, a city destined to become the great agricultural market for the farmers of the prairies and the plains.

The westward movement of agriculture was stimulated by Jefferson's genius in negotiating the Louisiana Purchase in 1803 for only $15 million. Subsequently, the purchase of Florida (1819), the annexation of Texas (1845), the Northwest Treaty (1846), the acquisition of Mexican territory (1848), and the purchase of Alaska (1862) provided the boundaries of almost all the present-day United States.

Farmers and ranchers continued to move west, with livestock and wheat leading the way in the North and cotton in the South. Corn and hogs followed wheat in many of the prairie states. Farther west, irrigation agriculture began in the United States in 1847, when Mormon pioneers settled Utah.

The new techniques and machines developed from about 1830 to 1860 were refined, improved, and widely adopted by the end of the 1860s. Substantially increased European demand for American agricultural products, combined with the shortage of human resources and the seemingly unlimited domestic demand brought about by the Civil War, encouraged farmers to boost production. Southern farmers had the problem of converting from cotton to food crops, mainly for local consumption. Northern and western farmers increased grain and meat production, both for export and domestic consumption. The result was an agricultural revolution, marked by the change from hand power to animal power for most farming operations and by a greater emphasis on producing for the commercial market. The development of a vast railroad system was also a significant factor contributing to improved transportation of agricultural products from food-producing to food-consuming areas.

The federal government became more directly involved in agricultural affairs in 1839, when Congress appropriated funds to the Commissioner of Patents to create a unit to collect statistics and investigate the promotion of agriculture. This move was the forerunner of the Department of Agriculture.

Under President Lincoln's administration in 1862, four laws were passed that were to influence agriculture substantially. The Homestead Act and the Transcontinental Railroad Act encouraged western settlement. The Homestead Act, which promoted the idea of the family farm (160 acres) and homestead ownership (if farmed for five years), was traditional with Lincoln's thinking. The Land-Grant College Act, sponsored by Senator Justin S. Morrill of Vermont, promoted agricultural education and became the major source of professional agricultural research and education through the system of land-grant universities, benefiting farmers as well as the rest of society. The fourth key law established the U.S. Department of Agriculture, providing more structure and direction for agricultural research and education.

Although science and technology were changing farming operations and agriculture was contributing by its exports to industrial growth, farmers seemed to benefit little from their improved efficiency. Northern and western farmers found that increasing production often resulted in lower commodity prices, without lower prices for machinery and other necessary inputs.

In the South, farmers had to overcome the destruction and disruption caused by the Civil War. For example, cotton farming took a different turn after the war. Many former slaves remained on cotton plantations as sharecroppers, while others migrated to towns and cities. Still others obtained small acreages and farmed on their own.

After the Civil War, the general economy also suffered increasingly from "boom" and "bust" cycles, periods of acute inflation followed by deflation. Industry continually sought higher tariffs, and farmers lower ones. Monetary policy was also in controversy between advocates of silver and gold. In response, farmers became increasingly concerned with the need for organizations operating in their behalf.

In 1867, Oliver Hudson Kelley organized the Patrons of Husbandry, or the National Grange, a fraternal order for farmers. It urged farmers to improve practices and organize marketing and purchasing cooperatives. In 1902, the Farmers Educational and Cooperative Union was organized, the forerunner of the present National Farmers Union. The American Farm Bureau Federation was established in 1920. While these three farmer organizations were the major ones, hundreds of others, all dedicated to solving one or more problems confronting farmers of that day, formed and disbanded in various parts of North America. Farmers were so severely discontented over their economic welfare that they were instrumental in offering a candidate, James B. Weaver of Iowa, for President of the United States on the Populist ticket in 1892. Weaver received over a million votes.

The Hatch Act of 1887 spurred agricultural research and development by establishing agricultural experiment stations in conjunction with the land-grant colleges in each state. The Reclamation Act of 1902 helped to develop the 16 arid western states by funding irrigation projects through the sale of public lands.

Many new inventions continued to add to the efficiency of production and marketing. A practical railroad refrigerator car was patented in 1869 that enabled the shipment of meat and perishable produce over long distances. A scientist at the University of Wisconsin, Stephen M. Babcock, perfected a test for measuring the butterfat content of milk.

Near the end of the nineteenth century, a number of inventors began working with the gasoline tractor. John Froelick of Iowa built the first of record that was an operating success in 1892. Machinery began replacing animal power at a more rapid rate. By 1910, with a supply of dependable mass-produced tractors coming on line, farmers had begun to adopt tractors on a widespread basis.

During the last 30 years of the nineteenth century, long-term expansion in agriculture continued, even though the percentage of American workers who were in farming declined from 53 percent in 1870 to 38 percent in 1900. In 1900, each U.S. farmer produced enough to supply seven persons with farm products.

In the latter years of the nineteenth century, farmers directed their complaints toward business monopolies and unfair competitive practices in railroading, meat packing, grain marketing, sugar refining, and certain other industries. These complaints resulted in the Interstate Commerce Act (regulating railroads) and the Sherman Antitrust, Clayton, and Federal Trade Commission Acts (controlling monopoly), which helped farmers gain better equity.

Although science and technology were capable of transforming American agriculture in the early 1900s, many farmers had not adopted the new techniques. Secretary of Agriculture James Wilson, a firm believer in the power of scientific research, took action. He sent Seaman A. Knapp to help cotton growers apply scientific information for controlling the boll weevil. Knapp encouraged farmers to participate by demonstrating control practices under the direction of trained agents. This lead to the establishment of the county agent system, with the first agricultural agent appointed in 1906. Eventually, under the leadership of Hoke Smith of Georgia and Asbury Lever of South Carolina, the Smith-Lever Act, approved May 8, 1914, established the nationwide county agent system.

Agricultural education at the high school level was also initiated under terms of the Smith-Hughes Act of 1916. Recognizing that technology and education alone would not cause improved practices to be adopted, Congress passed the National Farm Loan Act in 1916 to lend money to farmers to buy land on a long-term basis.

The problem of carrying scientific knowledge to the farmer was matched by the need to improve farm life. In 1908, President Theodore Roosevelt appointed the Country Life Commission, chaired by Liberty Hyde Bailey, to study the problem. Many of the commission's recommendations were carried out over the years. One was the establishment of the Bureau of Home Economics in 1923.

The period from the Spanish-American War to World War I was one of relative stability in agriculture. Agricultural productivity was increasing at a rate comparable to that of the growth in population. The healthy state of the general economy provided good domestic markets in addition to profitable export markets.

During World War I, farmers responded to appeals to increase grain and meat production for our allies and the armed forces. After the war, export demand fell and prices dropped sharply, but there was little that individual farmers could do except increase production even more in an attempt to meet rising costs. The agricultural economy entered a severe depression in the early 1920s, well ahead of the general depression of the 1930s.

As the agricultural depression deepened, farmers received less per unit for their increased production. George N. Peek, an implement manufacturer, called for equality for agriculture. This could be obtained, he said, by selling products for domestic consumption at a fair exchange value and surplus agricultural products at the world price. Legislation to this end was introduced by Senator Charles L. McNary of Oregon and Representative Gilbert N. Haugen of Iowa. The McNary-Haugen Bill passed twice and was vetoed both times by President Calvin Coolidge.

Help to farmers was extended with the Federal Farm Board, established by the Agricultural Marketing Act of 1929, aiding farmers through loans to cooperatives and price stabilization operations. However, the Great Depression of 1929 and lack of sufficient funds kept the board from achieving its goal of stabilizing market prices.

By 1929, the agricultural depression had extended into a vast economic depression, not only within the United States, but also throughout the world. While its causes are subject to debate, chronically low farm prices, high tariffs, unbounded speculation in the stock market, and mistakes in monetary and fiscal policies were paramount.

With the election of President Franklin D. Roosevelt and the adoption of his New Deal program, the federal government began a more active and direct role in agricultural and general economic affairs. The Agricultural Adjustment Acts of 1933 and 1936 and subsequent legislation were aimed at adjusting farm production to market demand and assuring farmers of fair prices.

Other legislation, prompted by the efforts of Hugh Hammond Bennett, tackled the problem of soil erosion. A rural electrification program was established on a permanent basis principally through the efforts of Senator George W. Norris of Nebraska. The Farm Credit Act of 1933 extended credit to farmer cooperatives and organized the production credit system. Farm tenancy, a chronic problem, was attacked with the Bankhead-Jones Farm Tenancy Act of 1937. Numerous other laws dealing with various farm problems were enacted.

Technological advances in production agriculture were not limited to farm machinery and equipment. Much of the increase in productivity and efficiency in animal and crop production can be attributed to the diligent efforts of animal and plant breeders and geneticists. Corn production illustrates their success well. From 1870 to 1940, there was no appreciable increase in average corn yields in the United States. During that 70-year period, average yields were from 27 to 30 bushels per acre. Then, during the 1940s, hybrid corn was adopted on a national scale, resulting in the first major sustained increase in corn production. Current yields are over triple what they were prior to corn hybridization. The average corn yield for the five years 1987 to 1991 was 109.6 bushels per acre.[1]

The abilities of U.S. farmers and the legislation of the 1930s were severely tested during World War II, when farmers were challenged with the slogan

---

[1] Richard Crabb, *The Hybrid Corn-Makers,* 2nd ed. (West Chicago: West Chicago Publishing Co., 1993), pp. 1–24.

"Food will win the war and write the peace." Farmers responded with a great outpouring of food and fiber.

The assured markets and high prices for farm products during World War II, combined with patriotic appeals and a shortage of labor, encouraged farmers to adopt all the skills and machines available to increase production. By 1945, each U.S. farmer produced enough to supply 15 persons with farm products, triple the output of farmers in 1870. This emphasis on increased production did not stop after the war. The result was a second agricultural revolution, which has continued.

Since World War II, the United States has had to contend with the somewhat fortunate problem of too much agricultural production rather than too little. This is unlike the situation in many countries around the world. Various techniques have been tried to control or adjust production to market demand, but most have fallen short of their announced objectives. As acreages were reduced, farmers added more fertilizer and used chemicals on their most productive land to expand output on the acreage allotted them. A variety of plans, schemes, and programs, among them the Soil Bank Act of 1956, were offered by amateurs and professionals alike, but the idea of effectively controlling farm production was politically repugnant to both farmers and legislators.

Considerable restructuring occurred in agriculture in the 1950s. The farm population in the United States decreased by almost 5 million persons from 1950 to 1958. Over 1.6 million farms went out of business during the 1950s. There was also rapid and widespread adoption of new production technologies.

In the late 1950s and early 1960s, some farmers began turning toward collective bargaining as a tool for securing economic improvement. The National Farmers Organization and the American Farm Bureau Federation each began farmer bargaining associations. Those of each group use different tactics. NFO's are based more on the pattern of labor union bargaining, while the Farm Bureau's are linked more closely with farmer cooperatives and voluntary efforts.

Conditions in the agricultural sector were relatively stable throughout the 1950s and 1960s. Growth in agricultural productivity and agribusiness was somewhat slow and steady. Price trends for major agricultural commodities were relatively flat, due largely to government involvement in the marketplace. Agricultural exports grew slowly from a post–World War II low of $2.8 billion to $4.0 billion in 1970.

The late 1960s and early 1970s brought forth a change in U.S. federal farm policy, namely, from one of enforced relative scarcity to one of encouraged abundance. The competitive market system was allowed to perform its job of allocating resources among competitive uses. The U.S. government positioned itself as a guarantor of "floor" prices below which farm prices would not be allowed to fall.

An unprecedented world demand for food helped boost this change in U.S. farm policy as world population growth tended to outstrip world food supplies. Devaluation of the U.S. dollar helped encourage U.S. exports and discourage imports into the United States. Also, new trade agreements with the Communist-bloc countries helped add to the demand for U.S. agricultural products.

Technical development in production agriculture through the 1970s continued trends observed in the 1950s and 1960s. Tractor size increased considerably, enabling much more rapid completion of field work. New herbicides, insecticides, and pesticides were developed and introduced.

During the 1970s and 1980s much effort was spent on fine-tuning agronomic and animal production systems. These decades saw the development of plant varieties for specific climatic zones, an increase in per acre plant populations, and the determination of efficient and specific fertilization rates. As needs were identified and adjustments made, agricultural productivity again began increasing.

The benefits of improved control over each farming operation were observed especially in the 1980s. Improvements in livestock production were even more marked than in crop production. Many technological developments came together to increase livestock productivity, including improved livestock-handling facilities, increased feed efficiency, better and uniform records to monitor individual animal performance, and improved disease prevention and control.

Perhaps the characteristic difference of the 1970s and 1980s in the agricultural sector was the dramatic rise and then fall of agricultural exports. Many factors unique to the 1970s lead to the increase in exports. Similarly, the decrease in exports in the 1980s was the result of a different, but specific, set of factors.

World demand for agricultural products rose dramatically due to increasing incomes and population in importing countries. Many of these importing countries used external credit to finance import purchases. The dollar depreciated in world markets during the 1970s, which resulted in lower prices for U.S. products in international trade. Finally, severe drought in many countries opened markets for U.S. agricultural products.

Domestically, the strong world demand for agricultural products and the accompanying high commodity prices spurred the use of credit to finance growth and expansion in agriculture. Many of the credit decisions were based on increasing collateral values, specifically land, rather than ability to cash flow.

Unfortunately, the strong demand for U.S. agricultural commodities was short-lived. A worldwide economic recession in the early 1980s resulted in a decrease in demand for agricultural products. The U.S. dollar appreciated relative to other currencies, driving up the cost of U.S. products on the world market. Many of the countries that had used debt to finance import purchases during the 1970s were experiencing credit difficulties that limited their ability to continue importing.

The downturn in agricultural exports between 1981 and 1986 precipitated a farm recession that was especially painful for farmers with highly leveraged operations, specifically those who had used credit to expand during the late 1970s. Farm prices declined generally and grain prices plummeted from 1981 to 1986. At the same time, the Federal Reserve System tightened the money supply to curb inflation, causing interest rates to increase. This caught highly levered operations in a difficult financial squeeze. Many failed and went out of business, along with many rural banks that had extended the credit. Declining land values compounded the difficulties as collateral values fell, turning good credit risks into poor credit risks and sometimes into business failures.

Agricultural exports again began to increase in 1987. As they did, farm prices began to rise, land values stabilized, and the unanticipated farm recession ended. The importance of strong export markets to the economic well-being of the agricultural sector had again been clearly established.[2]

Numerous interrelated factors have contributed to the large output and high productivity of American agriculture. They include (1) a large supply of land and water resources; (2) large investments for education that improve human skills and managerial abilities; (3) development and diffusion of new knowledge and technology; (4) complementary industrial development that supplies capital inputs for agriculture and nonfarm employment opportunities for people not needed in agriculture; (5) a structural organization of farm production and marketing that provides powerful economic incentives for farmers and marketing firms to increase output and productivity; and (6) public and private institutional services that help conserve and improve natu-

---

[2] Willard W. Cochrane, *The Development of American Agriculture: A Historical Analysis*, 2nd ed. (Minneapolis: University of Minnesota Press, 1993), pp. 150–164.

ral resources, increase the fund of knowledge about improved agricultural technology, and encourage capital formation and investments in agriculture.

Finally, the continuing accomplishments of U.S. agriculture may be summarized as follows:

1. A record-high capacity to produce food and fiber despite sharp declines in the total number of farms, farmers, and farm workers.

2. Greater yields per acre of land cultivated and greater total production of food and fiber.

3. An increase in the average size of farms.

4. Better qualities of plants and of the food they yield, and improved meats, dairy products, and fibers.

5. Improvements in the system of marketing and distribution, with accompanying benefits to consumers everywhere.

6. The discovery of alternative uses for traditional crops.

7. Lower prices of farm products as a result of changes and increased efficiency in agriculture.

8. A tremendous capacity to feed at least a major part of a hungry world.

9. The extension of electricity to rural areas, which has made possible higher standards of rural life and helped the farm mechanization process.

10. The development of agricultural research, extension, and education techniques on a vast scale available to all.

U.S. agriculture represents, according to most scholars of the subject, the greatest single achievement in the history of humankind's struggle for food, clothing, and shelter. So efficient has U.S. agriculture become that "one American farmer supplies enough food and fiber for 128 people—96 Americans and 32 citizens of other countries."[3] At the same time, "Americans spend less of their income on food than anyone else in the world—just 9 percent of total personal expenditures. . . ."[4]

---

[3] M. E. Ensminger in the Foreword to Hiram M. Drache's *History of U.S. Agriculture and Its Relevance to Today* (Danville, IL: Interstate Publishers, Inc., 1996), p. vii.

[4] Ibid.

Because American agriculture has been so successful within an economic system devoted to free enterprise, it is important that we compare that economic system with those in other parts of the world. We will make such comparisons in the next chapter.

Topics for Discussion

1. Discuss the economic conditions facing the early colonists in America.
2. Which four European nations played a significant part in the development of North America?
3. Discuss agricultural conditions and the development of the United States between the Revolutionary War and the Civil War.
4. Discuss the commercialization of U.S. agriculture after the Civil War.
5. Discuss the expanded role of the federal government in U.S. agriculture since World War I.
6. List the 10 major factors that summarize the growth and development of U.S. agriculture.

Problem Assignment

Prepare an essay on some historical aspect of U.S. agriculture of interest to you. (You may want to select a topic of local historical interest.)

Recommended Readings

Barger, Harold, and H. H. Landsberg. *American Agriculture, 1899–1939.* New York: National Bureau of Economic Research, Inc., 1942.

Benedict, Murray R. *Farm Policies of the U.S., 1790–1950.* New York: Twentieth Century Fund, 1953.

Cochrane, Willard W. *The Development of American Agriculture: A Historical Analysis,* 2nd ed. Minneapolis: University of Minnesota Press, 1993.

Crabb, Richard. *The Hybrid Corn-Makers,* 2nd ed. West Chicago: West Chicago Publishing Co., 1993.

Davis, John H. *A Concept of Agribusiness.* Cambridge, MA: Graduate School of Business Administration, Harvard University, 1957.

Drache, Hiram M. *History of U.S. Agriculture and Its Relevance to Today.* Danville, IL: Interstate Publishers, Inc., 1996.

Flint, Charles L. "A Hundred Years of Progress of American Agriculture," *Agricultural Science Review,* U.S. Department of Agriculture, Washington, DC, Second Quarter, 1973.

Peoples, Kenneth L., et al. *Anatomy of an American Agricultural Credit Crisis: Farm Debt in the 1980s.* Lanham, MD: Rowman & Littlefield Publishers, Inc., 1993.

Taylor, Henry C., and Anne H. Taylor. *The Story of Agricultural Economics in the United States, 1840–1932.* Ames: Iowa State University Press, 1974 reprint of 1952 ed.

3

Types of Economic Systems

An economic system is a set of relationships that organize human labor and natural resources to produce goods and services that support life. Economic systems are concerned with production, processing, and distribution. However, the forces that determine the manner in which these basic functions are performed differ between systems. The goals of each system are somewhat different. Economic organization, which is fashioned to achieve the goals, will also differ by system.

Earlier, economics was defined as the science of allocating scarce resources among different and competing choices and utilizing them to best satisfy human wants. The diversity and insatiability of human wants create competition for the limited resources available, and the strongest forces dominate when they are organized in such a way as to back up their wants with purchasing power. Wants plus purchasing power create demand.

Resources used to produce goods and services, including land, labor, capital, and management, are generally called the **factors of production,** or inputs. If productive resources were readily abundant and free for the taking, like the air we breathe, a study of economics would be unnecessary. But, because resources are scarce and in great demand, economic principles can provide guidance in resource allocation to attain maximum **utility** or satisfaction.

Ownership and control of factors of production serve as the foundation for any economic system. Usually ownership implies the freedom to use resources to the best advantage of the owner. In the classical economic sense, Adam Smith, an outstanding economist of the eighteenth century, expressed the concept that the welfare of the nation is best served when individuals operate for their own best interest. This concept of the "invisible hand," which served as the justification for **laissez-faire** development (noninterference by government, leaving coordination of an individual's wants to be controlled by the market), is the backbone of capitalism. Adam Smith maintained that entrepreneurs, if left free to seek their own fortunes, will be led as if by an invisible hand to promote the interest of society.

Owners of scarce resources have the potential of receiving payments for the use of their resources in the form of rents, wages, interest, and profits. Hence, owners are vitally concerned with making appropriate economic decisions regarding the best use of their resources. These decisions, however, will differ according to owners' individual goals. Maximizing profits may not be the basic goal of each individual, and more importantly, an attempt to maximize profits by one individual may work a hardship on another. Thus, some system of control, as well as rationing of resources, is essential in an economic system.

The functions of producing, distributing, and rationing goods and services are organized according to the basic philosophy of a society, the goals of political leaders, and/or the desires of individuals and groups having control over a significant stock of resources.

There are some 200 countries in the world.[1] With a diversity of cultures, different natural resource endowments, a wide range in education levels, and different problems of national security, it is understandable that economic organization differs between nations. Despite this, the characteristics of how an economy is organized are derived from four primary economic systems: capitalism, fascism, communism, and socialism. Remember, though, these systems are not entirely exclusive. Characteristic elements of one system may be found woven through the fabric of another. Some very important differences in economic organization, however, characterize each of these systems.

Capitalism

Capitalism is an economic system in which goods and services are produced for profit using privately owned capital goods and wage labor. Capitalists obtain the surplus product in the form of profits. Surplus product equals total product, the total amount of goods and services produced in an economy in a given period, less necessary product, the amount of goods and services required to maintain inputs in the production process.[2]

---

[1] The World Bank, *The World Bank Atlas* (Washington, DC: The World Bank, 1992).

[2] Samuel Bowles and Richard Edwards, *Understanding Capitalism: Competition, Command, and Change in the U.S. Economy*, 2nd ed. (New York: HarperCollins College, 1993), pp. 49–54.

Capitalism derives from the word "capital," which means produced goods that can be used to produce other goods and services. Machinery, tools, equipment, buildings, and mines are examples. Money is also referred to as capital, since it can be used to procure the elements of production, but it is not a capital good as such.

Ownership

Although capital refers to things used in the production process, capitalism implies a particular system of ownership and use of productive factors. Ownership may be vested in individuals or in a body of individuals organized into associations or corporations. The incentive to produce stems from the prospect of earning a profit; thus, privately held resources are organized to this end.

Free enterprise, on which organization of resources and production depends, is a vital ingredient of capitalism. Private property and individual control of resources are the mainstays of capitalism. Ownership and control of property convey the power to decide how, when, and where resources are used in production and consumption. Property ownership also permits wealth accumulation and encourages saving. Even death does not terminate property ownership. Through inheritance, ownership can be transferred to relatives, friends, or other specified entities.

By virtue of corporate ownership, vast holdings have developed in America. Large size, large volume of output, and the declining rural population are signs of the times. Commerce and industry are dominant. Although private owners have widespread interest, private ownership of capital accounts for only a small part of the total capital used in mass production of goods and services. Large corporations now dominate industrial production.

Economic Decision Making

Economic power is held primarily by resource owners, who make decisions that are generally directed toward (1) obtaining the greatest satisfaction at least cost, (2) maximizing income through efficient resource allocation, (3) keeping well informed, and (4) counteracting competition.

Capitalism is not easily divorced from its many related institutions of government, religion, finance, and law. These noneconomic societal institutions are so thoroughly blended with capitalism that to supplant capitalism with another system would require a substantial remodeling of all existing institutions.

Yet government regulation and government ownership of some resources discourage capitalism and thus conflict with the profit motive—the motivating force behind a capitalistic economy. Government also restricts competition by sometimes condoning certain monopolistic practices to restrict production. Some examples include price leadership, market dominance, basing point pricing, trade associations, market sharing agreements, intercorporate stock relationships, mergers, cartels, pools, exclusive production or selling contracts, production allocations, patents, price agreements, zone pricing, licensing, and patent pooling.

Critics maintain that under capitalism, as moral values succumb to materialism, human welfare is sacrificed for profit. Moreover, serious inequities of income are aggravated by windfall gains or unearned incomes obtained through inheritance. Critics also say that competition between enterprises leads to wasted resources and economic insecurity and that lack of balance in the economy is socially disruptive, leading to selfish individualism. Even employment is criticized as dependent on the whims of capitalists, who, critics claim, mercilessly exploit workers. Such criticisms are used by the disenchanted in their attempts to discredit the capitalistic organization.

Yet capitalist countries, for example, the United States and Japan, have the highest standards of living, the best levels of education, and productivity unequaled by countries having any other system.

In the United States, where most property is privately owned, freedom of enterprise, apart from government intervention and regulation, is limited only by individuals' abilities to organize resources for production. Workers are free to select their occupations, and consumers are free to buy whatever they can afford.

Competition maintains a balance between supply and demand through an automatic pricing mechanism that rewards the efficient and penalizes the inefficient. This might be likened to a self-policing system in which a minimum of government interference is necessary to conduct economic activities. The primary role of government is as referee. Political and governmental powers, theoretically, lie in the democratic vote and are limited by the Constitution, which also protects citizens' rights.

In a discussion of the relative merits of capitalism, two schools of thought emerge. Some contend that free enterprise enables individuals to make their own decisions, resulting in economic progress, as is evident in the United States. Alternatively, some contend that economic planning is necessary to achieve economic progress, attain full employment, and avoid capitalistic depressions.

Notable trends in economic planning in the United States since World War II include (1) greater power of the federal government, (2) reduced freedom of individuals, (3) greater control and regulation of industries, (4) more executive powers to government leaders, (5) more collective actions, (6) higher economic goals, and (7) numerous efforts to achieve full employment.

Fascism

Fascism is a centralized, autocratic national organization with intensely nationalistic policies exercising regimentation of industry, commerce, and finance. It forcibly suppresses opposition, censors criticism, and thus denies some freedoms to individuals.

Ownership

Under fascism, property is owned by individuals as under capitalism, but use of the property is rigidly controlled by the government. Hence, it may be stated that fascism is a controlled capitalist economy in which national direction is given to production and consumption. In addition to private property, fascism recognizes private enterprise, the profit motive, a free market, a free price system, and consumer choice—as long as these do not conflict with national interests.

Economic Decision Making

Although property is privately owned and business firms control production, the government maintains firm control over labor, employers, and consumers. Fascism seems to thrive under conditions of national emergency. Economic power rests with the government; production and consumption are directed toward accomplishing the goals of the state.

Fascism is characterized by a totalitarian state led by a single political party with a powerful leader. Nationalism is encouraged, and the individual is regarded as a servant of the state. Self-denial, government service, and general welfare are placed above capital gain.

Fascist Economies

The aims of a fascist economy are similar to those of mercantilist governments that favored the upper and middle classes instead of the laboring masses, the reverse of socialism and communism. Under fascism, production is nationally directed, with workers organized by occupation and industry into syndicates designed to promote national goals.

State security is given top priority, but goals are often extended to include (1) abolition of unearned incomes, (2) confiscation of war profits, (3) nationalization of trusts, (4) profit sharing, (5) pensions for the elderly, and (6) protection of public health.

Communism

Communism is a totalitarian system of government in which a single authoritarian party controls state-owned means of production with the professed aim of establishing a stateless society. It is a doctrine and program based upon revolutionary Marxist socialism subsequently developed by Lenin and the Bolshevik Party. Communists interpret history as a relentless class war that eventually results in the victory of the workers and the establishment of a workers' dictatorship. The state regulates all social, economic, and cultural activities through the agency of a single authoritarian party. The party leads the workers in communist countries in achieving the ultimate objectives of a classless society and the establishment of a world union of socialist republics.

Marxist socialism is another name for communism. The title is derived from its chief proponent, Karl Marx, who stated the doctrines that serve as the foundation of communism in his 1867 book, *Das Kapital*.

The views, aims, and tendencies of communism, however, were identified in the "Communist Manifesto," written nearly 20 years earlier in 1848. The manifesto, developed at a London assembly of communists of various nationalities, was designed to express the views of the Communist Party, which was already a significant power in many countries of Europe.

Based on a theory advocating elimination of private property, the communist philosophy implies a sense of socialism. "From each according to ability and to each according to need" is a basic communistic concept. How this philosophy is imposed upon society is subject to considerable criticism, especially when the transition from other systems is accomplished by violent revolution.

Political power is vested in a single party, with sovereign powers imposed by the government, as in a fascist-type organization. A major difference, however, is that private ownership of property is largely denied under communism but is recognized under fascism. Communism is a highly centralized form of government, and its tenets are fairly uniform as they extend nationwide and worldwide.

Ownership

Ownership of all goods, including consumer goods, is vested in the state. But goods and services are made available to each according to need, at least in theory.

Under communism, the government owns and controls property, presumably in trust for the workers, who, according to communist philosophy, are the common owners and beneficiaries of the entire economic system. The Communist Party generally maintains control over the economic system. The fact that relatively little, if any, democratic control is exercised by the workers is explained as being only a temporary situation. Theoretically, control will be turned over to the workers as soon as they become experienced and capable of exercising administrative control. When this final step occurs, a socialist economy is promised.

Economic Decision Making

Economic activity is couched in planning for growth. Requirements are anticipated, and output is planned to meet those requirements.

A central planning agency collects information to assess needs and then evaluates availability of resources to satisfy those needs, along with the necessary production and investment decisions required. Various government agencies are then advised regarding production and distribution plans. Thus, every enterprise receives orders on what and when to produce, from whom to purchase inputs, how much to pay labor, and to whom to sell, as well as details concerning profits, costs, capital repairs, and staff organization.

Instead of market demand being registered by price, as in a capitalistic economy, the central planners decide what, when, where, and how production is to take place. The complexity of such decision making is directly proportional to the size of the country, its population, the variety of its natural resources, and the diversity of its agricultural and industrial enterprises.

Agriculture is one sector notoriously unresponsive to centralized planning. The shortcomings of agriculture are highlighted by underinvestment, which leads to shortages of machinery, fertilizers, farm buildings, and livestock.

Peasant farmers typically lack incentives to produce because no mechanism is in place to reward superior productivity, not even a fair share of gross farm income. Instead, peasant farmers prefer to devote more effort to their own private activities, such as vegetable gardening or small-scale livestock production. Authorities have attempted to discourage this by insisting that more labor time be devoted to the collective or by taxing private production.

In the former Soviet Union, about two-thirds of the workers were engaged in agriculture, with most of the farming done on collective units of two types:

1. *State farms*—operated as state enterprises under the Ministry of State Farms. The government owned the land, agricultural machinery, livestock, etc. Labor was hired on a piecework basis. Workers belonged to trade unions and had paid vacation rights. A farm director was responsible for farm operation.

2. *Collective farms*—each about 2,500 acres in size with about 200 peasant families. Mechanization was a prime objective. Machine tractor stations owned and operated the machines for collective farms. Each family generally provided for its own needs and was allowed a farmstead of ½ to 2½ acres on the collective. Tools to be used on the farmstead belonged to the peasant. Only 1 percent of the cultivated area was operated by private owners.

After deductions of all expenses (machines, seed, rent, insurance, cultural fees, farm improvements, and administration and managerial costs), the balance was distributed among the workers according to the number of workday units per respective worker. The more skilled tasks were rated more highly in workday units.

Allocation of materials and equipment was rather clumsy. Farms accepted machinery and supplies sent to them by the authorities, whether or not they were needed or had been ordered. Moreover, large-size farm units often were beyond the ability of farm managers to organize for efficient operation.

Profit as a criterion for performance was not deemed to be inconsistent with Marxism. But, instead of profit going to the management, it went to the state.

With an increasing emphasis assigned to profits within the former Soviet Union, a new role seems to be emerging for prices. However, many decisions are still linked to political priorities in the allocation of scarce resources rather than to prices.

Under communism, each firm works with two profit figures: planned and unplanned. The planned profit is that resulting from the planned procedures. The unplanned profit is realized when the plant manager is able to economize on inputs, thus reducing costs more than anticipated.

Planned profits could easily be changed to planned losses by a mere change in price, and the state sometimes sets a low price to stimulate consumption, even though the products may be produced at a loss.

Economic Functions

Under communism, some costs are ignored in accounting calculations. Although interest on short-term working capital and rent on some buildings are taken into account, interest on long-term capital and rent or depletion costs on land and minerals are not, mainly because the latter are incompatible with the Marxist concept that labor is the sole source of value.

Profits under communism play a very minor role, for the important goal is the fulfillment of production targets according to the central plan. Bonuses to employees and managers who achieve production quotas provide a reward and presumably an incentive to produce according to plan and even to exceed the planned output.

Profits in a given enterprise accrue to the state and are not generally available to the plant for reinvestment. Investment capital comes through appropriations, usually without regard to the profit record of the enterprise.

Since firms are not required to promote the sale of their products, they need not worry about quality or grading of the products. As a result, efforts are directed toward a common product with as little variety as possible, because too much variety disrupts mass production schedules. A common size steel bar, a common size truck, and a common style coat are examples of the uniformity that results from mass production goals.

In a communistic setting, capital and land are virtually free goods and are used extravagantly on massive projects. As industrial maturity is attained, consumers become more discriminating in their choices, and mass production leads to a buildup of surplus inventories. Under the system, reduced production means the managers must forfeit their bonuses. Thus, they continue to produce without regard to consumer demand.

Innovation is also discouraged under communism, for it presents an element of risk to production managers. Unless a new product or process is given priority by the planning committee, it will not be produced.

In summary, communism differs from capitalism in the following ways:

1. Communism discourages individual initiative because capital for development cannot be freely marshaled from investment sources. It must be garnered through state officials.

2. Production without competition, under communism, leads to complacency and gross inefficiency.

3. Communism, characterized by a lack of integrated market communications, compounds the problems of inventory control and gearing supply to demand.

4. Lack of authority to delegate power and allocate capital under communism shackles the managers and postpones adjustments necessary to improve efficiency.

To clear up some misunderstandings, it is important to note that most Westerners do not know that communism often permits private ownership of homes, furniture, cars, and tools and allows the use of individual plots of ground for gardens or small farming operations. Nor do they know that differential wages exist and that the ultimate aim of communism is to build a socialist state.

On the negative side, communism still lacks agricultural and industrial efficiency. In general, the quality of goods is very poor. Social motivation is questionable, and both social and economic inequalities persist.

Socialism

Socialism is a political and economic theory of social organization based on collective or government ownership and administration of the essential elements needed for production and distribution of goods. In this system, economic controls are maintained by the people, and the economic benefits are distributed equally among them.

Theoretically, socialism implies equal distribution of benefits among members of society, but another version has been suggested to recognize incentives and to reward those having ambition. The alternate socialist philosophy seems to say, "From each according to desire for remuneration, and compensation to each according to productivity."[3]

Under socialism, public service and usefulness substitute for private profit. Furthermore, a democratic organization permeates all socialist programs, implying government response to the popular will of an enlightened citizenry.

---

[3] William N. Loucks and William G. Whitney, *Comparative Economic Systems*, 9th ed. (Lanham, MD: University Press of America, 1985), pp. 172 and 182.

Political Controls

Socialism implies self-government, with little centralized machinery or discipline. General policies of a national scope are relatively few in number because regional or local societies are more democratically oriented, recognizing that populations are segregated regionally.

Socialists, as opposed to communists, feel that transition from capitalism to socialism can be accomplished peacefully. The chief functions under socialism are economic. The legislative branch is most important, with the executive and judicial branches subservient to it.

Bureaucracy poses a threat in socialism, as it does in other economic systems. Strong altruistic feelings on the part of leaders are essential for effective socialism.

Ownership

Under socialism, property is owned by society for the benefit of society and for public service. The intensity of ownership may vary, as was true under communism. State ownership may be limited to public utilities and large-scale industries.[4] Resources of a local nature are owned by local society groups, small industries by local cooperative associations, and personal goods by individuals, including land and tools used for personal use. If additional labor is required, individual enterprises can no longer be regarded as being operated solely for personal benefit.

Economic Decision Making

Decisions are made by and for the masses, not by or for individuals. Savings are planned to serve the socialized economy, with taxation as the primary technique used to accumulate additional capital.

Management of a socialist economy is enormously complex. Most of the managerial machinery must be created and tested on an experimental basis. Special agencies and keen insights as to current and future needs of society are needed. Demands, supplies, prices, substitutes, new inventions, new processes, etc., must be anticipated. In capitalism, economic adjustments are predi-

---

[4] Ibid., p. 170.

cated on competition, but in socialism, there is no adequate substitute. Government management boards make the decisions.

With the profit motive either eliminated or severely limited under socialism, potential higher wages are the only economic incentive. Workers are allowed differential payment, based on individual productivity, considering both quantity and quality of output. Noneconomic rewards are also available to workers. Income must be derived from productive effort and not merely from ownership or speculation in ownership.

Economic planning is of prime importance in socialist economies, for enterprises must fit into the total sphere of action acceptable to society. Under this system it is quite difficult to avoid infringement of choice of occupations and consumption. Without price guides, it is almost impossible to sense people's wants. Planning and executing the plan require authority to manipulate prices and impose taxes as needed. Through these techniques economic activity is regulated, with resources shifted according to plan. Planning agencies are typically organized along geographical and functional lines.

To avoid stagnation or economic relapse, socialists propose to adopt modern machinery and techniques according to plan, thus improving output and reducing costs. Furthermore, economic security is offered to researchers and inventors to free them for creative efforts, with special awards for significant accomplishments.

Labor organizations are encouraged to protect worker interests, but unions are of the industrial type rather than craft unions. Stability of industry and employment is to be achieved by equitable distribution of purchasing power.

One serious problem confronting socialists is that of changing from an established order, like capitalism, to socialism. Property ownership is the most serious obstacle. Modern socialists, however, have developed a series of steps designed to make the transition without necessitating a revolution or wholesale confiscation of property. Nationalization would occur in the following sequence: banking functions, public utilities, natural resource industries (except agriculture), manufacturing industries, monopolies, and finally agriculture and handicrafts.

Weaknesses of Socialism

The socialized order often rebels at decisions made by the management organization, much as a child rebels against parental authority.

Because of government monopoly, socialism lacks the competitive forces to determine prices and efficiency in production that would be undertaken by many sellers. Planned arbitrary actions are frequently too sluggish and are resisted by society. In contrast, automatic responses to market forces are usually more readily accepted by society.

Summary

It must be emphasized that the philosophy of economic systems where political control is democratically determined is the philosophy of the society. Otherwise, the political leadership determines economic organization and policy. In practice, existing systems are composites of economic organizations engaged in production and consumption. A pure system, according to dictionary definition, is rarely, if ever, found. A synopsis of the characteristics of each of the four primary economic systems is presented in Table 3–1.

Table 3–1

Brief Characteristics of Several Economic Systems

Items	Capitalism	Fascism	Communism	Socialism
1. Goals	Profit	Power and profit	Classless society	Public service
2. Ownership of property				
Natural resources	Individual	Individual	State	Society
Homesteads	Individual	Individual	Some individual plots	Individual
Personal tools and equipment	Individual	Individual	Individual	Individual
3. Use of property (decision making)	Individual	Government	State	Society
4. Economic planning:				
Control of production	Individual	Government	State	Control boards
Control of consumption	Individual	Government	State	Individual
5. Political decisions	Democratic	Government	Communist Party	Democratic
6. Labor	Competitive	Workers' syndicates to promote nationalism	Trade unions for state goals	Industrial unions Differential wages

Topics for Discussion

1. What is the derivation of the following terms: capitalism, fascism, communism, socialism?
2. How does ownership of resources differ between capitalism and fascism? Between communism and socialism?
3. What are the basic economic goals of the four economic systems?
4. Who makes the economic decisions in each of the four systems?
5. What are the desirable aspects of each of the four systems? The undesirable aspects of each?
6. What systems are most closely tied to political forces?
7. What was *Das Kapital*? The "Communist Manifesto"?
8. Differentiate between state farms and collective farms under communism.
9. What is the difference between modern socialism and communism?

Problem Assignment

In an essay discuss your preferred economic system, including an explanation of features that appeal to you. At the same time explain any weaknesses recognized in your selected system, and present your arguments for counteracting them.

Recommended Readings

Barkema, Alan, Mark Drabenstott, and Karl Skold. "Agriculture in the Former Soviet Union: The Long Road Ahead," *Economic Review,* 77(4):79–85 (1992).

Bastiat, Frederic. *The Law.* New York: Irvington-on-Hudson, 1950.

Bowles, Samuel, and Richard Edwards. *Understanding Capitalism: Competition, Command, and Change in the U.S. Economy,* 2nd ed. New York: HarperCollins College, 1993.

Bromley, Daniel W. "Revitalizing the Russian Food System: Markets in Theory and Practice," *Choices,* 8(4):4–8 (1993).

Colander, David C. *Economics,* 2nd ed. Chicago: Richard D. Irwin, Inc., 1995. Ch. 3 and 39.

Edwards, William, and Dennis Judd. "Playing the Market Economy Game in Russia and Ukraine," *Choices,* 9(1):33–35 (1994).

Friedman, Milton. *Capitalism and Freedom.* Chicago: The University of Chicago Press, 1963; with a new preface, 1981.

Grossman, Gregory. *Economic Systems,* 2nd ed. Englewood Cliffs, NJ: Prentice-Hall, Inc., 1974.

Halm, George N. *Economic Systems: A Comparative Analysis.* New York: Holt, Rinehart & Winston, Inc., 1960.

Lee, Dwight R., and Richard B. McKenzie. *Failure and Progress: The Bright Side of the Dismal Science.* Washington, DC: The Cato Institute, 1993.

Schumpeter, Joseph A. *Capitalism, Socialism, and Democracy.* Magnolia, MA: Peter Smith Publishers, Inc., 1983.

Varga, Eugen. *Two Systems: Socialist Economy and Capitalist Economy.* Trans. by R. Page Arnot. New York: Greenwood Press Publishers, 1968 reprint of 1939 ed.

Viner, Jacob. *Essays on the Intellectual History of Economics,* Douglas A. Irwin (ed.). Princeton, NJ: Princeton University Press, 1991.

Von Mises, Ludwig. *Socialism: An Economic and Sociological Analysis.* New Haven, CN: Yale University Press, 1951.

4

Types of Business Organization in a Capitalistic Economy

How resources are organized and used in a productive enterprise usually reflects an economic organization that denotes a certain type of ownership, assumes management responsibilities, and bears economic risks. The purpose of this chapter is to discuss types of business organization used to facilitate production in a capitalistic system, particularly that of the United States.

The four main types of business organization are sole proprietorships, partnerships, corporations, and cooperatives. Although sole proprietorships are most numerous, corporations carry on the bulk of the business in industries outside agriculture.

The business firm is the basic production unit in a capitalistic economy. Decisions concerning type of output, methods of production, level of employment, product distribution, and pricing are made within the firm.

In a capitalistic economy, no government planning board predetermines output levels, sets product prices, or establishes employment levels. These basic decisions are made by the millions of independent business firms in the United States. Thus, activity in the American economy is the sum total of decisions made by all business firms.

The U.S. economy is dynamic and has changed dramatically in the past century. Before the Civil War, most business firms were relatively small and individually owned. Now, many business firms are large and complex conglomerate corporations. Although this trend has also been occurring in agriculture, most farms are still privately owned and operated.

There are about 20.1 million business firms, excluding farm proprietorships, in the United States.[1] Of these, 73.7 percent are sole proprietorships, 7.8

---

[1] U.S. Bureau of the Census, *Statistical Abstract of the United States: 1994* (Washington, DC: U.S. Department of Commerce, 1994).

percent are partnerships, and 18.5 percent are corporations, including cooperatives. Individually owned businesses are more numerous in every industrial group. The difference between the number of corporations and the number of sole proprietorships is least in manufacturing and wholesale trade. Also, corporations exceed the number of partnerships in every industrial group except mining and agriculture, forestry, and fishery services. The number of corporations and the number of partnerships are about equal in those industrial groups.

Of the agricultural service industries in the United States, just under 60 percent are sole proprietorships, with the remainder split evenly between partnerships and corporations. Sole proprietorships are the dominant form of business organization for U.S. farm operations (86 percent), followed by partnerships (10 percent), and corporations (4 percent).

Sole Proprietorships

The **sole proprietorship** is the most common type of business firm in agriculture, as well as in retail trade, service industries, and many professions. The structure is relatively simple: an individual owner supplies the working capital, directs or manages the firm, and receives all profits or bears all losses. The individual who owns and controls the business determines, to a great extent, how long the business lasts. Should the owner want to retire or close the business for some other reason, this can be readily accomplished by fulfilling existing business obligations and then stopping business operations.

The firm and its owner are one and the same. Since assets of the owner are assets of the firm, the firm's creditors may claim at least some of the owner's personal possessions—automobiles, boat, collections, etc.—if the firm is unable to pay existing debt obligations. Thus, liability for obligations of the firm extends beyond business assets to include personal assets.

Sole proprietorships are normally small, with limited available capital. However, their economic efficiency generally compares favorably with larger firms. This is true of farming and of service industries such as barbering, dentistry, and medicine. Because control and profits belong only to the owner, sole proprietorship is an attractive economic organization. The unattractive feature is exposure of personal property, in addition to business assets, to satisfy creditors' claims. Advantages and disadvantages of sole proprietorships and other types of business organization are presented in Table 4–1.

Table 4-1

Advantages and Disadvantages of Certain Types of Business Organization

A. Sole Proprietorship

Advantages:

1. Low start-up costs
2. Freedom from regulation
3. Owner in direct control
4. Minimal working capital requirements
5. Tax advantages to small owner
6. All profits to owner

Disadvantages:

1. Unlimited liability
2. Lack of continuity
3. Difficult to raise capital

B. Partnership

Advantages:

1. Ease of formation
2. Low start-up costs
3. Additional sources of risk capital
4. Broader management base
5. Possible tax advantages
6. Limited outside regulation

Disadvantages:

1. Unlimited liability
2. Lack of continuity
3. Divided authority
4. Hard to find suitable partners
5. Difficulty in raising large amounts of capital

C. Corporation

Advantages:

1. Limited liability
2. Specialized management
3. Ownership transferable
4. Continuous existence
5. Legal entity
6. Possible tax advantages
7. Easier to raise capital

Disadvantages:

1. Closely regulated
2. More expensive to organize
3. Charter restrictions
4. Double taxation
5. Management more complicated

D. Cooperative

Advantages:

1. Limited liability
2. Specialized management
3. Continuous existence
4. Legal entity
5. Substantial tax advantages
6. Easier to raise capital
7. Enjoys certain antitrust and regulatory exemptions

Disadvantages:

1. Incorporating statutes quite restrictive
2. Cooperation among members difficult to achieve
3. Slow in organizing and getting started
4. Members fail to recognize their ownership responsibilities
5. Business community resentment against cooperatives

Partnerships

A **partnership** is an association of two or more persons to carry on a business for profit as co-owners. This arrangement provides access to more capital and allows the firm to be larger. Two basic types of partnerships exist, general partnerships and limited partnerships.

In a *general partnership,* each partner is involved with both ownership and management of the firm. Each partner has unlimited liability for both his or her own business obligations as well as for those of the other partners. Like sole proprietors, partners are responsible to the extent of their personal and business assets.

A *limited partnership* is a partnership formed by two or more persons, with one or more general partners and one or more limited partners. A limited partner contributes only a fixed amount of capital to the partnership. Limited partners are not bound by the obligations of the partnership beyond their investment in the partnership, except in special cases. Limited partners may contribute money or other property to the partnership, but they may not contribute personal services.

There are several major differences between the limited and general partners in a limited partnership. Unless otherwise noted, laws applicable to general partnerships also apply to limited partnerships. The liability of limited partners does not extend beyond their individual, agreed-upon contributions. General partners have unlimited liability. Limited partners who assume control of the business or participate in its management may be declared general partners and made subject to the same liability as the general partners.

Partnerships have other disadvantages. Like sole proprietorships, partnerships have limited life. A partnership is usually terminated if a general partner dies or withdraws from the business. Provisions can be made to ensure business continuity by requiring the partnership, or the surviving partners, to purchase the interest of the deceased or withdrawing partner. Limited partnerships specify ways in which limited partners may transfer their interest to other limited partners without disrupting business operations.[2]

In addition to an improved capacity over that of sole proprietorships to secure funds, partnerships permit specialization of labor. Since more than one person is involved, partners may have the flexibility to devote attention to

---

[2] Barry, Peter J., Paul N. Ellinger, John A. Hopkin, and C. B. Baker, *Financial Management in Agriculture,* 5th ed. (Danville, IL: Interstate Publishers, Inc., 1995), pp. 572–574.

areas in which they have comparative advantage. Areas of specialization could include production, financing, advertising, or marketing.

Corporations

The **corporation** as a form of business organization was devised to make possible the accumulation of large sums of money needed to finance massive business enterprises. The corporation also implies a certain degree of separation of control from the owners. As a legal entity authorized by the state, a corporation owns its own assets and must meet its own liabilities. Thus, it is a separate entity from its personal owners, the stockholders. Stockholders are not responsible for corporate debts beyond the amount of their individual investment in the corporation, nor can they claim corporate assets while the business is in operation. The limited liability has much greater appeal to investors than the total liability associated with sole proprietorships and general partnerships. Much capital can thus be accumulated to finance large-scale corporate firms.

The corporate business structure has certain disadvantages. First, the federal government and most state governments tax corporate incomes at fairly high rates. Moreover, stockholders must also pay income taxes on personal income received as dividends. Had they invested the same capital in a proprietorship or partnership, they would have had to pay only one income tax on owner income. Second, periodical financial status reports not required of other types of business organization must be prepared. Third, certain expenses, including legal fees, are incurred when incorporating a firm. These added costs and public regulatory features discourage some businesses from incorporating.

Although the owners are the decision makers in a proprietary corporation, relatively few in fact play an active part in the operation of the business. The board of directors, elected by the stockholders, determines business policies and hires managers to execute these policies.

Some owners (stockholders) know very little about their company and are concerned only about the dividends received on money invested. Although invited to attend stockholder meetings and vote on corporate matters, few actually attend. If they do attend, they do not share equally in the right to vote. The number of shares owned determines the number of votes a stockholder has. Thus, voting power and control rest with the larger investors.

A typical procedure for electing directors of a large corporation begins with an announcement of an annual meeting for that purpose. Stockholders receive notices of the meeting and proxy forms on which they may designate their choices and assign their votes to one of the officials of the corporation. If a proxy form is signed and returned, the designated official may vote that stockholder's shares. If a proxy is not returned, the shares are not voted unless the stockholder attends the meeting.

Recognizing that thousands of shareholders in a large corporation own so few shares individually and have a very limited knowledge of the workings of the corporation, it is not surprising that actual control of the corporation lies with major stockholders, officers, and directors.

Management (i.e., the officers and the directors) usually dominates the affairs of a corporation. This situation may exist even with the management owning a relatively small proportion of the total stock. The power derives from the fact that management often controls the voting machinery. So, although control of a corporation is legally in the hands of its stockholders, in many large corporations, it is actually in the hands of management.

Laws governing incorporation of a business vary from state to state. Invariably, however, requests for incorporation are made to the state official who supervises the granting of corporate charters. The request begins with preparation of official forms on which the name of the corporation, the purposes of the corporation, the length of existence, and the names and addresses of persons who are to serve as initial directors are designated. After the charter is granted, the stockholders must meet to complete the incorporation process. At this meeting, corporate bylaws are adopted and a board of directors is elected. This board then elects the officers, who will have charge of business operations.

Bylaws for the corporation usually provide details about (1) location of the corporation's principal office; (2) stockholder meetings—date and place, quorum requirements, voting privileges; (3) directors—number, method of electing, time and place of director meetings, quorum requirements; (4) officers—method of selecting, duties, terms of office, salaries; (5) stock—kind of certificates to be issued, transfer and control, right of directors to declare dividends; (6) how to amend.

Subchapter S Corporations

The Subchapter S corporation, distinguished from the regular or Subchapter C corporation, was authorized in 1958 to benefit small businesses, including those in agriculture. To qualify, a firm must satisfy each of the following conditions:[3]

1. It must have only one class of stock outstanding.

2. It must have no more than 35 initial stockholders, all of whom must be individuals or estates (another corporation cannot be a stockholder).

3. It must have consent of all stockholders to the choice.

The Subchapter S corporation was enacted with the objective of enabling enterprises "to select a form of business organization desired, without the necessity of taking into account major differences in tax consequences." It provides a form of business organization that is taxed differently from regular corporations. In fact, Subchapter S corporations are taxed as partnerships, so they pay no income tax. The popularity of the Subchapter S corporation stems largely from the fact that using this type of organization avoids "double taxation" of corporate income without loss of such corporate advantages as limited liability and favorable tax treatment of "fringe benefits" to shareholder-employees.

The Subchapter S corporation is well adapted to a family-sized farm where the present owner wishes to provide for the continuation of the farm upon his or her death without having to divide the acreage and assets among the heirs.

A competent lawyer and accountant should be retained when establishing such a corporation.

Cooperatives

A variation from the regular corporation is the **cooperative**, which is designed to give the corporation more democratic control. Instead of the one share–one vote arrangement prevailing in an ordinary corporation, the cooperative is organized on the principle that each member has one vote regardless of the share of ownership. However, some states permit voting on the basis of patronage, or business done with the cooperative.[4]

---

[3] Ibid., pp. 576–579.

[4] Roy, Ewell P., *Cooperatives: Development, Principles and Management*, 4th ed. (Danville, IL: Interstate Publishers, Inc., 1981), Ch. 1.

A cooperative is usually organized to provide goods and services for its own members rather than for the general public. This form of business organization is made possible through special laws that establish certain limitations as to the nature of the operation, prescribe maximum returns to invested capital, and stipulate that the distribution of net earnings be prorated to patrons on a patronage basis.

A summary of characteristics of the four types of business organization is presented in Table 4–2.

Table 4–2

Four Methods of Doing Business Under the Free Enterprise System

Phase	Sole Proprietorship	Partnership	Corporation	Cooperative
1. Why is the firm operated?	To buy, sell, or produce goods and services	To buy, sell, or produce goods and services	To buy, sell, or produce goods and services	To buy, sell, or produce goods and services
2. To whom are the goods and services rendered?	The public or non-owner customers	The public or non-owner customers	The public; incidentally to stockholders	Primarily its own members
3. How is the firm started?	Decision of individual	Agreement between associates who become partners	Organization by associates who become stockholder-investors	Organization by associates who become owner-members
4. How does the firm become legal?	By the owner's attaining legal age and controlling the business	By contract between two or more individuals, preferably written and recorded	Usually incorporated under general laws giving corporations great freedom in their operations	Usually incorporated under special laws requiring operation according to co-op principles
5. Is a charter required, and if so, by whom?	No charter required	No charter required; partnership contract may be recorded	Charter required; only states can charter; no federal charter	Charter required; most cooperatives chartered by states; some federally chartered
6. How does one get into the firm?	By starting or buying the business	By consent of the partners and a new agreement	By buying stock	By meeting the qualifications for membership, obtaining approval of the board, and doing business with the association

(Continued)

Table 4-2 (Continued)

Phase	Sole Proprietorship	Partnership	Corporation	Cooperative
7. Who controls the firm, selects the manager, and makes policy decisions?	The individual	The partners, by agreement	The board of directors elected by the stockholder-investors	The board of directors elected by the member-patrons
8. How is voting done?	None necessary	Informal agreement; sometimes by vote of partners	Usually one vote for each share of common stock	Usually one member, one vote; sometimes by patronage
9. Who owns the business?	The individual	Two or more individuals	The stockholders	The member-patrons
10. What do the owners put into the business?	The individual puts in capital and personal effort	Each partner may put in capital or personal effort, or both	Capital is supplied by investors seeking profits	The members do business through the cooperative and put in funds or leave retains
11. What returns can be received on the money invested?	Unlimited	Unlimited	Unlimited	Usually limited to a maximum of 8% or as prescribed by state law
12. How may net earnings be used?	As desired by individuals	As agreed upon by partners	As dividends to stockholders, or as reserves, or both	Prorated to patrons on patronage basis
13. What are the owner's(s') liabilities?	All property of individual, except legal exemptions	All property of all partners, except legal exemptions	Assets of the corporation	Assets of the cooperative
14. How may the business be ended?	Death, disability, bankruptcy, or retirement of owner	Death of any partner, bankruptcy, or decision to dissolve	Bankruptcy or legal dissolution of company	Bankruptcy or legal dissolution of cooperative
15. How are net earnings taxed?	As an individual	As individuals	At regular corporate tax rates	As a partnership

Source: Ewell P. Roy, *Cooperatives: Development, Principles and Management,* 4th ed. (Danville, IL: Interstate Publishers, Inc., 1981), pp. 14–16.

Trusts

Although no data are available on the extent of the use of trusts as a form of organization for farm businesses, a substantial amount of farm land is held in trust. Use of a trust as a landowner and lessor of land to tenants is relatively common.

As an artificial being, a trust is similar to a corporation. A corporation, however, is created under a charter approved by the state and operates within the state corporation law. A trust functions largely within rules specified in the trust instrument. Approval by the state is generally not required, although in some states periodic reporting to the local court is necessary unless waived.

The trust instrument specifies, at a minimum, the name of the trustee, the identity of income beneficiaries and of individuals to receive principal and any accumulated income upon termination of the trust, and the property to be transferred to the trust. The trustee, who normally is an adult individual or a bank or trust company, manages the property, invests funds, handles the accounting and bookkeeping, distributes income as specified by the trust instrument, and, in general, functions as manager of the trust. Because the law imposes a relatively high standard of care (fiduciary duty) on trustees, they tend to be somewhat conservative in decision making unless a broad grant of power and authority is clearly specified in the trust instrument.

Topics for Discussion

1. Discuss advantages and disadvantages of sole proprietorships, partnerships, regular and Subchapter S corporations, and cooperatives.
2. Name at least one kind of business enterprise best adapted to each of the above.
3. What do you think is the future for family farm corporations?

Problem Assignment

Visit and interview the operator or operators of one of the types of business organization discussed in this chapter. Prepare a report evaluating their success with this particular type of business organization.

Recommended Readings

Barry, Peter J., Paul N. Ellinger, John A. Hopkin, and C. B. Baker. *Financial Management in Agriculture,* 5th ed. Danville, IL: Interstate Publishers, Inc., 1995. Ch. 20.

Boehlje, Michael D., and Vernon R. Eidman. *Farm Management.* New York: John Wiley & Sons, Inc., 1984. Ch. 9.

Harl, Neil. *Agricultural Law* (15 vols.). New York: Matthew Bender & Co., Inc. 1981; updates.

Lee, W. F., M. D. Boehlje, A. G. Nelson, and W. G. Murray. *Agricultural Finance,* 8th ed. Ames: Iowa State University Press, 1988. Ch. 5 and 16.

Luening, Robert A., Richard M. Klemme, and William P. Mortenson. *The Farm Management Handbook,* 7th ed. Danville, IL: Interstate Publishers, Inc., 1991.

Roy, Ewell P. *Cooperatives: Development, Principles and Management,* 4th ed. Danville, IL: Interstate Publishers, Inc., 1981 (out of print).

Thomas, K. H., and M. D. Boehlje. *Farm Business Arrangements: Which One for You?* North Central Publication 50. Ames: College of Agriculture, Iowa State University, 1982.

5

The U.S. Monetary System and Fiscal Policy

This chapter explains briefly the U.S. monetary system, with emphasis on the sources and purposes of money, the functions of banks and the Federal Reserve System, and monetary policy and fiscal policy as they relate to the national economy. There are also brief discussions of inflation, deflation, and devaluation, three popular economic terms often heard in the media and in casual conversations but seldom fully understood.

Sources and Purposes of Money

Throughout time people have acquired possessions of one sort or another. People have recognized that it is possible to give up one possession in exchange for another. Early explorers traded beads and other goods for pelts. Later, settlers exchanged items of food, clothing, and shelter to survive better in a sparsely settled wilderness. But, exchanging commodities is an awkward way of obtaining articles that are needed or getting rid of those held in excess of personal needs.

Exchanging goods for goods is called **barter.** Its limitations have led to the development of a common medium of exchange, known as **money**, whose chief characteristics are acceptability, stability, durability, transportability, and divisibility. Historically, it is believed that China was the first to use coins as money. India was next to adopt the use of coins. Coin usage then spread to the Middle East. Egypt and Greece minted gold and silver coins with such artistic skill that they were in great demand in international trade channels of the world.[1]

Gold and silver, because of their natural beauty, were more readily accepted than other metals. Not only were they used as coins, but also as ornaments. As ornaments, these valuable metals could be carried upon the person with a feeling of greater security than would be normally associated with a pouch full of coins.

The use of coins spread rapidly through Europe with the expansion of the Roman Empire. Silver for Roman coins was obtained mainly from Spanish mines. With the fall of the Roman Empire, a new source of silver was developed in the mines of Joachimsthuler in Bohemia. The resulting coins were called "Joachimsthaler," later abbreviated to "thaler" and then corrupted to "dollar."

Early colonists of North America had little use for money from their home countries. Grain, fish, and furs were the common exchange in New England. In the southern colonies, especially in Virginia, tobacco was used as a medium of exchange. However, powder and bullets were regarded as legal tender in all the colonies.

In 1651, Massachusetts authorized the first mint in this country.[2] From this mint came the first American coins: copper and silver pence and shillings bearing crudely stamped pictures of willow, oak, and pine trees. Several additional states established their own mints after the American Revolution. These operated until the United States Mint was established by act of Congress in 1792.

The nation's first cents and half-cents were struck in 1793 from copper and were about the size of our current half-dollars and quarters. A profile of Liberty, surrounded by stars, decorated one side, while the other bore the inscription as to denomination enclosed by a wreath.

In 1794, the silver half-dime, half-dollar, and dollar were added to our coinage. Then, in 1795 and 1796, the eagle, half-eagle, and quarter eagle were introduced. At that time the eagle was a gold coin, and the dollar a silver coin. Also, in 1796, the first quarters and ten-cent pieces appeared.

The silver dollar of 1840 took on a new design of Liberty, and in 1849 Congress authorized coinage of a double eagle and a gold dollar. In 1851, silver three-cent pieces were minted, but because they were so small and thin,

---

[1] *An Introduction to the History of Coinage and Currency in the United States* (St. Louis: Federal Reserve Bank, 1953), p. 3.

[2] Ibid., p. 4.

they proved impractical. In 1853, a three-dollar gold piece was put into circulation. Foreign coins, especially the Spanish silver dollar, were regarded as legal tender in the United States until 1857.

Additional developments in U.S. coinage include:

1856 Flying Eagle penny made of copper and nickel
1859 Indian Head penny
1864 Bronze two-cent piece with the motto: "In God We Trust"
1866 Five-cent piece
1875 Twenty-cent piece
1883 Liberty Head nickel
1909 Lincoln Head penny

Several times during our history, coin circulation became so limited that some businesses issued tokens redeemable in legal currency when used in transactions with the issuing firms. During the depression of 1834, these were dubbed "Jackson hard time cents." In 1863, during the Civil War, substitute coins in large circulation were called "Civil War tokens."

In addition to coins, our monetary system includes paper currency. The paper itself has very little intrinsic worth, but its value is sustained by faith—a faith in the government and the financial institutions that it empowers to carry on money exchange. Paper money can be likened to a ticket or a token that can be exchanged for coins or other commodities. But the instrument even more important than coins or currency in our monetary system is "checkbook money." In terms of their dollar volume, most transactions in the United States are paid for by check, much more so than by coin and currency. Before considering the role of the checkbook, however, let's look at the history of paper money.

As early as 1690, with American soldiers clamoring for their wages, the Massachusetts Bay Colony found it necessary to issue paper money. Other colonies and the Continental Congress also issued paper money. So much was issued that paper money could no longer be redeemed in coin and it became worthless. This led to the expression "not worth a Continental."

About the time of the War of 1812 and on up to the Civil War, the government issued interest-bearing treasury notes to accumulate money to meet expenses. However, these were not intended to circulate as currency. The first paper money designed for general circulation by the U.S. government was authorized by the Acts of July 17 and August 5, 1861.

Nevertheless, in the early 1800s banks of the various states issued their own bank notes, many of which were unsecured. In 1838, New York passed a free banking law that gave rise to many locally owned independent state banks throughout the territory. Some, which were developed in communities still regarded as wilderness areas, were called "wildcat banks." Many of these subsequently failed, leaving depositors holding worthless bank notes.

One way the government could obtain resources during the Civil War was to print fiat money. Most of the paper money the North printed was United States notes, commonly called "greenbacks." The greenbacks, which were considered temporary, were supposed to be redeemed within about two years after the war, but they became permanent government-issue currency.[3]

In 1863, a national banking and currency system was established by Congress. Banks chartered under this law were private concerns authorized to issue currency only to 90 percent of the par value of their holdings of government bonds, which, with a 5 percent redemption fund, were deposited as security in the federal Treasury.

During the Civil War, seceded southern states and the Confederate government issued millions of dollars in paper money. However, the value of this money soon dwindled to nothing because of the Confederate defeat and the return to national unity.

In 1878, silver certificates were first issued, and in 1879, treasury notes were made redeemable in gold. The Federal Reserve System, which entered the picture in 1913, empowered Federal Reserve Banks to discount paper for member banks, engage in banking operations, and issue Federal Reserve notes. These notes were liabilities of both the issuing Federal Reserve Banks and the U.S. government.

Soon after the nationwide bank holiday of 1933, gold coin, gold bullion, and gold certificates were withdrawn from circulation. In recent years, expanded credit card usage has reduced the need for ready cash.

The functions of money, in the final analysis, are to (1) serve as a medium of exchange, (2) provide a standard of value, and (3) serve as a store of wealth.

Functions of Banks

Banks are accepted today as commonplace business establishments, yet banking, as we know it, is of comparatively recent origin. The Bible mentions religious temples being used as places of business by money changers. There

---

[3] Richard H. Timberlake, *Monetary Policy in the United States: An Intellectual and Institutional History* (Chicago: The University of Chicago Press, 1993), p. 86.

is evidence also that stone tablets were used as promissory notes. But the first banking service worth mentioning was that of serving as custodian of funds. Today, many people put savings in a bank to earn interest, but the Bank of Amsterdam, founded in 1609, charged customers for storing the gold they deposited. Similarly, goldsmiths accepted gold and coins for safekeeping and issued receipts for the amount of the deposits. The receipts then could be exchanged in much the same way as bank checks are used today.

The goldsmiths discovered that part of the money left with them could be loaned to other persons in return for promise of repayment. In fact, the borrower was often willing to accept receipts for the amount borrowed instead of the actual cash because the receipts were acceptable as money. This extension of responsibility from that of custodian to that of lender marked the beginning of our commercial banking system. Modern banks, however, do far more than merely act as custodians of funds and lenders of money. As part of their many services, they now handle investments, trust accounts, savings, collections, credit cards, travelers checks, and loan accounts and provide counseling and supervisory services.

Banks may be chartered by the national government or by the states. Less than half the banks now existing are members of the Federal Reserve System, but nonmember banks are small.

The Federal Reserve System

The American monetary system before World War I was structurally unsound. It was subject to periodic monetary stringency, which at times resulted in collapse—bank failures, widespread bankruptcy, and general economic depression. This state of affairs, which finally became recognized as intolerable as a result of the Panic of 1907, led to the establishment of a "central bank," the Federal Reserve System.

The Federal Reserve System, often called the "Fed," was designed to provide the nation with a supply of money and credit appropriate to its needs over time. Also, it was designed to be ready, in the event of a developing liquidity shortage, to serve as a "lender of last resort," able to create new money quickly. The system was not intended to manage or control the money supply but rather to respond to needs of the productive units of agriculture, commerce, and industry within the rules of the international gold standard.[4]

---

[4] Robert A. Degen, *The American Monetary System: A Concise Survey of Its Evolution Since 1896* (New York: Free Press, 1987), p. 199.

The Federal Reserve System, established by the Federal Reserve Act of 1913, consists primarily of three entities: the Board of Governors (Federal Reserve Board) in Washington, the 12 regional Federal Reserve Banks, and the many member banks.

The Federal Reserve Board consists of seven members appointed by the President of the United States and confirmed by the Senate. Appointments are for 14-year terms that are staggered to assure continuity of responsibilities and policies. It is this group, separated from political pressures, with each member selected from a different district, that establishes monetary policy and directs activities of the system.

The Federal Reserve Act also provides for an Advisory Council to meet several times a year to review monetary affairs and advise the Board of Governors. This council is composed of elected representatives, one from each of the 12 regional Federal Reserve Banks.

The 12 regional Federal Reserve Banks are located in financial centers representing each of 12 geographic regions of the nation. The capital stock in these banks is owned by the member banks in each district. However, the member banks do not control the regional banks. This control rests with the Board of Governors, and ultimately with Congress.

A board of directors for each regional bank, consisting of nine members, six elected by member banks and three appointed by the Board of Governors, sets operating policy. All segments of the economy are represented on the board. Three of the six elected by member banks represent business, commerce, and agriculture. The other three are bankers, each representing a group of large, medium, and small banks, respectively. The three appointed by the Board of Governors cannot be bankers but must represent the general public.

The Federal Reserve Banks provide other banks with currency and coin as needed and make corresponding charges against the member banks' deposits. The Federal Reserve Banks also ensure that money in circulation is in good condition. Worn or mutilated currency is taken out of circulation and sent back to the mint for replacement.

A more important role of the Federal Reserve Banks, however, is their banking service to the U.S. government. Government agencies have deposit accounts with Federal Reserve Banks. When the government borrows money, the U.S. Treasury normally sells promises-to-pay in the money market. These take the form of treasury bills, payable in 90 days; certificates of deposit, payable in 1 year or less; treasury notes, payable in 5 years or less; or treasury bonds, payable after 5 years. Savings bonds sold to individuals also represent government borrowing.

Like the federal government, the Board of Governors usually concentrates on national matters, while the 12 district banks tend to specialize in matters of regional importance. Furthermore, the Federal Reserve has four principal functions:

1. It is a bank serving other banks.

2. It is the U.S. government's bank.

3. It supervises member banks to help them stay safe and strong.

4. It manages the nation's money supply.

In a sense, the Federal Reserve is a wholesaler, while local commercial banks are retailers of banking services. As a service to other banks, the Fed supplies the currency. It also acts as a clearinghouse. Within its clearinghouse function, checks issued on the various banks arrive at the check collection departments of the Federal Reserve Banks. There they are sorted, and deposits are credited to banks making the deposits at the same time that corresponding charges are made against banks on which the checks were drawn.

In its second function, the Federal Reserve keeps the government's checking account and helps to handle the government's borrowing of funds.

The third function, supervision of member banks, results in a set of regulations telling banks what they can and cannot do. The Federal Reserve also examines bank records at least once a year to ensure that they are properly managed, and it controls credit buying.

The fourth function, managing the money supply, is probably the most important. The objective is to maintain a fair balance between the supply of money and credit in relation to the supply of goods and services.

There are three principal ways to obtain more spending money: earn it, take it out of savings, or borrow it. The Federal Reserve operates primarily through the medium of borrowed money.

Money for bank loans comes from two sources: from savings deposits and from commercial banks that have the power to create "checkbook" money. The Federal Reserve has very little control over savings, but it does have control over the amount of money commercial banks can lend, thus regulating the amount of "checkbook" money they create.

The Fed uses three main techniques to influence the amount of "created" money. First, the Fed may change the reserve requirement imposed on banks. If banks are required to maintain a 20 percent reserve, they must have $1 on reserve for every $5 in deposit accounts. Reducing the requirement to 10 percent would require banks to have only $1 on reserve for each $10 on deposit.

By lowering the required reserve ratio, the amount of reserves that member banks are required to maintain in their reserve accounts is lowered correspondingly, and these released reserves are available to expand "checkbook" money. A reduction of reserve requirements from 20 percent to 19 percent, for example, could release as much as $1 billion in excess reserves. Assuming that the total reserve deposits of member banks amounted to $100 billion, a 20 percent reserve requirement means $20 billion would be on legal reserve. If the reserve requirement were reduced to 19 percent, the legal reserve would be $19 billion, leaving $1 billion in excess reserves.

With a reserve requirement of 20 percent the banking system can expand the supply of money fivefold, as follows:

Banks	New Deposits	New Loans	Reserves
#1	$1,000	$ 800	$ 200
#2	800	640	160
#3	640	512	128
#4	512	410	102
etc.			
Totals	$5,000	$4,000	$1,000

Reserve requirements vary according to class of bank and types of deposits. Also, some leakage in the system occurs when money is hoarded or circulated without being deposited into the banking system; furthermore, some agencies are not covered by the Fed.

The second technique is to vary the "discount rate," the rate of interest that the Federal Reserve charges for loans to commercial banks. If it is desired to limit growth in money supply, the discount rate is increased. This means the banks must pay more for money borrowed from the Federal Reserve Banks. In turn, banks will raise the interest charged the public on mortgages, operating loans, automobiles, etc. This sequence of higher rates discourages borrowing, slowing the economy.

The third technique is for the Federal Reserve to enter the open market to buy or sell government securities. To stimulate a sluggish economy, the Federal Reserve buys back government securities and issues checks in payment. The checks go to the deposit accounts of member banks, increasing bank reserves and enabling the banking system to extend more credit. In contrast, the sale of government securities by the Federal Reserve decreases deposit reserves of other banks, reducing their lending power.

In principle, then, the Federal Reserve can influence the money supply created by bank loans. The Fed can slow growth in the money supply by (1) raising reserve requirements, (2) raising the discount rate, and (3) selling securities. Or, the Fed can accelerate growth by (1) lowering reserve requirements, (2) reducing the discount rate, and (3) buying securities on the open market.

The Federal Reserve System's primary efforts are directed toward exercising control over the total supply of "checkbook" money. When inflation threatens, the Fed tightens the money supply. When the economy is sluggish and needs a boost, it augments the supply. These measures alone, however, may not be adequate. Other economic influences must be recognized. Continued government spending, for example, may counteract efforts of the Federal Reserve System to control inflation. Therefore, increased taxation is imposed to take some of the extra money out of circulation. Also, the Federal Reserve can do very little about pressures of labor unions and industrial giants to boost wages and prices upward.

Fiscal Policy

In addition to monetary manipulations, fiscal policy and administrative measures have been used in attempts to regulate economic activity and thus foster economic stability. Administrative measures, such as rationing, price control, and allocation of resources by a central government agency, have proven generally unacceptable in our society and very disruptive to our economic system; hence, they will not be considered further at this point. Fiscal policy, however, is much in evidence and deserving of considerable attention, not only by economics students, but by all responsible citizens in our society.

Fiscal policy embraces governmental tax and expenditure programs designed to affect private business, consumer expenditures, and the economy of the nation as a whole. It is generally aimed at achieving full employment, price stability, and gradual economic growth. In accomplishing these goals, fiscal programs and techniques are constantly in use to correct inflationary or deflationary situations. In this respect, fiscal policy may be said to have built-in stabilizers. These include:

1. Our progressive income-tax rate structure, which causes government tax receipts to rise or fall in keeping with the rise or fall of personal and corporate incomes.

2. The provision for increased welfare payment transfers and unemployment compensation as the economy slows down and net national product declines.

3. Federal aid programs designed to stabilize incomes among disadvantaged economic sectors.

Although these built-in stabilizers are designed to reduce economic fluctuations, they sometimes need additional help. One technique is to expand or cut back on public works and other expenditures. A second is to adjust entitlement programs, primarily welfare, social security, and unemployment payments. And a third is to change the tax rates—imposing a surtax, for example.

In a democratic system, changes in tax rates or changes in designation of income brackets require congressional action. Accordingly, changes in fiscal policy are subject to the laborious and slow-moving process of arriving at a congressional decision. Not only is it difficult and time-consuming to enact a corrective measure, but it is equally hard to reverse the action once the change has served its purpose and a return to the original policy is desired.

Although enactment of tax legislation is distressingly time-consuming, government expenditures are subject to faster manipulation. Since funds are appropriated by the government on a project-by-project basis, according to fiscal budgeting procedures, Congress exercises more flexibility and can cut or expand certain program appropriations. This is not to imply, however, that budget adjustments are easily achieved. Usually, it is easier for Congress to approve an increase in program funding than to initiate a cutback. Each program develops its own special interest group, and members of Congress are generally prevailed upon to support programs rather than to undermine them. In this respect, members of Congress are responsive to public opinion and they must regard each program on the basis of its merit and in terms of national need. Moreover, the federal budget shows the expenditure plan for the government and the anticipated surplus or deficit position. Fiscal policy is generally not intent on maintaining an annually balanced budget. An ideal philosophy is to maintain a balanced budget during periods of full employment while setting aside surpluses as a reserve to balance deficits that can be expected during periods of recession. The government has fallen far short of that ideal, and the national debt continues to increase.

Two views seem to prevail regarding the national debt. On one hand, public debt is considered in the same light as private debt. Continued spending in excess of collected revenues constitutes a shifting of the burden to future generations and ultimately leads to bankruptcy. On the other hand, the national debt is treated rather lightly, with the philosophy that the debt reflects financial claims both owned and owed by the citizens. In order to spend more than is received, the government must borrow money, which is generally done by selling government bonds or other obligations to banks and to the public. Interest on these bonds is then paid from revenues collected from taxpayers. Thus, the interest on the debt consists of an internal transfer of funds from taxpayers to bondholders, who may, in the final analysis, be the same people. Debts owed to international investors outside the country, however, would not be self-liquidating.

There is currently much discussion in Congress about a constitutional amendment to require a blanced budget. Regaining control of the national debt will be a monumental task, but with interest payments on that debt increasing rapidly, many feel our policy makers must curb government spending. The balanced budget amendment has been proposed as one way to require the federal government to spend less than it receives. The key to success in balancing the federal budget will be courage and conviction among the policy makers as they make the decisions to limit spending.

In summary, it can be said that fiscal policy is concerned with the taxing and spending goals of our government. The three main objectives are full employment, price stability, and economic growth. Moreover, built-in stabilizers are intended to keep the economy on an even keel. As disturbing forces develop, however, discretionary actions may be necessary to bring the economy back on course. These actions are regarded by economists as efforts to "fine-tune" the economy.

Inflation and Deflation

Money by itself cannot feed, clothe, or shelter us. But, with the value money represents, we can obtain necessary goods and services. Thus, there is a flow of money in exchange for a flow of goods and services. If these flows continue with a recognized balance, or stability, over time, it can be said that the relationship of money to goods is satisfactory and that the economy is working well.

Inflation occurs when the money supply expands while the supply of goods and services remains the same or expands at a slower rate, thus resulting in an increase in the prices of goods and services. As prices increase, money reduces in value because it buys less. For example, you may pay $50 for a pair of shoes that cost $25 some years ago or $200 for a coat that once cost $100. The dollars used to buy these articles today are worth only half as much as before; twice as many dollars are needed to have the same purchasing power. As prices rise, the dollar's purchasing power falls. The **rate of inflation**, then, is the percentage change in the price level from one year to the next.

The economic welfare of each citizen is tied to the nation's fiscal policy and monetary policy, specifically regarding inflation. Savings and life insurance are seriously diminished by inflation. Investments that grow in value as inflation grows are the best hedge against inflation. These include ownership of homes, properties, businesses, stocks, commodities, and valuable metals.

What are the causes of inflation? There are several contributing factors. No one alone can bear the total blame. Businesses often blame rising costs for the higher prices of their goods and services. Laborers clamor for higher wages because they see costs of things they buy constantly increasing. The ultimate result is a *cost-push* situation, which forces prices upward.

Recall the earlier definition of "inflation": an expansion of the money supply without a compensating increase in the supply of goods and services. Credit increases the money supply. For example, a woman borrows $1,000 from her bank. The bank does not give her cash but rather credits her checking account with $1,000, against which she can write checks. The bank has thus created or added $1,000 to the money supply. As the money is repaid to the bank, the deposit will be erased and the money supply reduced accordingly. Credit card usage has the same impact.

Although commercial banks can add to the money supply, they are not the only inflation culprit. Excessive government spending is also to blame. When the government spends more than it recovers through taxation, a deficit occurs. That deficit must be covered by borrowing. If the government raises money by selling securities to the banks, that adds to the money supply. The banks pay for the securities by crediting the government's "checking account," which in turn is paid out to meet government expenses.

Government deficits, to be sure, need not add to the money supply. If government securities are sold to the people or to investors instead of to banks, the money supply is not really affected, because savings are used to buy securities and, in this process, money is simply being transferred from one group to another.

Nevertheless, in time of war or other economic crisis, the government is called upon to spend more than it can gather through normal taxation. It then leans upon its unique ability to create money. In the past it resorted to the printing press and turned out new money. This led to serious financial difficulties for many people through depreciation of old currency. Now, to get money, the government must operate through the Federal Reserve System, issuing bonds in exchange for the fresh supplies of credit. This is more sophisticated than the old method, but the results are quite similar. Through this process, the government avails itself of a large volume of fresh money, which is distributed to various government agencies to be spent for the military, housing, education, entitlement programs, and other purposes. These added deficit dollars are essentially the same as newly printed dollars. They do not represent goods that have been produced, yet they can buy goods just as other dollars do. This additional money supply increases the demand for more goods, encouraging prices to rise. This is referred to as *demand-pull* inflation (usually caused by too much government spending).

A share of the blame for inflation must also go to labor unions and businesses that demand unreasonable wages and profits. Labor unions insisting on pay raises that exceed the increased output per hour of labor attempt to get something for nothing, with consumers paying for the raises in the form of higher prices. Likewise, businesses taking unrealistically high profits push prices up. Both these influences lead to cost-push inflation.

Controls on inflation must inevitably focus on government spending, wage increases, and pricing policies of large industries. Reducing government spending and increasing taxes tend to reduce the money supply, bringing it into balance with the supply of goods. Holding wages and prices at a stable level or in line with productivity is intended to allow the supply of goods to catch up to the supply of money in circulation.

The reverse of inflation is called **deflation.** Instead of a rise in consumer spending, prices, money supply, and credit, there is a sharp decrease in consumer spending, a declining price level, a tight money supply, and reduction in credit.

Devaluation

A reduction in the value of a country's currency in relation to another currency is called **devaluation.** Limiting this discussion to the American dollar, it can be stated that the dollar is recognized as good international currency. As long as it retains its value, or normal purchasing power, international traders remain content. If, because of inflation, the dollar loses value, confidence in the ability of the United States to redeem dollar holdings in goods at normal prices becomes shaken.

A declared devaluation of 10 percent would mean the value of a dollar would be reduced 10 percent. Imported goods formerly bought for $1.00 would now cost $1.10. On the other hand, U.S. exports would now cost 10 percent less, only $.90 for a good formerly priced at $1.00. In effect, devaluation of the U.S. dollar results in lower prices for U.S. goods and services exported to other nations and in higher prices for goods and services imported from abroad.

In small countries that have limited international trade, a devaluation of currency may occur frequently with little repercussion in international finance circles, but it is quite another thing for a nation like the United States to consider frequent devaluation. In fact, it is theorized that devaluation is a relative thing and that after the initial upheaval, international currencies would be adjusted to arrive at the same relative position that prevailed originally.

Coordination of Monetary Policy and Fiscal Policy

Monetary policy and fiscal policy need to be well coordinated for a healthy economy. If monetary authorities are disposed to control inflation, fiscal authorities should not try to undo their work, and vice versa. Fiscal policy that involves raising taxes is not likely to be attractive to politicians and thus is usually enacted long after need has been demonstrated. Repeal of tax legislation, however, is likely to occur ahead of need. Unbalanced budgets are also a fiscal plague during periods of high economic activity. Budget-deficit psychology is commonplace and viewed as acceptable, regardless of the level of economic activity.

Monetary policy, on the other hand, has less political orientation than fiscal policy. Monetary policy is more apt to be timely, although its overall effectiveness, in comparison with fiscal policy, is of some debate. Unfortunately, over-reliance on monetary policy becomes necessary when fiscal policy fails, such as incurring super-deficits in a fully employed economy or fighting wars without increasing taxes. Thus, monetary policy sometimes is called upon to rectify mistakes of fiscal policy, which compounds the economic trouble rather than lessening it.

It appears that greater rapidity in adjusting fiscal policy is needed. Some have suggested, for example, that federal income tax rates be made variable, tied to economic indicators or indexes, to keep the economy on a more stable course with less inflation or deflation (Table 5–1).

One suggestion regarding monetary policy would be to maintain a stable rate of growth in the money supply (say, 4 percent) instead of swinging from zero growth to growth rates of as high as 12 percent. A stable growth rate would reduce economic instability and allow the rate of inflation to be moderated. Businesses and consumers would know more what to expect of monetary policy authorities.

Consult Chapter 26 for further discussion of monetary policy and fiscal policy.

Table 5-1

Counter-Cyclical Policy Actions

Type of Policy Instrument	Inflation	Deflation
Monetary Policy:	Restraint	Ease
Open-market operations	Sell securities	Buy securities
Discount rate	Raise	Lower
Reserve requirements	Raise	Lower
Selective measures	Strengthen	Relax
Fiscal Policy:		
Expenditures	Restrain	Expand
Revenue	Rise	Decline
Automatic response	Rise	Decline
Tax rates	Increase	Decrease
Budget	Surplus	Deficit
Debt Management:		
Maturity of debt	Lengthen	Shorten
Types of securities:		
Short-term	Restrict sales	Expand sales
Long-term	Expand sales	Restrict sales

Topics for Discussion

1. When is barter more important than money?
2. Why is it necessary to have both coins and paper money in our monetary system?
3. What is the purpose of the Federal Reserve System?
4. What are the principal functions of the Federal Reserve System?
5. Differentiate between fiscal policy and monetary policy.
6. Why is fiscal policy more difficult to change than monetary policy?
7. What is inflation? Who gains from it, and who loses?

8. What is deflation? How does it relate to the general price level?
9. How does devaluation affect the domestic economy?
10. Discuss the need for closer coordination of monetary policy and fiscal policy.

Problem Assignment

Prepare a report discussing the announced goals of U.S. monetary or fiscal policy and the techniques used to achieve those goals.

Recommended Readings

Colander, David C. *Economics,* 2nd ed. Chicago: Richard D. Irwin, Inc., 1995. Ch. 11, 13, 14, 15.

Degen, Robert A. *The American Monetary System: A Concise Survey of Its Evolution Since 1896.* New York: Free Press, 1987.

Eisner, Robert. "Good and Bad Deficits: Views of a Liberal Keynesian," *Choices,* 8(1):6, 9 (1993).

Figgie, Harry E., Jr. "Ammunition for the Deficit War: A Doomsayer Speaks Out," *Choices,* 8(1):5, 8 (1993).

Golob, John E. "Does Inflation Uncertainty Increase with Inflation?" *Economic Review,* 79(3):27–38 (1994).

Hall, Robert E., and John B. Taylor. *Macroeconomics: Theory, Performance, and Practice,* 3rd ed. New York: W. W. Norton & Co., Inc. 1991.

Meltzer, Allan H. "The Deficit: A Monetarist's Perspective," *Choices,* 8(1):7–9 (1993).

Schaub, James D., and Daniel A. Sumner. "The Deficit and Agriculture," *Choices,* 8(1):10–11, 32 (1993).

Skousen, Mark. *Economics on Trial: Lies, Myths, and Realities.* Burr Ridge, IL: Irwin Professional Publishing, 1991.

Timberlake, Richard H. *Monetary Policy in the United States: An Intellectual and Institutional History.* Chicago: The University of Chicago Press. 1993.

6

Macro-profile of the U.S. Economy

To understand the macro-aspects of the U.S. economy better, we begin with the term **gross national product** (GNP). GNP is the total dollar value of goods and services produced in this country during a given period of time, usually one year. GNP is a dollar measure of the value of all goods and services sold to their final purchaser. A closely related term, **gross domestic product** (GDP), omits net earnings from the rest of the world and measures only output produced by factors in the United States.

The value of final items is recorded in the GNP and GDP, not the intermediary stages or sales of products or services to first buyers. For example, the value of a $500 suit sold at retail is recorded as $500, from which, of course, the cotton or wool producer, assembler, textile manufacturer, wholesaler, distributor, and retailer must share. Likewise, for services rendered, the value is recorded at their final point of transfer. Such services include laundry and dry cleaning; repairs; personal care; medical, dental, and legal services; and many others.

The Bureau of Economic Analysis in the U.S. Department of Commerce monitors the GNP and GDP of the United States.

Obviously, GDP will vary from year to year because of at least two factors: (1) variation in the volume of goods and services offered for final sale and (2) variation in prices paid at the final purchase for goods and services. Increases in GDP from one year to the next must be *deflated,* in terms of changes in the value of the dollar, to obtain the "net" increase or change in the GDP. Sometimes an increase in the GDP may reflect only a change in the price level and not an actual increase in output of goods and services.

GDP can also be expressed on a per capita basis, representing average productivity per person. However, this average does not imply that there is an equitable distribution of productivity, hence income, among the population.

68 / ECONOMICS: APPLICATIONS TO AGRICULTURE AND AGRIBUSINESS

Productivity and income may be concentrated in the top segment of society while very little economic wealth may be present in the lowest segment.

Consequently, the GDP has limitations. It does not measure nonmarket goods and services, improved product quality, distribution equity, or objective pursuit of leisure.

Purchasers of the Gross Domestic Product

GDP represents the dollar value of all goods and services purchased by final consumers. These consumers can be classified into four principal categories: (1) individual consumers, (2) business investors, (3) government (local, state, and federal), and (4) foreign purchasers. Expenditures made by each of these four groups are about as follows in accounting for the GDP:

Category	Classification of Expenditures	Estimated Percentage of GDP
1. Individual consumers	Personal consumption, all items	67.6
2. Business investors	Gross private domestic investment	15.1
3. Government	Government purchases of goods and services	19.0
4. Foreign purchasers	Net exports of goods and services (exports-imports)	–1.7
	Total	100.0

On average, about 68 percent of total annual GDP is accounted for by personal consumption expenditures, slightly over 15 percent by gross private domestic investment, and 19 percent by government. Imports have exceeded exports, reducing GDP by 1.7 percent.

Individual Consumer Purchases

The 67.6 percent of GDP accounted for by individual consumer purchases may be further classified as follows:

Item	Percent of Personal Consumption
Food	14.8
Housing	13.4
Medical care	11.8
Household operation	5.7
Motor vehicles and parts	5.4
Clothing and shoes	5.2
Furniture and household equipment	4.6
Transportation	3.5
Gasoline and oil	2.5
Other	33.1
Total	100.0

Business Investor Purchases

Gross private domestic investment represents 15.1 percent of the GDP. Expenditures within the gross private domestic investment category are divided about as follows:

Item		Percent of Gross Private Domestic Investment
Consumption of fixed capital		74.0
Nonresidential investment		13.3
Structures	5.5	
Producers' durable equipment	7.8	
Residential		11.1
Change in business inventories		1.6
Total		100.0

Taken together, nonresidential structures and the residential category constitute new construction, including new housing, factory buildings, shopping centers, and public utility plants. Producers' durable equipment includes such items as machinery and equipment, machine tools, trucks, locomotives, etc. Business inventories consist of raw materials and semi-finished and finished goods in inventory at industrial, farm, wholesale, and retail firms.

Government Purchases

Purchases by local, state, and federal governments normally represent about 19 percent of the total annual GDP. These purchases are divided about as follows:

Item	Percent of Government Purchases
Federal government	41.7
National defense	31.0
Other	10.7
State and local	58.3
Total	100.0

Net Exports of Goods and Services

Goods and services imported into the United States have exceeded goods and services exported, resulting in a slight reduction of 1.7 percent in GDP.

Receivers of Final Expenditures: The National Income

The expenditures of the four groups of consumers just discussed become the income of other groups in the U.S. economy. Expenditures are always offset by becoming income to other parties. Thus, expenditures will equal income, and vice versa. National income (NI) represents the annual earnings derived from the production of final goods and services.

The components of national income are as follows:

		Estimated Percentage of	
Component	Description	NI	GDP
Compensation of employees	Wages, salaries, and benefits paid to all employees	73.6	59.5
Proprietors' income	Earnings of sole proprietors and partnerships, including independent professional persons and farmers	8.2	6.6
Corporate profits	Earnings of corporations (before taxes)	8.3	6.7
Net interest	Net interest payments to individuals from all sources except government	10.1	8.2
Rental income of persons	Net rental income	–0.2	–0.2
	Total	100.0	80.8

Compensation of Employees

The 73.6 percent of national income represented by wages, salaries, and benefits paid annually to all employees in the United States may be further analyzed as follows:

	Estimated Percentage of	
Item	Compensation of Employees	NI
Commodity-producing industries	22.9	16.8
Distributive industries	19.4	14.3
Service industries	25.4	18.7
Government	15.6	11.5
Supplements to wages and salaries	16.7	12.3
Total	100.0	73.6

Supplements to wages and salaries include employer contributions for social insurance and to private pension, health, and welfare funds.

Proprietors' Income

Proprietors' income accounts for 8.2 percent of the total annual national income and is comprised as follows:

	Estimated Percentage of	
Item	Proprietors' Income	NI
Farm net income	10.6	0.9
Nonfarm net income	89.4	7.3
Total	100.0	8.2

These earnings are gross receipts minus expenses, with capital consumption adjustments.

Corporate Profits

Corporate profits before taxes represent 8.3 percent of national income. They are comprised as follows:

	Estimated Percentage of	
Item	Corporate Profits	NI
Profits tax liability	37.5	3.1
Dividends	38.0	3.2
Undistributed profits	20.6	1.7
Inventory valuation adjustment	−3.9	−0.3
Capital consumption adjustment	7.7	0.6
Total	100.0	8.3

Net Interest

Net interest income represents 10.1 percent of national income, which for the most part covers interest payments made to individuals from all sources except government. It includes items such as the interest paid on various savings investments in banks, savings and loan associations, credit unions, and other financial institutions.

Net Rental Income

Net rental income, which consists of receipts minus expenses, is actually a slight draw on national income of about 0.2 percent. Three kinds of rental income are included: (1) cash rent from house, apartment, and condominium rentals; (2) royalty income from patents, copyrights, and rights to natural resources; and (3) imputed rent, or the value of housing services to homeowners owning their own residences.

Residual Between National Income and GDP

National income is about 80 percent of GDP. In 1991, GDP was $5,677.5 billion, while national income was $4,544.2 billion. A logical question at this point is, What comprises the remaining 20 percent of the GDP?

Let's begin the explanation by stating that both the national income and the gross domestic product measure the output of final goods and services. They do so, however, from different points of view. National income, as we have seen, measures it from the earnings viewpoint; gross domestic product, from the market value or expenditure viewpoint.

National income is the total of the annual earnings derived from the production of these final goods and services. Gross domestic product, in its measurement of the market value of final goods and services, includes national income plus two main non-income components: (1) consumption of fixed capital and (2) indirect business taxes. The first, consumption of fixed capital, represents about 11 percent of GDP, just over half of the residual between national income and GDP. The second, indirect business taxes, represents 8.4 percent of GDP, just under half of the residual between national income and GDP. There are other non-income components—business transfer payments, for example—but they are of minor importance.

Consumption of Fixed Capital

The first of these non-income cost components is consumption of fixed capital, consisting of (1) depreciation charges on business assets and (2) depreciation charges on owner-occupied dwellings.

Depreciation represents a decline in the value of an asset brought about by wear and tear or by obsolescence. Businesses should regularly (i.e., each year) set aside funds to have adequate monies available for the replacement, renewal, or repair of depreciated assets as they are used up or become obsolete in the production process. The amount set aside in each period is the amount of depreciation charged in that period. Since these assets usually last a number of years, the depreciation charges enable the businesses to match the amount of the machinery and equipment used up in a production period to the production period in which it was used. The sum total of these annual depreciation allowances becomes part of consumption of fixed capital in the GDP.

Depreciation charges on owner-occupied dwellings are based on the assumption that home ownership is a business that produces housing services and therefore is entitled also to depreciation allowances.

Indirect Business Taxes

The second non-income cost component, which accounts for just under half the difference between gross domestic product and national income, is indirect business taxes.

Within the value of final goods as we buy them are certain taxes that the producer has paid to the government before offering the items for sale. For example, the makers of automobiles have paid a federal tax on each car before it is offered for sale. Such taxes are sometimes called "excise taxes." The excise tax stamp affixed to a bottle of liquor indicates that the tax has already been paid by the manufacturer. This item of indirect business taxes includes, among other things, sales taxes paid by consumers and property taxes. On the other hand, income taxes paid by corporations would be classified as direct business taxes.

Recapitulation

Thus far, we have indicated that the gross domestic product may be adjusted for changes in the price level, in which case it becomes known as the "real" or "deflated" domestic product.

National income comprises about 80 percent of the GDP, with the majority of the remaining 20 percent being divided between consumption of fixed capital and indirect business taxes.

Gross national product (GNP) equals GDP, plus receipts of income by U.S. residents from the rest of the world, minus payments of income earned in the United States by foreign residents. If consumption of fixed capital alone is subtracted from the GNP, we obtain the net national product (NNP).

Personal Income

Personal income can be derived by making the following series of adjustments to national income:

Subtracting

- corporate profits, with inventory valuation and capital consumption adjustments
- net interest
- contributions for social insurance
- wage accruals less disbursements

Then *adding*

- personal interest and dividend income
- government and business transfer payments to persons

Specific sources of personal income include:

Source		Percent of Personal Income
Wage and salary disbursements		58.2
Commodity-producing industries	15.3	
Distributive industries	13.4	
Service industries	18.3	
Government	11.2	
Other labor income		6.0
Proprietor's income		7.6
Farm	0.7	
Nonfarm	6.9	
Net rental income		−0.2
Personal dividend income		2.8
Personal interest income		14.5
Transfer payments		16.0
Personal contributions for social insurance		−4.9
Total		100.0

Transfer payments include, among others, old-age, survivors', disability, and health insurance benefits; government unemployment insurance benefits; veterans' benefits; government employees' retirement benefits; and aid to families with dependent children. These transfer payments are received by individuals from the government, for which no services are rendered.

Disposable Personal Income

The personal income accruing to U.S. individuals is not fully disposable or spendable because personal income tax and nontax payments, about 12.8 percent of personal income, must be deducted. Just over 76 percent of these personal income tax and nontax payments go to the federal government, and the remainder goes to state and local governments. In addition, just over 4 percent of personal income is saved, while about 2.3 percent of personal income is allocated to interest payments by persons. The remainder of personal income is available for personal consumption expenditures, which involve all types of goods and services and represent what is termed *consumer spending* or *purchasing power*. This is equivalent annually to about 68.5 percent of GDP and 85.6 percent of NI.

Macroeconomic Equation

The macroeconomic flow equation may be stated as follows: **(1)** GNP – Consumption of Fixed Capital (CFC) = Net National Product (NNP); **(2)** NNP – Indirect Business Taxes (IBT) + Subsidies, less current surpluses of government enterprises (S) – Business Transfer Payments (BTP) = National Income (NI); **(3)** NI – Corporate Profits (CP) – Net Interest (I) – Social Insurance Contributions (SIC) + Government Transfer Payments (GTP) + Personal Interest Paid (PIP) + Personal Dividends Paid (PDP) + BTP = Personal Income (PI); **(4)** PI – Personal Taxes (PT) = Disposable Personal Income (DPI); **(5)** DPI – Personal Savings (PS) = Personal Consumer Expenditures (PCE); and **(6)** PCE + Gross Private Domestic Investment (GPDI) + Government Purchases (GP) + Net Exports over Imports (NE – I) = GNP.

U.S. Farmers and the GDP

Just over 15 percent of the total GDP can be attributed to the food and fiber system. Of this amount, processing and marketing of food and fiber account for about 12 percent; farm supply industries account for 1.5 percent; and agricultural production accounts for 1.4 percent.

As business investors, farmers play an important part in private domestic investment in terms of new construction, durable equipment purchases, and product inventories held on farms. Gross private investment by farmers may comprise between 12 and 16 percent of total private domestic investment.

A most important role of the agricultural sector is in the net exports of goods and services. The agricultural sector traditionally contributes toward a trade surplus, with the value of agricultural exports exceeding the value of agricultural imports.

Net farm income comprises only about 1 percent of the national income, while farm population comprises about 2 percent of the total population. There are many reasons for the farmers' relatively small share, including the following: (1) Farmers are in a most competitive position among themselves, which tends to drive their net incomes down relative to other economic groups who are not as competitive. (2) Farmers have difficulty adjusting supplies to market demand, which means that a relatively small oversupply will drive prices to low levels, reducing net farm income. (3) The market for agricultural products is relatively inelastic with regards to price and income. (4) Most productive resources used in farming have few, if any, alternative uses. As a result, these resources remain in production much longer than would otherwise be the case. These points are discussed in greater detail in subsequent chapters.

Indicators of Agricultural Performance

Three overall indicators attest to agriculture's important role in U.S. economic growth during the first eight decades of the 1900s: (1) the percentage of the gross national product accounted for by agricultural products, (2) the percentage of national wealth required to meet the nation's food and fiber needs, and (3) the percentage of the total labor force employed in agriculture.

In 1900, the agricultural component accounted for 23 percent of the gross national product; however, in recent years it has accounted for about 1 percent.

The percentage of national productive wealth represented by agricultural assets gives a similar picture. The nation's total wealth in 1900 is estimated to have been $163 billion in 1929 prices. Of that amount, about 27 percent consisted of farm lands, buildings, and crop and livestock inventories. In more recent years, only about 5 percent of national productive wealth is accounted for by agricultural assets.

Even more significant has been the decline in the farm population and in the percentage of the nation's labor force employed in agriculture. The rural farm population declined from 35 percent of the total in 1910 to about 2 percent in more recent years. The percentage of the labor force employed in agriculture, a good index of economic growth, dropped from 38 percent in 1900 to just over 2 percent in more recent years.

As shown in Table 6-1, there is generally an inverse relationship between GDP per capita and the share of agriculture in total GDP. Nations whose agricultural sector comprises a smaller share of GDP, generally those with a smaller percentage of farm population, have considerably higher GDP per capita. As fewer people are required to feed, clothe, and house the population, relatively more people can engage in other productive endeavors, providing a higher standard of living for everyone. Developing nations should, therefore, give priority to agricultural development before they can make substantial advances in other sectors.

Table 6–1

Relationship Between GDP per Capita and Share of Agriculture in GDP

Country	GDP per Capita	Share of Agriculture in GDP
United States	22,560	2
Sweden	25,490	3
France	20,600	3
Australia	16,590	3
Canada	21,260	4
Mexico	2,870	9
Brazil	2,920	10
Thailand	1,580	12
Colombia	1,280	16
Honduras	570	23
Pakistan	400	26
Ethiopia	120	42
Mali	280	44
Mozambique	70	65

Source: The World Bank, *The World Bank Atlas* (Washington, DC: The World Bank, 1992), pp. 18–19.

U.S. agriculture's apparent declining share of the GDP should not be cause for alarm. Rather, it is a sign that with an efficient agricultural economy, more resources are available for deployment to production of other goods and services, thus providing a higher standard of living for all citizens.

Moving Toward a Larger GDP

It is expected that within the next few years, the GDP of the United States will reach $6 trillion annually. Some of the steps that have been suggested to insure that this growth in the GDP is attained and exceeded are as follows:

1. *Increase the education, training, and retraining opportunities for all U.S. citizens* to enable them to become more productive in their jobs and occupations. This will result in more goods and services from a given amount of effort expended. Increased human productivity may also come from better schools, colleges, vocational training, and on-the-job training. Special problem areas that may need more attention include reducing school dropout rates, providing job aids for persons with disabilities, and retraining persons with obsolete jobs and skills. A better-trained work force would lead to a greater supply of capable business managers, technicians, and entrepreneurs who have ability to develop resources and exploit technology effectively.

2. *Expand job opportunities* by developing new enterprises, industries, plants, and offices; expanding research and development; creating more part-time jobs for students, older persons, and persons with disabilities; and relocating jobs from congested to more sparsely settled areas.

3. *Increase competition in the free enterprise economy* to allow for greater mobility and flexibility in the use of resources and the exchange of ideas. This may call for attacks on monopoly, in both business and labor; lower tariffs; and full and open release of knowledge obtained through public funds.

4. *Decrease the dependence of people on public welfare and other entitlement programs* by increasing reliance on jobs and developing a better partnership between government and private enterprise in terms of social security, pensions, and retirement plans.

5. *Improve the efficiency in government operations at the local, state, and federal levels* through better long-range planning, consolidation of overlapping agencies and programs, wise use of tax funds according to priority budgets and needs, and a better division of tax receipts among various levels of government.

6. *Strengthen the family unit for social and economic improvement.* Emphasis on the importance of the family unit as the basis for a properly functioning society, with corresponding measures of support enhancing and protecting the family, would, in general, improve society. This would lead to less unemployment and welfare; more creative expression; and fewer consumer problems, including those of space, pollution, health, and education. The result could be a higher rate of savings, which would sustain a higher rate of domestic investment for producing goods and services.

Inflation, Unemployment, and the Gross National Product

Policies designed to control inflation and/or unemployment or to expand the "real" gross national product may be viewed in terms of Figure 6–1. At 95 percent employment, which is considered here as full employment, the potential GNP is realized. This requires an aggregate demand of the level A. If aggregate demand rises to level B, inflation results, because the economy is already at its full potential; thus, inflation will be of the amount P_0 to P_1. If aggregate demand falls to level C, employment will fall from 95 percent to 90 percent, and the full potential of the economy will not be realized by that amount of difference between 90 and 95 percent, with the latter being considered as full employment.

When unemployment increases simultaneously with inflation, it is likely that unemployment results not only from a deficiency in aggregate demand but also from "structural" or "institutional" causes. Businesses, instead of hiring new workers, tend to pay overtime to the existing work force and thus save on pension funds, life and health insurance, etc. Also, some persons who are unemployed do not have the necessary skills to fill available jobs. In addition, welfare payments and food stamps, among other entitlements, tend to discourage job seeking and job acceptance, maintaining the unemployed pool. Another possibility is that data on unemployment may be in error or exaggerated.

Potential GNP and the inflation and unemployment gaps

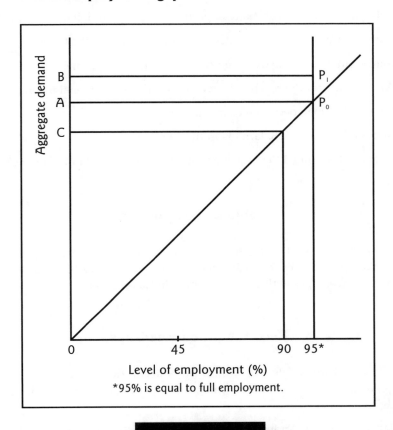

Figure 6–1

In another situation, when price inflation occurs while aggregate demand is below the potential GNP, the cause may be cost-push inflation, with organized labor elevating wage rates faster than worker output or productivity.

Another reason for unemployment is that business firms reduce output instead of reducing prices when economic activity slows down. This is due in part to the way nonagricultural firms compete among themselves; that is, they fail to compete on price. Labor unions, likewise, would rather hold wage levels and take unemployment rather than reduce wage levels and maintain employment. Agriculture, on the other hand, takes lower prices and maintains or even increases output.

Topics for Discussion

1. Define the terms "gross national product" and "gross domestic product."
2. Name and discuss the four categories of purchasers of the GDP.
3. Name and discuss the five components of "national income."
4. What comprises the residual between the GDP and national income?
5. Discuss the "macroeconomic equation."
6. Discuss the role of U.S. farmers in the GDP.
7. How can the United States achieve an even larger GDP in the future?

Problem Assignment

Obtain the latest data on the GDP for the United States. Calculate the appropriate percentages for each category and subcategory of the GDP. Compare your percentages with those presented in this chapter.

Recommended Readings

Barry, Peter J., Paul N. Ellinger, John A. Hopkin, and C. B. Baker. *Financial Management in Agriculture,* 5th ed. Danville, IL: Interstate Publishers, Inc., 1995. Ch. 18.

Bradley, Edward, Jay Anderson, and Warren Trock. "Outlook for U.S. Agriculture Under Alternative Macroeconomics Policy Scenarios." WREP 102. Department of Agricultural Economics, University of Wyoming, 1986.

Colander, David C. *Economics,* 2nd ed. Chicago: Richard D. Irwin, Inc., 1995. Ch. 8, 9, 10, 11.

Gwartney, James D., and Richard L. Stroup. *Economics: Private and Public Choices,* 7th ed. Orlando, FL: The Dryden Press, 1995. Ch. 5, 6, 7.

Hildreth, R. J., Kenneth R. Krause, and Paul E. Nelson, Jr. "Organization and Control of the U.S. Food and Fiber Sector," *American Journal of Agricultural Economics,* 55(5):851–859 (1973).

Kitchen, John, and David Orden. *Effects of Fiscal Policy on Agriculture and the Rural Economy.* AGES 9131. Washington, DC: Economic Research Service, USDA, 1991.

Keynes, J. M. *The General Theory of Employment, Interest and Money.* New York: Harcourt, Brace & Co., 1965.

PART TWO

7 Human Resources in Agriculture

8 Natural Resources

9 Capital Resources

10 Selected Characteristics of U.S. Farming Operations

7

Human Resources in Agriculture

American society continues to become more and more urbanized and decidedly less rural. This trend alone has important implications regarding housing, education, employment, and family life, among other considerations.

In discussing human resources, we refer to different sectors of the economy classified for our purposes as follows:

1. *Farm,* which includes rural residents actively engaged in farming and selling $1,000 or more of agricultural products annually.

2. *Nonmetro,* which includes rural residents engaged in, and obtaining the major source of their income from, nonfarming pursuits. Rural residents include people living in towns and cities with populations of less than 2,500.

3. *Urban,* which includes all residents of towns and cities with populations of 2,500 or more. Urban residents may be further classified according to those living in smaller towns and cities with populations of less than 100,000 but more than 2,500 and those living in metropolitan centers with populations of 100,000 or more. Suburban residents may be persons living in concentrated nonmetro areas that are unincorporated or perhaps in suburban towns of less than 2,500 that surround large cities.

People who live in large cities and their suburbs and in towns of at least 2,500 residents account for about 75 percent of the total population. Rural people in the open country or in towns with less than 2,500 residents numbered about 61.6 million in 1990. The farm population comprises less than 2 percent of the total U.S. population, compared with over 3 percent in the 1970s.

About 25 of every 100 persons in the United States are rural residents. Of these 25 rural persons, only about 2 are farm residents, while 23 are nonmetro residents. Although it is expected that the number of farm residents will continue to decline, future losses in farm population will be small compared to

U.S., rural, and rural farm populations, 1900–1988

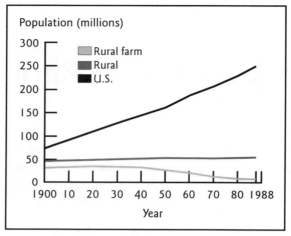

Source: Economic Research Service, USDA

Figure 7–1

those observed in recent decades. Since farm population represents such a small proportion of the total rural population, future changes in farm population will only slightly affect overall rural population trends, except in predominantly agricultural areas.

Why have these trends in residence taken place? First, technological advances, specifically increased mechanization, greatly reduced the work force required in agriculture. Some of those displaced stay in rural areas but work in other jobs, while many, especially rural youth, small farm owners, and minority farm workers, leave for urban areas. Second, the consolidation of small farms into larger units means fewer farms and a reduction in the number of persons farming. Third, agribusinesses located in small towns and industry in larger cities employ more and more rural people in nonfarm work. Fourth, relatively few children of farm families are staying in farming because of limited opportunities. They either stay in rural areas as nonfarm workers or migrate to urban areas. Thus, the farm population continues to decline while the nonmetro and urban population continues to increase.

These trends pose certain social and economic problems. First, farm workers released because of technological advances are often ill-prepared to work at nonfarm jobs. Many are middle-aged or elderly, poorly educated, and unskilled. About 55 percent of hired farm workers in 1990 had less than 11 years of education.

Second, rural residents commuting to city employment need and demand certain services (water, gas, sewerage, fire protection, schools, roads, etc.) but contribute relatively little toward financing these services. At the same time, farm residents are reluctant to tax themselves sufficiently to provide such services. Zoning problems in rural areas also arise as restaurants and drinking establishments, service stations, and other strip developments are built indiscriminately in areas leading into the cities.

Third, agribusinesses must employ well-trained technical and sales people, but rural farm areas are often unable to provide them; thus, nonfarm residents are recruited for agribusiness jobs.

Fourth, many children from farm families cannot be accommodated on their home farms, so they seek nonfarm work. Unfortunately, small rural schools, in many cases, have been unable to train youngsters adequately for nonfarm vocations. Farm youth often face limited opportunities, either in rural areas at low-paying jobs or in urban areas at a decided disadvantage in competing with others for jobs.

Those who remain as farm residents and farm workers face a more competitive role as agriculture continues to become more technical, mechanized, and business-oriented. Farming will require increasingly large amounts of fixed capital, production credit, and technical ability. Corporate farms and contract farming operations controlled by corporations are becoming more prominent as technology and efficiency call for large, integrated operations. The number of corporate farms, other than family-held corporations, increased almost 30 percent from 1987 to 1992. Farming as a "way of life" becomes more a "way of making money." The traditional concept of farming and farm life gives way rapidly to business and technical efficiency.

Let's briefly discuss some of the socioeconomic characteristics of rural people to determine some of the changes that are occurring and their consequences.

Characteristics of Rural People

Age

The nonmetro population contains a higher proportion of children and older persons and a lower proportion of young adults between the ages of 20

88 / ECONOMICS: APPLICATIONS TO AGRICULTURE AND AGRIBUSINESS

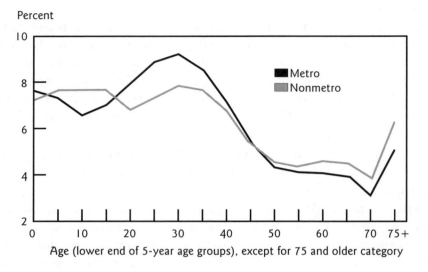

Figure 7-2

and 46 than the metro population. Active farmers are generally older than nonmetro workers, in general, because many young adults have migrated to urban areas. The average age of farm operators was 53.3 years in 1992, up from 52 years in 1987. Older farmers remain in agriculture due to either personal preferences or lack of better alternatives.

Increases in life expectancy over the past 50 years and the aging of the large population segment born in the 1920s increased the proportion of elderly between 1970 and 1990. The percentage of the population over 75 rose dramatically, especially in nonmetro areas. Retirement migration to nonmetro areas, coupled with historically high levels of nonmetro outmigration of young adults and their children, placed a higher proportion of older people in nonmetro locations. The percentage of nonmetro population over age 65 was 15 percent in 1990, compared to 12 percent in metro areas.

Gender

From 1970 to 1990, participation in the labor force by women increased in all regions of the country and at all levels of urban influence. In contrast, participation by men fell in all regions and at most levels of urban influence. The proportion of nonmetro women working or looking for work increased from 37.3 percent in 1970 to 51.8 percent in 1990. Labor force participation for metro women increased from 42.6 percent to 58.2 percent over the same period.

From 1970 to 1990, participation by nonmetro men in the labor force fell slightly, from 72 to 69.6 percent. The fall can be attributed partially to earlier retirement and partially to the fact that men laid off during middle age have a difficult time finding new jobs. Labor force participation for men in metro areas showed a similar decline over the period 1970 to 1990, falling from 78 to 75.9 percent.

In the farm population, there are 112.7 males for every 100 females. In contrast, for the total resident population in the United States, there are only 95.2 males for every 100 females. Young farm women, who tend to leave farms sooner than young men, have a relatively better chance of finding work and adjust to urban living better than farm men. Conversely, there are more job opportunities for rural males than for females. For example, males make up 83 percent of hired farm workers.

Just under 8 percent of all farm operators in 1992 were female. However, the percentage of female farm operators has been increasing over the last several decades. Only 5.4 percent of farm operators were female in 1970.

Race

In 1990, 8.7 million nonmetro residents belonged to one of four minority groups: Blacks, Hispanics, Asians, and Native Americans. Blacks made up close to two-thirds of the minority nonmetro population in 1980, but their share declined as other groups grew much faster during the 1980s. Minorities constituted only 14 percent of the total nonmetro population in 1980, but they accounted for 50 percent of the people added during the 1980s.

In 1992, about 2.3 percent of all farm operators were minorities, down slightly from 2.4 percent in 1982. Blacks constituted about 43.3 percent of minority farm operators in 1992.

Blacks constitute about 12 percent of the total population of the United States. Of the Black population, only 14.4 percent live in nonmetro areas. Only 2 percent of Blacks live in farming-dependent counties.

There were only about 19,000 Black farm operators in the United States in 1992—about 2 percent as many as in 1920, when there were more than 900,000 Black farmers. After 1920, Black families migrated at a dramatic rate out of farming. Many sought a better life in cities, while many were pushed out by the near demise of cotton and tobacco tenant farming. Increased mechanization in agriculture, particularly in the South, also caused many Blacks to migrate to urban areas, where they comprise an increasing proportion of the population.

Minorities in farm areas and unskilled farm workers generally have been particularly disadvantaged by the federal minimum wage law, which has fostered mechanization, causing their layoffs. They have also been affected by federal farm support programs, which, for some commodities, have restricted farm production, reducing labor requirements. It is likely that some federal farm and wage policies have actually contributed to the mass departure of minorities, Blacks in particular, from rural areas.

Dependents

Historically, rural couples have had more children than urban couples. For example, in 1970, there was an average of 3.41 children per woman for women ages 34 to 44 in rural areas. For the same age group in urban areas, there was an average of 2.87 children per woman. In rural areas, the higher fertility sustained population size despite substantial outmigration.

However, the reproductive difference has declined substantially. In 1990, the average number of children per woman of age 35 to 44 in rural areas decreased to 2.27, compared to 1.90 in urban areas. With rural fertility converging toward the national average, population stability will be more difficult to sustain and the rural population will age more rapidly than in the past. Rural couples marry at an earlier age than urban ones, but that age has been increasing over time, mirroring patterns observed in metro areas. The relatively large number of older people in rural areas means more dependents must be supported by the productive adults. Not only must the productive adults rear and educate more children, but they must also be responsible for their elders and must shoulder this financial responsibility with less income than their urban counterparts.

The decreased fertility observed in nonmetro areas could reduce the pressure for young people to migrate from rural areas in future years. Less outmigration of youth could then result in more stable residential patterns and kinship ties. Less outmigration would also enable rural communities to capture the benefits of educating their youth rather than losing them to urban areas.

Often, a rural community educates a rural child through high school, only to have the youth leave the area upon graduation. The urban area, on the other hand, receives the rural emigrant at essentially no expense, ready to work, pay taxes, etc. This is perhaps one reason urban areas should be interested in helping rural areas bear part of the cost of educating rural children. The better educated and trained the rural children are, the better urban or rural citizens they shall make.

Education

Educational levels increased significantly between 1960 and 1990. The gap between the percentages of the population ages 25 and older with at least a high school diploma in nonmetro and metro areas narrowed over that period. However, the metro-nonmetro gap in the percentage of the population with at least a four-year college degree widened. About 13 percent of the nonmetro population 25 years and older has a college degree, up from 5.1 percent in 1960. In contrast, 22.5 percent of the same age group in the metro population has a college degree, up from 8.5 percent in 1960. This widening gap is partially due to the migration of better-educated nonmetro residents to metro areas to find jobs that more fully use their skills.

High school completion rates show the opposite trend from college completion rates, with the gap between metro and nonmetro rates narrowing between 1960 and 1990. The percentage of the nonmetro population ages 25 and up with a high school diploma increased from 34 percent in 1960 to just over 69 percent in 1990, a 35.2 percentage point increase. The percentage of the metro population, same age group, with a high school diploma increased from 43.3 percent in 1960 to 77 percent in 1990, an increase of 33.7 percentage points.

The patterns discussed above were also observed in farming-dependent counties. The percentage of the people ages 25 and up who completed high school increased 19.5 percentage points (from 36.9 to 56.4 percent) from 1970 to 1990, compared to an increase of 13.2 percentage points (from 41.7 to 54.9 percent) for the United States as a whole. However, the percentage of the population ages 25 and over in farming-dependent counties with a college degree increased only 5.5 percentage points (from 5.3 to 10.8 percent) compared to an increase of 9.6 percentage points (from 10.7 to 20.3 percent) for the United States as a whole.

The education level of hired farm workers is still lacking. While 75.2 percent of the U.S. population over age 25 has at least a high school education, only 44 percent of hired farm workers have completed high school. Almost 33 percent of hired farm workers have less than nine years of education.

The expense of leaving home to go to college is a serious obstacle to many young people, especially from lower-income groups living in rural areas. As we have already discussed, recent data indicate that the gap in higher education between metro and nonmetro areas is widening, not lessening. This could be explained in part by the increasing availability of city colleges and university branches and the mounting costs of going to college, which discourage rural residents with lower incomes.

These educational differences would not be of great concern if farming was still uncomplicated and labor intensive. But it is not. Farming requires and will continue to require individuals with good education if the farm economy is to prosper and serve the total U.S. economy efficiently.

Migration

Migration is undertaken primarily by young adults seeking to further their educations and/or enter the labor force. Thus, net migration (the number of immigrants to an area less the number of outmigrants) is closely tied to changing employment opportunities.

The number of persons moving out of nonmetro areas exceeded those moving in by more than 500,000 during the 1980s. This net migration caused the 1990 nonmetro population to drop 1 percent lower than what it would have been otherwise. The effect of this net migration loss varied at the county level. Many retirement-destination counties actually had increases in their population, while farming-and mining-dependent counties had significant population losses through outmigration.

It is recognized that migration from farm to nonfarm residence has been a necessary and logical process that must occur because there are more rural people than are actually needed in farming. However, it is not necessary that this migration lead mainly to large urban areas. Migration could, and perhaps should, be channeled more toward nonmetro areas, into smaller towns and cities. The main obstacle to this movement has been a lack of job opportunities in rural areas. However, with improved telecommunication technology, businesses may have the flexibility to locate in smaller, relatively remote towns. With the improved ability to communicate, other aspects of life-style environments become more important, including the quality of schools, the character of wide-open spaces, etc. Movements such as these will serve to improve the quality of life in rural areas as well through increased opportunities for rural people.

Employment

Labor force participation rates and number of persons in the labor force increased in nonmetro and metro areas between 1980 and 1990. These increases were due to more women working or looking for work. Women's labor force involvement accelerated in the 1980s as more women sought careers outside the home and also worked to maintain overall living standards. The largest increase in labor force participation was among women with children.

Metro labor force participation rates have exceeded nonmetro rates for men and women for decades. In part, relatively more people over age 65 and relatively more discouraged workers who think they cannot find jobs have kept nonmetro labor participation rates below metro rates. At 54.1 percent in 1970, the nonmetro labor force participation rate was 5.3 percentage points below the metro rate. The gap had widened to 6.3 percentage points in 1990, when the nonmetro labor participation rate was 60.4 percent. The small increase in the gap was partially due to a larger drop in nonmetro labor participation by men. A major contributor to the overall difference, though, is the higher proportion of retirees in nonmetro areas.

In nonmetro counties dependent on farming, the labor force participation rate increased from 52.6 percent in 1970 to 59.3 percent in 1990. The percentage of men in the labor force in these counties fell from 72 percent in 1970 to 69.4 percent in 1990. In contrast, the percentage of women increased dramatically, from 34.4 percent in 1970 to almost 50 percent in 1990.

Total employment increased almost 27 percent in farming-dependent counties from 1970 to 1990. The increase stems from growth in manufacturing and in service industries. Employment in farming and agriculture-related jobs fell almost 19 percent from 1970 to 1990 in farming-dependent counties.

Agriculture provides employment to many people. About 50 percent of the hours worked on farms are by farm operators, 34 percent by hired farm workers, and the remainder by unpaid farm workers—farm operator family members, for example.

The relative importance of hired farm workers has increased over the last 40 years. More mechanization and other technological improvements, together with greater off-farm employment opportunities, have reduced the number of people working on farms. However, the decline in operators and family members has been greater than the decline in hired farm workers. Hired farm workers made up only about 25 percent of the total farm employment in the 1950s, compared to about 34 percent in the 1980s.

The amount of hired labor required varies by size of farm. Larger farms will be more likely to need additional labor beyond what the operators and their families can provide. Labor requirements also vary by the commodity produced. Production of some crops—corn and soybeans, for example—is largely mechanized, lowering the need for hired labor. In contrast, some crops are labor intensive, requiring hand cultivation and harvesting—for example, fruits, vegetables, and other horticultural crops.

Underemployment is also an issue in rural communities. "Underemployment" refers to the relationship between a person's potential capacity for work and work actually done. Farmers and farm workers are not often employed year-round in their farm work. In general, specialized row-crop farming, especially of horticultural crops, leads to more underemployment than livestock farming. Many farm workers, both operators and hired workers, are not employed consistently during the year, although they may work long hours during peak periods, such as at harvest. For underemployed farm workers, part-time jobs are actively sought to supplement incomes.

The social and economic problems of migrant farm workers are especially distressing. Because of the nature of their work and the economic circumstances surrounding their employment, migrant farm workers lack economic security; have frequent moving costs; have poor schools, housing, recreation, and health care; and are often ineligible for certain economic benefits provided other workers.

Incomes

Through the late 1980s, rural median household income was about 25 percent below the urban income level. Rural median household income, which was $24,691 in 1991, has stayed between $24,500 and $24,700 (in 1991 dollars) since the late 1980s. The gap between rural and urban median incomes is wider than it was in the late 1970s and early 1980s. On a net income basis, however, the disparity is not quite so pronounced, because rural families may produce some of their food and may have lower housing costs and lower property taxes, among other favorable considerations.

Sources of income for average farm operator household

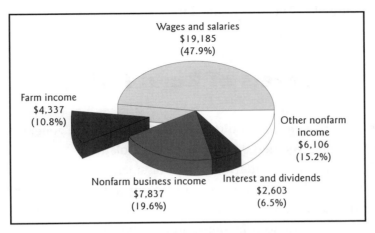

Source: *Agriculture Fact Book—1994*, USDA

Figure 7–3

Like many other U.S. households, farm operator households receive income from a variety of sources, one of which is farming. The 1992 average income of farm operator households was $40,068, comparable to the average U.S. household income. About 90 percent of all farm operator households receive some income from off-farm sources, with many operators spending most of their work effort in off-farm occupations. Off-farm income includes earned income, such as wages and salaries from off-farm jobs and net income from nonfarm businesses, and unearned income, such as interest and dividends, royalties, annuities, social security, Medicare, and income from other nonfarm sources.

Why is net farm income lower than other incomes for given units of productive resources such as land, labor, and capital? Several reasons are suggested: (1) Agriculture, because of its competitive nature, provides less income per work unit than other industries not as competitive. (2) Though improving, the work force in agriculture is not as well educated or trained. (3) Agriculture cannot provide full, year-round employment because of its seasonal, biological nature. (4) Farm prices, the principal determinant of farm income, are more variable than nonfarm prices and wages. (5) Farm family income, though comparable to nonfarm family income, has to be divided among more household members, resulting in lower per capita incomes. For example, if a farm family and a nonfarm family earn $40,000 each, on a per capita basis the farm family would probably average $10,000 (four in the family) and the nonfarm family $13,333 (three in the family).

Levels of Rural Living

Levels of rural living are influenced by at least three general factors: (1) income, (2) community inertia, and (3) relative isolation.

Insufficient income is often the result of small, uneconomic size of farm units; unproductive land resources; and lack of farm business ability on the part of the operator. Thus, insufficient income is associated with lower levels of rural living, as evidenced in substandard education, health, housing, and other services.

Community inertia is a reflection of conservatism and individuality, characteristic of traditional farmers. Thus, community inertia and organizational inexperience lead to a general lack of community planning, political action, and group effort in achieving higher levels of living by providing better schools, utilities, services, and the like.

Relative isolation leads to higher costs of public services, such as highways, schools, water, electricity, telephone, and waste disposal. Costs of sparsity arise because of high transport costs to serve a given number of people and a lower tax base available to finance needed services.

Rural Poverty

Poverty is as much a fact of life in rural America as it is in inner cities. Families are classified as being at, above, or below the poverty level by using the poverty index developed in 1964 and revised in 1969 and 1980. The poverty index is based solely on money income and does not reflect the fact that many low-income persons received noncash benefits, such as food stamps, Medicaid, and public housing.

The poverty rate in rural areas has been higher than in urban areas since 1959, when collection of poverty data using the official government definition began. In the early 1990s, the rural poverty rate was 16.1 percent, significantly higher than the 13.7 percent urban poverty rate.

Poverty rates, 1959–1991
Nonmetro-metro poverty differential narrowed over time, especially during the 1960s.

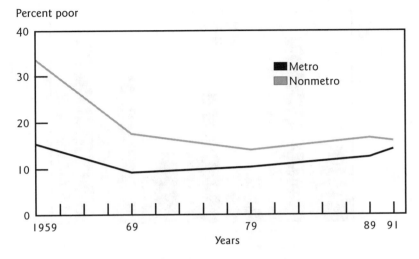

Source: *Agriculture Fact Book*—1994, USDA

Figure 7-4

Rural poverty is concentrated mainly in the Appalachian and Piedmont regions, the Ozarks, the Mississippi Delta, and the southwestern United States. Poverty in these areas is associated with small one-crop farms; elderly operators with a low level of education and training; large families; lack of year-round employment; poorer soils in some cases; a high proportion of seasonal farm labor; and minority racial and ethnic groups.

Rural poverty is often unobserved by the general public, since residents in poverty are dispersed and less visible, not concentrated as in inner cities. Politically, the rural poor are less effective because it is difficult for them to assemble, organize, and express their views. In contrast, the urban poor are in a better position to express their views to political groups. Hard-core rural poverty is likely to persist because the rural poor have little mobility, are strongly rooted to their rural institutions, and are not well equipped with education, occupational skills, or motivation to better themselves.

Numerous state and federal programs have been established to help the poor, including welfare assistance; Aid to Families with Dependent Children (AFDC); Medicare and Medicaid; food stamps; Women, Infants, Children

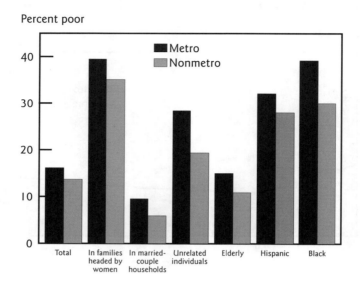

Poverty rates by population group, 1991
Nonmetro residence increased poverty risk for all groups.

Source: *Agriculture Fact Book*—1994, USDA

Figure 7–5

(WIC); school breakfast and lunch programs; and a variety of housing loan programs.

Rural Schools

Expenditures for education per rural student are generally lower than expenditures for education per urban student. This results mainly from the lack of rural revenue sources to support better education. There is an increasing tendency in many states to equalize the school tax burdens by channeling funds obtained statewide into certain poorer rural areas. In this way, urban areas share in the cost of educating rural children, many of whom will eventually migrate to urban areas, anyway.

Improving rural education is a difficult task. Because of high transportation costs and high fixed costs per pupil for operating school plants, less of the tax revenue is left for teachers' salaries. As better rural school teachers retire or change professions, it becomes difficult to recruit younger educators to teach in rural areas. As rural schools weaken, it then becomes harder to upgrade them, because substantial tax increases are required. Consolidation of rural schools is one answer; more state and federal support is another. Reorganization of the curriculum is also required if rural children are to be prepared for an increasingly urban society.

Beyond high school, rural youth encounter other educational problems. Urban youth can attend city colleges and branch universities, but rural youth are often at a disadvantage from the standpoints of both distance and money. Also, rural high schools may not have emphasized a college preparatory curriculum or even a preparatory curriculum for vocational nonfarm employment.

A critical need for rural youth is for vocational schools where skilled trades and service occupations may be learned. Fortunately, vocational schools are most often supported by state and federal funds, thus easing the tax burden on local school districts.

Rural Health

Health services and medical facilities in rural areas are often neither as comprehensive nor as good as those in urban areas. For example, 1,307 areas were identified in 1988 as lacking adequate health care personnel. The number of medical doctors in the United States increased by 76 percent between 1970 and 1988, from 268,000 to 472,000. However, the ratio of doctors to total population is still significantly lower in rural areas compared to urban areas. There are almost 230 physicians for every 100,000 persons in metro areas, compared to only 100 physicians per 100,000 persons in nonmetro areas.

Medical specialists, who depend on high population concentrations for clientele, rarely locate in rural areas. More than half of all physicians in the United States are specialists. Of the 243,000 specialists in 1988, only 22,000 were located in nonmetro areas.

Rural people, in general, carry less health insurance than urban residents, which tends to restrict their use of medical facilities. Federal- and state-sponsored medical aid benefits and the federal Medicare program for older people have helped improve rural health standards somewhat.

Prospects for more physicians, nurses, dentists, and medical specialists locating in rural areas are not too promising. Rural areas are likely less attractive because of lower income potential and a lack of modern medical facilities. It is not unusual for rural residents to commute 100 miles to receive expert medical attention.

Increased state and federal support for medical facilities and nursing homes has improved this aspect of rural health care, but adequately staffing these facilities remains a problem. The National Health Service Corps (NHSC) was created in the early 1970s to offer scholarships to medical students and other health care trainees who agreed to work in rural areas or inner cities identified as having a shortage of health care providers. Funding for this program was increased in the early 1990s in an attempt to meet the needs of underserved areas.

Rural Housing

Farm housing is generally less adequate than nonmetro and urban housing, with the exception of housing in some inner cities. For example, about 56 percent of all occupied housing units in the United States that lack complete plumbing are found in rural areas, but less than 24 percent of occupied housing units are located in rural areas.

About 2.7 percent of housing units in rural areas lack complete plumbing facilities, compared to only 0.5 percent in urban areas. Similarly, 2.1 percent of housing units in rural areas lack complete kitchen facilities, compared to 0.7 percent in urban areas. The situation in rural farm housing is slightly better than for rural housing in general, with 1.9 and 1.2 percent of rural farm homes lacking complete plumbing and kitchen facilities, respectively.

The southern United States contains more of the rural housing deficiencies than any other region of the country. This is due in part to the milder climate and the higher density of low-income families.

There is evidence, however, that rural housing is improving relatively faster than urban housing. Rural people have more equity in their homesteads, which allows them greater borrowing capacity to improve their homes. But some farmers tend to divert funds that could go to family housing improvements into more and better farm machinery, more land, and other farm improvements. Farmers may reason that such investments are more productive than family housing improvements, which are principally consumptive.

A number of USDA rural housing and community development programs have been initiated to improve the rural housing situation. For example, through the Direct Rural Housing Loan program, loans are made to low-income families to purchase, build, repair, or refinance single-family homes in rural areas. Through the Farm Labor Housing Loans and Grants program, loans and grants are available for housing for migrant, seasonal, and year-round farm workers. Assistance is also provided through this program for essential support services, such as day care, laundries, and small medical clinics.

Rural Utility Services

Rural electrification programs have brought electricity to nearly all rural areas and farm homes. Natural and butane gas distribution has also expanded in most rural areas. These services have greatly enhanced farm output and levels of living in rural areas.

Telephone service to rural and urban areas has been expanded and improved considerably over the last several decades. About 7.1 percent of occupied rural homes have no telephones, compared to 4.7 percent of urban homes. Only about 3.5 percent of rural farm homes do not have telephones.

Central water supplies are still lacking in many rural areas due mainly to the high cost of developing water facilities. About 46.6 percent of rural homes have water supplied by a public system or a private company, and 44 percent by individual drilled wells. Fewer rural farm homes (21.4 percent) have water service provided by a public system or a private company, while relatively more have drilled wells (63.7 percent). Financing of central water supplies in rural areas by the USDA is helping in this regard. Improving central water supplies should help to reduce risks from fire in rural areas and thereby lessen insurance costs.

Rural Churches

Churches in the more rural areas are experiencing a decline in memberships and revenues as farm population declines. Often, small churches and religious denominations merge. In nonmetro areas less dependent on farming and mining, churches are faring relatively better because of population stability and higher family incomes. Denominations that are regional, national, or international in scope often subsidize poorer, rural churches.

It appears logical to expect fewer but larger rural churches consisting of merged congregations and denominations. Some rural areas pool resources to build a united community church with adequate worship, recreational, and social facilities for use by several denominations. This appears to be working satisfactorily in many areas.

Rural Communication

Rural areas have enjoyed much improvement in radio and television communication. All rural areas have access to radio communication, and 95 percent of all rural residents have television sets. Hilly and isolated rural areas are being reached with cable or satellite television.

Newspaper deliveries to rural areas have improved. Larger urban dailies now reach out to rural areas and are supplemented by fewer but more diverse weekly newspapers.

Farmers increasingly subscribe to trade journals that have helped to improve farming technology. Popular magazines also have more rural subscriptions despite the loss in population. Library service in rural areas has also improved through the development of branch libraries and the use of bookmobiles.

Rural Politics

In the past, farmers have had a strong hand in shaping their own political and economic destiny. Now the farm population comprises only a small part of our total population, with the proportion shrinking every year. This means that under our democratic system, farmers have less potential political influence than they had in earlier times. Therefore, it becomes even more important that farmers understand democratic and political processes and use their farm organizations to attain political effectiveness.

Rural Government

Units of government in rural counties face several problems: (1) reapportionment, especially when towns and cities are expanding in population while rural areas are declining in population; (2) an eroding tax base, where retail sales and incomes are declining, leaving the property tax as the main source of tax funds; and (3) demand for increased services brought on by a higher level of expectations for better roads, schools, hospitals, etc.

Suggested solutions might include (1) consolidating local governmental units to reduce administrative overhead costs; (2) restructuring the level and kind of services provided through public funds to eliminate less important services; and (3) achieving a better balance between local, state, and federal tax revenues. For example, some of the local tax funds for education might be replaced with state and federal funds, permitting local governments to use these tax revenues for other needs and purposes.

It is evident that, in general, rural areas are overstocked with governmental units that need to be consolidated for greater efficiency and economy.

Rural-Urban Differences

It is apparent that rural and urban societies are converging into one society, but with the urban having a relatively greater impact on rural society than vice versa. Thus, it is expected that purely rural values and creeds will disintegrate and be absorbed by urban values and creeds. However, this convergence of values and creeds may modify present urban society, giving it new values and creeds. Greater interdependence between peoples in terms of education, employment, travel, politics, and communication, among other facets, will produce new sets of values and beliefs. People will rely less on local community roots and more on their jobs or professions with a regional, national, or international scope.

Topics for Discussion

1. Discuss the trends associated with characteristics of rural people.
2. Discuss the trend in levels of rural living.
3. Why is rural poverty so extensive?
4. Are rural-urban differences converging or diverging?

Problem Assignment

Analyze the impact of particular technological advancements (mechanical or biotechnological) on human resources in agriculture. Give careful consideration to all aspects: social, economic, and political.

Recommended Readings

Brown, David, and Linda Swanson (eds.). *Population Change and the Future of Rural America.* Staff Report No. AGES 9324. Washington, DC: Economic Research Service, USDA, 1993.

Economic Research Service. *Agricultural and Rural Economic and Social Indicators.* Agriculture Information Bulletin No. 667. Washington, DC: Economic Research Service, USDA, 1993.

Economic Research Service. *Rural Conditions and Trends.* Washington, DC: Economic Research Service, USDA. Published quarterly.

Evans, Martha. "Number of Rural Doctors Has Increased," *Farmline,* 13(8):15–16 (August 1992).

Hoppe, Robert A. (ed.). *The Family Support Act: How Will It Work in Rural America?* Rural Development Research Report No. 83. Washington, DC: Economic Research Service, USDA, 1993.

Martinez, Doug. "Poverty a Persistent Problem in Rural America," *Farmline,* 14(3):15–16 (March 1993).

Office of Communications. *Agriculture Fact Book—1994.* Washington, DC: U.S. Department of Agriculture, 1994. Pp. 24–54.

Smith, Deborah T. (ed.). *Americans in Agriculture: Portraits of Diversity—1990 Yearbook of Agriculture.* Washington, DC: U.S. Department of Agriculture, 1990.

8

Natural Resources

Natural resources include land, forests, water, minerals, and air. The nature and extent of the use of these natural resources comprise an integral part of the subject matter of agricultural economics.

Land

Land is immobile and indestructible, although its fertility is destructible. As a resource, land is unique because immobility gives it location value, which may increase or decrease depending upon the nature of development that occurs. Thus, local influences greatly affect the use and value of particular plots of land.

Land itself is indestructible as far as a particular plot of ground is concerned. It can be flooded, plowed up, or burned over, but the plot as space remains in ownership. However, its fertility as an agricultural resource is destructible. But even its loss of fertility for certain crops does not preclude land from being used in other ways, such as for grazing, timber, residential lots, or commercial purposes.

Land Uses

The total land resource of the United States encompasses 3,618,770 square miles, or about 2.3 billion acres. Of that, 365 million acres are in Alaska and 4 million acres are in Hawaii. The discussion in this chapter focuses on only the 48 contiguous states, since Alaska has very little cropland and the primary crops grown in Hawaii are not grown elsewhere in the United States.

Grassland pasture and range comprises the largest area among the major land uses in the 48 contiguous states, accounting for 31 percent in 1987 (Table 8–1). However, grassland pasture and range has declined steadily since the mid-1960s. Grazed forest land has also declined. Grazing lands provide necessary forage for dairy cattle, beef cattle, horses and mules, sheep, goats, and hogs. Grazing for beef and dairy cattle and for sheep is the most important.

Forest land serves multiple uses, including timber production, wildlife habitat, recreation, waterflow origination, and livestock grazing. At 29 percent, forest-use land comprised the second largest area among major land uses in 1987. However, the land area devoted to forest use declined from 32 percent in 1945. All land with forest cover comprises an even larger area, almost 601 million acres in 1987. However, many areas of forested land are designated to special uses, such as parks, wilderness areas, and wildlife areas, that prohibit normal forestry uses, in terms of timber and other wood products.

Table 8–1

Major Uses of Land in the Contiguous United States, 1945–1987

Land Use	1945	1949	1954	1959	1964	1969	1974	1978	1982	1987
	Million acres[1]									
Cropland	450.7	477.8	465.3	457.5	443.8	471.7	464.7	470.5	468.9	463.6
Cropland used for crops[2]	363.2	382.9	380.5	358.4	334.8	332.8	361.2	368.4	382.6	330.7
Cropland idled	40.1	25.6	18.7	33.6	51.6	50.7	20.8	26.0	21.3	68.0
Cropland used for pasture	47.4	69.3	66.1	65.4	57.4	88.2	82.7	76.1	65.0	64.9
Grassland pasture and range[3]	659.5	631.1	632.4	630.1	636.5	601.0	595.2	584.3	594.3	588.8
Forest-use land[4]	601.7	605.6	615.4	610.9	611.8	602.8	598.5	583.1	567.2	558.2
Forest land grazed	345.0	319.5	301.3	243.6	223.8	197.5	178.9	171.3	157.5	154.6
Forest land not grazed	256.7	286.1	314.1	367.3	388.0	405.3	419.6	411.8	409.7	403.6
Special use areas:[5]										
Urban land	15.0	18.3	18.6	27.1	29.2	30.8	34.6	44.2	49.6	55.9
Other special use areas	85.0	87.0	91.6	97.3	115.3	112.3	113.4	123.0	127.3	135.3
Miscellaneous other land[6]	93.4	84.0	80.5	78.9	63.0	78.4	90.6	91.9	88.5	93.9
Total land, 48 states[7]	1,905.4	1,903.8	1,903.8	1,901.8	1,899.6	1,897.0	1,897.0	1,897.0	1,895.7	1,895.7

[1] Distribution may not add to totals due to rounding.
[2] Includes cropland harvested, crop failure, and cultivated summer fallow. Estimates based on data from the U.S. Department of Agriculture, National Agricultural Statistics Service, and the U.S. Department of Commerce.
[3] Other grassland pasture and nonforested range (excludes cropland used only for pasture and grazed forest land).
[4] Excludes forest land in parks and other special uses of land.
[5] Includes land in rural transportation areas, rural parks, wildlife areas, defense and industrial areas, farmsteads, farm roads and lanes, and urban areas.
[6] Includes land in miscellaneous areas not inventoried and areas of little surface use, such as marshes, open swamps, bare rock areas, desert, and tundra.
[7] Totals differ over time due to remeasurement of the U.S. land area.

Source: Economic Research Service, USDA.

About 35.4 percent of all forest-use lands are in the Mountain and Pacific States; 34.2 percent in Appalachia, the Southeast, and the Delta States; and 12.4 percent in the Northeast. Over 62 percent of grazed forest land is in the Mountain and Pacific States.

About 25 percent of the land area in the 48 contiguous states was cropland in 1987. Total cropland has not changed much since the 1940s. Because of changes in cropland idled and used for pasture, greater variation has occurred in cropland used for crops. Cropland idled is influenced by federal crop programs. Cropland used for grazing varies considerably, depending largely on prevailing and anticipated market conditions.

Corn (grain and silage), wheat, soybeans, and hay accounted for over 80 percent of all harvested crop acres in 1993 (Figure 8–1). Additional "principal crops" included sorghum, oats, barley, rice, rye, and cotton, among others, and accounted for 15 percent of the area of crops harvested. Acres of fruits, vegetables, nuts, melons, and all other crops accounted for just over 4 percent of the area of crops harvested in 1993.

Special uses of land include urban areas, highways and roads, railroads, airports in rural areas, parks, wilderness and wildlife areas, and national defense and industrial areas. Although urban land increased 273 percent from 1945 to 1987, it still accounts for less than 3 percent of total land area. The other special land uses listed accounted for 7 percent of land in 1987.

About 75 percent of the U.S. population lives in urban areas and on only 3 percent of the nation's total land area. Despite dramatic increases in urban land

Harvested crops, 1993

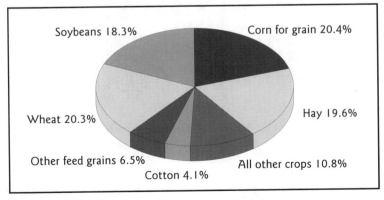

Source: Economic Research Service, USDA

Figure 8–1

area (from 15 to 55.9 million acres between 1945 and 1987), percentage decreases in rural land area are small because of the size of the rural land area. It is estimated that converting land to urban uses will not affect the level of agricultural production and the ability to meet food and fiber demand into the next century.

Land converted to urban uses comes from several different major land uses (Figure 8–2). During the 1970s, for example, 37 percent of new urban development came from cropland and pasture, 24 percent from forest land, and 23 percent from rangeland. Only about 15 percent of the acres converted to urban uses were considered prime farm land.

Wilderness and wildlife areas and other related land uses are a major part of the special uses category. Congress has set aside an area the size of Alabama, some 34.6 million acres, as wilderness. There are also areas designated as national monument areas, national recreation areas, national game refuges and wildlife preserves, national wild and scenic rivers, national primitive areas, and national volcanic monument areas.

Urbanized land, by prior land use, fast-growth counties, 1970–1980

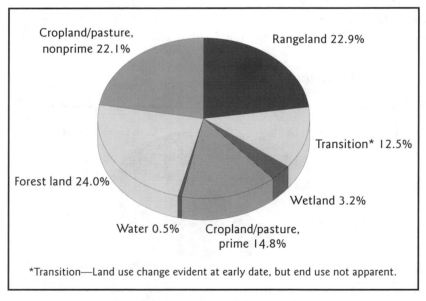

*Transition—Land use change evident at early date, but end use not apparent.

Source: Economic Research Service, USDA

Figure 8–2

Land Ownership

About 70 percent (1.3 billion acres) of the land area in the 48 contiguous states is privately owned (Figure 8–3). Most of the country's crop, range, timber, and pasture production is on private land. Another 21 percent of the land is federally owned. It is used for timber, grazing, mineral development, recreation, watershed development, and wildlife habitat. (About half the land in 11 western states and almost all the land in Alaska is federally owned.) State and local governments own about 6 percent, with American Indians owning the remaining 3 percent of the land area.

Water Resources

There is no actual overall shortage of water in the United States. In some areas, of course, water is relatively scarce and can be obtained only at great expense. In other areas, water is overly abundant, and considerable expense is incurred in controlling the surplus. Another water problem is its distribution throughout the year with varying rainfall periods.

The United States receives average annual precipitation of 30 inches, but variation extends from a little over 7 inches in the Southwest to almost 70 inches along the Gulf Coast.

Water needs change over time. Population growth in the Southwest and West, for example, has greatly increased urban water demand. At the same time, competition for water supplies is increasing for nonconsumptive (instream) uses, like recreation and riparian habitat.

Expanding water supplies, usually through dam construction and groundwater pumping, has traditionally been the response to increased demand for water to meet both urban and agricultural needs. However, new opportunities to increase supplies in that manner are limited because of a lack of suitable project sites, tight fiscal constraints, and increased public concern over the environmental effects of such projects. As a result, with demand increasing but with limited potential to increase supply, reallocating existing supplies, at least to some extent, will be necessary.

Agriculture is the largest water user. Growing demand for municipal and instream uses will result in reduced water use in agriculture. Changes in water allocation will affect irrigated farming and rural communities the most. There were over 279,000 farms irrigating almost 49.5 million acres of crop and pasture land in 1992.

110 / ECONOMICS: APPLICATIONS TO AGRICULTURE AND AGRIBUSINESS

U.S. land ownership, 1987

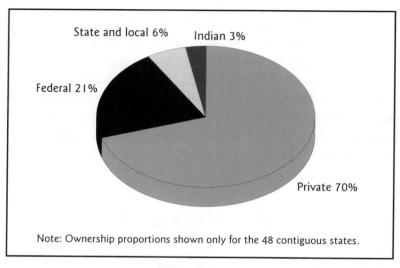

Source: Economic Research Service, USDA

Figure 8–3

Trends in farm landowners, farms, and farm land, 1900–1987

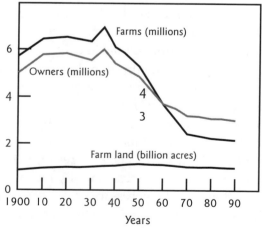

Source: Economic Research Service, USDA

Figure 8–4

Water consumption in irrigation and other uses, 1990, by region*

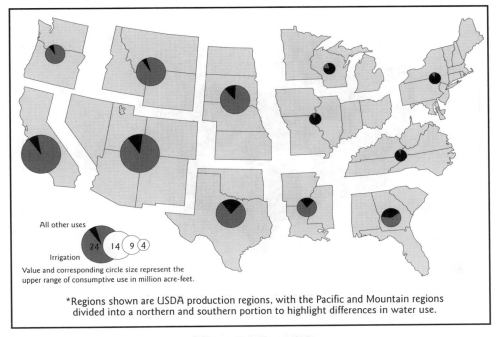

Source: Economic Research Service, USDA

Figure 8–5

Water Quality

Water quality is a serious problem. As urban areas grew and we became more industrialized, there was much dumping of untreated or inadequately treated municipal and industrial wastes into rivers and streams, endangering the quality and purity of water. Agricultural uses, which result in soil erosion and water runoff containing harmful chemicals and animal wastes, also affect water quality. Runoff from cropland contributes much of the sediment and the nutrients reaching freshwater systems. Water conservation and management have become much more important as the long-term debilitating effects of pollution have been recognized. Federal legislation, such as the Clean Water Act, addresses this need.

Monitoring and controlling agricultural sources of pollution can be difficult, because it is almost impossible to determine an exact source of water pollution, such as a particular farm. Therefore, the main focus of efforts to

Agricultural sources of surface water pollution

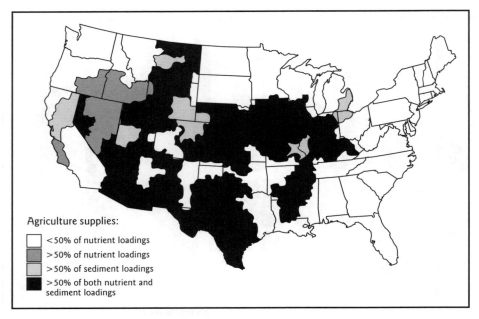

Source: Economic Research Service, USDA

Figure 8–6

address the water pollution problem has been on *point sources* of pollution, such as sewage treatment plants and factories. Much more attention will now be focused on reducing pollution from *nonpoint sources*, such as agriculture.

The economics of controlling water pollution appears to favor some type of effluent fee levied on each unit of waste discharged, with the amount of the fee dependent upon what damage is being done to others. Under this plan of pollution control, firms would try to limit discharges and/or discharge less harmful pollutants. However, one difficulty would be the determination of equitable fees to levy.

Water Rights

The subject of water rights is a critical one to farm operators, especially those who depend on water for irrigation.

Basically, two doctrines govern the use of surface water in the United States. The prior appropriation doctrine controls the use of such water in states west of the 100th meridian and in Mississippi, and the reasonable-use riparian doctrine governs the use of surface water in states east of the 100th meridian other than Mississippi, with the exception of Iowa, which adopted a system of water-use permits in 1957.

Under the doctrine of prior appropriation, the first person to put the water to beneficial use has the right to continue using the same amount for the same purpose at the same place. This system requires a state agency to administer and record water rights, which carry with them the right to use a definite quantity of water. Under the doctrine of prior appropriation, water rights are lost by nonuse, and during periods of drought, senior (older) rights have priority over junior (more recent) rights.

The reasonable-use riparian doctrine means that every proprietor of land on a bank of a river (a riparian owner) has equal rights to the water that flows in the stream adjacent to that land. A riparian owner has no property in the water itself but has a simple usufruct while it passes the property. The riparian owner has no right to detain or give new direction to the water and must return it to its proper channel before it leaves the property. The law requires that a riparian owner must use only a reasonable amount of water and must not use the water to the injury or annoyance of neighbors. Reasonable use is a question of fact for a jury whose verdict is final. Therefore, reasonable use cannot be defined exactly. It depends upon the relevant conditions and the sense of judgment of the jury. What is considered reasonable use in one locality might not be considered reasonable use in another.

The subject of water rights is very complex and often contradictory, since state and federal laws vary.

Mineral Resources

The primary mineral resources of importance to agriculture are petroleum, salt, sulfur, nitrates, phosphates, potash, and lime. These resources are exhaustible and not reproducible. Simply knowing of their existence is not enough; hence, their location and development are paramount.

Petroleum deposits in the United States are found principally in Alaska, Texas, California, Louisiana, Oklahoma, and Kansas. A barrel of crude oil yields about 43 percent gasoline, 37 percent fuel oil, 4 percent kerosene, 2 percent lubricants, and 14 percent a variety of other products. Even though energy consumption in agriculture has declined since the late 1970s, farmers are major consumers of petroleum products. They will likely continue to be as farm mechanization continues. Petroleum imports will remain important.

Refined petroleum fuels, such as butane and propane gas, are important to agriculture in weed control, fuel, irrigation pumps, heating, and other farm and home uses. Natural gas, which is found with or near petroleum deposits, comes primarily from Louisiana and Texas. It is used for heating, cooling, household uses, and various other purposes on farms.

The principal fertilizer ingredients are nitrates, phosphates, and potash. Phosphate rock, which provides phosphorus fertilizer material, is found in North and South Carolina, Tennessee, and Florida. Potash is found in New Mexico and California, with imports from Canada. Liming materials come mainly from Indiana, Pennsylvania, Wisconsin, Missouri, and Texas. Their use is to treat acid soils and render them more alkaline.

Sulfur is found primarily in Texas and Louisiana. It is used in formulation of pesticides and fertilizers and in making of paper pulp.

Salt is found principally in Louisiana, Michigan, Texas, New York, and Ohio. Its use in animal feeds and feeding is of prime importance. Other uses are in meat packing and food processing.

Air Resource

Pure, fresh air is no longer totally free but has become a scarce resource in many areas because of vehicle exhaust fumes, trash incineration, industrial smoke, and other pollutants. Efforts to clean up the air give rise to a search for methods that are the most economical in yielding pure, fresh air.

It is estimated that 90 percent of the urban population lives in areas with air pollution problems. Of course, rural areas are also affected. Polluted air harms the health and well-being not only of humans but also of animals and plants. Agriculture suffers huge financial losses each year from air pollution but is now itself being examined with increasing vigilance as a source of pollution. Controlled fire has always been used by farmers for the preservation of food, for the destruction of pests and diseases, and for the disposal of wastes. Disposal of wastes—including straw, stubble, tree pruning waste, dead trees, and brush cleared from rangeland—produces smoke, odors, dust, and airborne particulate matter that are increasingly objectionable (but not necessarily harmful) to city dwellers as they continue to relocate in rural areas. It is therefore important that there be continued surveillance of air pollution damage to agriculture, as well as measurement of amount and effects of agriculturally produced pollution.

There is a cost in purifying air, but also one in failing to do so. The job of purifying air cannot be confined to local governmental units, because impure air moves from one area to another and, hence, from one local government authority to another. State and federal control of air pollution seems more appropriate, for polluted New York City air may move into New Jersey, and vice versa. City, county, and state air pollution control districts are becoming more prominent and necessary.

Selected Aspects of Farm Land Use

Value of Farm Land

U.S. farm real estate, the primary source of collateral for the farm business, accounts for nearly 75 percent of the value of all farm assets. Farm real estate values are reliable indicators of the general economic condition of the agricultural sector.

The average per-acre value of farm real estate in the 48 contiguous states was $744 in 1994. Average farm land prices were highest in the Northeast, with average per-acre values ranging from $1,081 in Maine to $4,686 in Connecticut, $4,840 in New Jersey, and $5,334 in Rhode Island. Outside the Northeast, farm land values were highest in Florida, with an average value per acre of $2,205, and in California, with an average value per acre of $1,722. The average value per acre in the productive Corn Belt states was $1,285.

General factors that give value to farm land include its productivity in terms of supplying food and fiber products and the expectations such productivity generates for future cash income from farming; pride and security in owning farm land; location of the farm land in relation to roads, utilities, urban centers, etc.; population and economic growth; and the fact that it is a resource with a fixed supply in relation to a variable but increasing demand.

The value of farm land rose continually each year from the mid-1940s through the early 1980s. Factors contributing to this sustained increase in land value include the following:

1. As they adopt new technology, many farmers seek to lower their per unit production costs by increasing their landholdings, a process that bids up prices of connecting land tracts.

2. Nonfarmers often seek farm land as an investment, as a hedge against inflation, or purely for speculation.

3. Businesses and industries moving into the fringe of urban areas acquire land for present and future uses.
4. The number of available acres per person shrinks each year as population increases but total land area remains the same.
5. Income from nonfarm sources supplements farmers' ability to buy land.
6. Monetary benefits of government price supports and acreage allotment programs are capitalized into land values. For example, if one expects cotton prices to be supported by the federal government at a certain level for a reasonable length of time, then that person will be willing to pay more for cotton land than if cotton prices were set in a free market. Therefore, the seller of cotton land obtains a price that covers not only the value of land in producing cotton but a windfall gain resulting from government price supports for cotton.

Purchasing Farm Land

It is difficult to buy good-quality farm land in the desired quantity and at the desired price. The basic question is, What can an operator afford to pay for farm land? Obviously, if farm land prices rise but farm product prices do not, the ability of a farmer to pay for land decreases. This could be especially critical for young producers entering farming.

In many cases, farm land is overpriced relative to income produced from that land. However, when land is appreciating in value, capital gains from that land may be realized. It is difficult, however, to generalize about farm land, since each plot is unique. Factors important in purchasing farm land are (1) the cost of the land, (2) applicable acreage allotments, (3) enterprise production costs, (4) future prices for farm output, (5) expected yields, (6) extra labor and machinery needed, (7) expected rate of return on the investment in land, and (8) prospects for inflation or deflation.

One method of arriving at prices to pay for land is to use the following formula:

$$\text{Value of land per acre} = \frac{\text{Annual net return to land per acre}}{\text{Current market interest rate per year}}$$

Suppose land income is $80 per acre per year and the going rate of interest is 10 percent. By dividing $80 by 0.10, the resulting value of that land is $800 per acre.

Farm land by method of acquisition, 1987

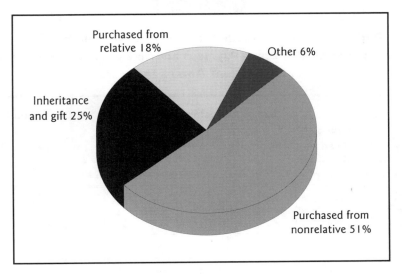

Source: Economic Research Service, USDA

Figure 8–7

Data presented in Table 8–2 may help determine the feasibility of buying farm land. Assume a farmer is considering the purchase of a 50-acre tract of land adjacent to her present farm at $800 per acre, or $40,000. She knows she can borrow the money to purchase this land at an interest rate of 10 percent. Assume she plans to make annual payments. She has asked you to help her decide whether the purchase is financially feasible.

Based on her farm records, she is confident that she will be able to generate net returns (all operating expenses and fixed expenses, except for the land payment, have been subtracted) of $80 per acre, or $4,000 total, which she can use to service the land note. This total amount divided by the amount paid for the land, which equals 10 percent (4,000/40,000) in our example, is the size of the annual payment in cents per dollar paid for the land. How long will it take to pay for this land?

The repayment length will be influenced, of course, by the size of the down payment. Based on the assumptions of our example, this farmer would not be able to pay the note off at all if she borrowed the entire $40,000 (no down payment) and had only $.10 per dollar paid for the land, or $4,000, to use to make payments.

Table 8–2

Approximate Years Needed to Pay Off Farm Land Purchase in Relation to Size of Down Payment and Size of Annual Repayment at 10 Percent Annual Interest

Buyer's Equity	Amount of Loan	Size of Annual Payments in Cents per Dollar Paid for Land								
		.06	.07	.08	.09	.10	.11	.12	.13	.14
Per dollar paid for land		*Years needed to pay off at 10 percent*[1]								
$ .00	$1.00	x	x	x	x	x	25	19	15	13
.10	.90	x	x	x	x	24	18	15	12	11
.20	.80	x	x	x	23	17	14	12	10	9
.25	.75	x	x	29	19	15	12	10	9	8
.30	.70	x	x	22	16	13	11	9	8	8
.35	.65	x	28	18	13	11	9	8	7	7
.40	.60	x	20	15	12	10	8	7	7	6
.50	.50	19	13	10	9	7	6	6	5	5

[1] x denotes that the rate of total payment is equal to, or less than, the interest rate; in such cases it is impossible to pay off the principal without supplementing with cash from other sources.

Now, let's assume she can make a 20 percent down payment. In this scenario she pays $8,000 down ($40,000 ∗ 0.20) and finances the remaining $32,000. She still can make payments equal to $.10 per dollar paid for the land, or $4,000. Now how long will it take to pay for the land? Look at the third row in Table 8–2 to find out. At 20 percent down and an annual payment equal to $.10 per dollar paid for the land, it will take 17 years to repay. If our farmer could put 50 percent down, it would take only 7 years to repay the land note.

Factors associated with changes in farm land prices include (1) the level of mortgage interest rates, (2) the level of net farm incomes, (3) optimism or pessimism of buyers, (4) amount of down payment required, (5) level of property taxes, (6) level of government payments to farmers, (7) expected capital gains, (8) desire for farm growth and expansion, (9) nonfarm population density, and (10) technological advances.

Renting Farm Land

In many cases, it is financially advantageous to rent farm land rather than to purchase it. Rental charges generally reflect more closely the value of farm land in terms of its economic productivity, while land sale prices also reflect such factors as location, urbanization, mineral rights, etc. Farm producers may not be able to afford payments for land reflecting both agricultural and urban values, but they will usually be able to meet rental payments based on agricultural productivity. In determining cash rental rates on land, customs (the way agreements are typically reached in a given area) play a large part. Aside from using customary rental rates, a technique for deciding upon a per acre rental rate is to use this equation:

Rental Rate = Current Land Value * Interest Rate on Alternative Investment

For example, if land is selling for $1,000 an acre and the going rate of interest on an alternative investment is 8 percent, the rental rate per acre should be in the vicinity of $80 ($1,000 * 0.08). The property taxes per acre should also be included.

Taxing Land

The property tax is one of the oldest sources of revenue used by governments. In the early days of the United States, land was the main source of income and wealth. Since it was visible and immovable, land could not escape taxation. As the nation prospered, income and wealth in other forms were created, and land lost its uniqueness as a tax base.

Taxation of real property (land and improvements) is the largest source of revenues for most local and county governments. For state governments, the sales tax is the major source of revenue, while the income tax (individual and corporate) is relied upon more heavily by the federal government (Table 8–3). Generally speaking, sales and income taxes generate revenues as the economy expands, while the property tax is less flexible as a revenue source. Moreover, the property tax is not always distributed equitably, because of the tendency to overtax poor land and undertax fertile land.

Property taxes are calculated by multiplying the "assessed" or "taxable" value of the property by the appropriate tax rate. Rates are expressed in mils (10 mils equal 1 cent) per dollar of assessed valuation. Mil levies can be, and frequently are, translated into dollar terms and expressed as dollars per $1,000 of assessed valuation.

Table 8–3

Sources of Tax Revenue for Federal, State, and Local Governments

Sources	Government		
	Federal	State	Local
	—Percent—		
Propery tax	0	2	75
Individual income	73	32	5
Corporation income	15	7	1
Sales[1]	9	49	15
Other[2]	3	10	4
Total	100	100	100

[1]Includes custom duties; general sales tax revenue; and selective sales tax revenue, which includes taxes on motor fuel, alcoholic beverages, tobacco products, and public utilities.
[2]Includes motor vehicle licenses, operators' licenses, and death and gift taxes.

Taxes levied on farm real estate vary throughout the country, depending on land values and revenue needs. The highest property taxes are in the Northeast, followed by the Pacific States and the Midwest. The lowest real estate taxes prevail in the South, with the exception of Florida, and in the Mountain States. The southern states are prone to use other forms of taxation for revenue; the Mountain States have low land values, generating low tax revenues.

As the rural nonfarm population increases relative to the rural farm population, public services grow in demand. The usual procedure is to pass bond issues secured by the property tax. Thus, farmers may find their property taxes increased greatly without necessarily enjoying these added public services. One benefit to farm landowners, however, comes from higher property values, but these higher values often are taken into account in assessing and taxing farm land.

In some states, property taxes on farm land have become exorbitant because of assessments based, not on farm value, but on market value. In other states, farm property taxes are ridiculously low, with most farm landowners not paying any property taxes. Liberal homestead exemptions, inequitable assessment practices, and political favors are the main causes.

If farm land is heavily taxed, owners are forced to intensify farming, sell the land, or shift it into nonfarm uses. A more equitable procedure is to tax farm land in terms of its agricultural value as long as it remains in that use. On the other hand, land that is taxed little or not at all is subject to idleness and speculation, with owners waiting for price rises so they can sell out at an opportune time. Also, land taxes that are relatively low shift the burden of taxation to nonowners of land.

One serious limitation of the property tax is its inflexibility in relation to farm prices and incomes. Whether farm prices and incomes are high or low, the tax stays the same. In times of low prices, farmers are at a serious disadvantage. Other taxes, such as sales and income, are levied in relation to expenditures, salaries, and earnings; hence, they are relatively less burdensome in times of distress.

Rapid increases in property taxes intensify the pressure for tax relief, especially for farmers located on the fringes of growing cities. Taxes per acre levied on farms in metropolitan areas are, in general, much higher than taxes per acre levied on farms in counties adjacent to metropolitan areas. Likewise, taxes per acre levied on farms in counties adjacent to metropolitan areas are much higher than taxes per acre levied on farms in rural counties.

High tax levels have given rise to a number of proposals for property tax relief. The major type of relief proposed for land on the urban fringe is some form of preferential assessment. A number of states now have such laws.

Preferential assessment may take the form of assessing farm land on its "use" value, not its "market" value. As long as land stays in an agricultural use, it is taxed as such, regardless of its value for other uses. Other plans designed to ease property taxes on farm land include tax deferral. This suspends or defers part of property taxes until land passes into nonfarm uses, at which time the deferred taxes become due and payable.

In regard to forest and mineral lands, a severance tax usually applies as products are harvested and sold. Although property taxes may also apply, the lands are usually assessed at much lower rates than those for conventional farm land. The severance tax policy is intended to encourage conservation by easing the pressure for rapid exploitation to avoid paying property taxes over a long period.

Land Planning, Zoning, Eminent Domain, and Easements

Planning for land use involves a comprehensive master blueprint made by competent personnel to decide the optimum utilization of a given land area. Successful planning consists of citizen involvement, public information and education, and competent technical staffs or committees. After land use plans have been approved or accepted by the area or community, they can be used to guide zoning ordinances, industrial and housing developments, recreational uses, etc.

Zoning is a legal method employed by government under its police power to restrict the conduct of certain activities in specified areas. Zoning is intended to ensure orderly growth and development, preferably according to a master plan. Without a plan and accompanying regulations, haphazard mixing of incompatible land uses occurs, which disrupts, and possibly depresses, property values. Zoning based on economically optimal land use within a long-term plan will prevent blight and minimize related problems. Zoning tends to maintain the land for purposes for which it is logically suited and, when enacted, is not retroactive but applies only to future land use.

The usual zoning procedure begins with a governmental unit announcing a zoning plan, holding public hearings, and adopting ordinances specifying permissive uses for land. Zones might be designated "A-1," "A-2," etc., with "A" meaning residential, "1" single-family occupancy, "2" multiple-family occupancy (apartments), etc. Business and industrial uses might be in Zone B. Zone C might permit row crops, while Zone D might allow livestock, dairy, and poultry enterprises. This pattern of zoning is termed "exclusive," while "cumulative" zoning indicates that Zone A uses are permitted also in Zone B and Zones A and B uses are permitted also in Zone C, etc. In more recent years, exclusive zoning has gained in importance relative to cumulative zoning. Appeals by property owners to zoning commissions and courts of law regarding classification of their property are permissible. Exceptions may be granted when justified.

The acid test of a zoning ordinance is the reasonableness of its provisions. Zoning must be rational, guided by common sense; must promote the public interest; and must protect the property owner from willful and unjust regulations. Farmers should approach zoning with a positive rather than a negative attitude, make a thorough study of the purposes and objectives of community planning and zoning, and encourage planning boards and/or zoning commissions to include someone thoroughly familiar with farming.

Rural areas need zoning as much as urban areas. As the countryside becomes inhabited with rural nonfarm people, land use problems accelerate because of the development of business and recreational facilities—some desirable, some not. Farmers, as a rule, have resisted zoning, in the belief that zoning would discriminate against them. This need not be so if proper efforts are made to establish sound zoning plans and procedures before zoning becomes an after-the-fact proposition under which nuisances get zoned in instead of zoned out.

Another element in the use of land and property is **eminent domain,** a right assumed by public governing bodies to condemn private property, with compensation, for public use. Land may be taken under eminent domain for roads, schools, recreation, and other purposes. A governing body usually decides what public projects are needed; plans are then drawn and sites selected. Landowners thus affected are offered certain sums for their property. If owners feel that the amounts offered are inadequate, they can resort to the courts and have the matter settled by a judge. After this procedure, private property becomes public property.

Certain problems arise out of eminent domain: (1) an increasing number of governing bodies are being given property condemnation rights; (2) the federal urban renewal programs have accelerated eminent domain proceedings, as have the interstate highway system and airport and school construction programs, among other things; (3) corruption among certain civil servants in condemnation proceedings has occurred; (4) small landowners have not received adequate compensation for their property; and (5) compensation for property often fails to meet other costs involved in relocating farming operations.

The use of **easements** is another technique for land use control. A governmental unit or a franchiser purchases the development rights to land, but not the land itself, for a fixed number of years. During this period, the landowner cannot develop the land for any purposes other than those agreeable to the easement purchaser. One difficulty in easement procedures is establishing a value for easement rights.

Soil Conservation

Conservation is the effort expended to maintain the land's productive capacity, while development is the effort expended to increase its productive capacity.

Before 1935, conserving soils in the United States did not have a high priority in national economic life. Land was viewed as abundant, and virgin soils were very fertile; thus, exploitative practices were commonplace. The dry years of the early 1930s, with the resulting "dust bowls," reminded the nation of this neglect and waste. Consequently, the passage of soil conservation legislation at that time was made much easier. Through the Soil Conservation Service (now called the Natural Resources Conservation Service) and other agencies, including soil conservation districts, national programs of soil and water conservation were put into effect. These programs were varied. They consisted, among other things, of land shaping; cover crops; crop rotations;

small dams, ponds, and other water control devices; farm mapping; soil surveys; liming programs; soil testing and fertilizer recommendations; reforestation; drainage programs; and land leveling.

In terms of federal land policy, it is necessary to distinguish between conserving land and developing land. The former is more necessary than the latter, considering the issue of agricultural capacity now prevalent. Even in conserving land, mistakes in policy are made. For example, many of the federal assistance programs for soil conservation consist of payments for practices that add to farm productivity that the federal government subsequently may seek to restrict with production controls. In other words, the government may make payments for conflicting purposes—one to expand output and another to restrict it—even to the same farmer. Payments for efforts to prevent serious soil erosion, floods, etc., are not in question here, but payments for practices that increase productivity in the short run are sometimes questionable. It would appear more appropriate for farmers to make these latter investments privately, based on their own cost-returns analysis. Such short-term investments are probably more output-increasing than they are soil-conserving, although they are both. However, society has a legitimate role in expending funds for conserving soils even though land may be privately owned, because the national interest extends far beyond the life span of any one farmer. Loss of soil fertility not only reduces productivity but also magnifies the problem of silting and impairs water management.

Another inconsistency in federal soil conservation programs is that farm tenancy, which causes some of our soil management problems, is not regulated or controlled by the federal government. In fact, some government price support programs may cause land to be overtilled by farmers who want to expand output to obtain high support prices and higher net incomes. Thus, tenants often mine the soil as much as possible to obtain greater returns above the rentals paid to landlords, many of whom are absentee. Land that should remain in grass or timber is cultivated. Therefore, a prime need in federal soil conservation policy is a master plan coordinating all programs into a clear, unified, long-range soil management policy. Without such a comprehensive and coordinated policy, it is likely that a part of federal expenditures on soil conservation is being wasted or misused. Federal programs designed to foster resource conservation are discussed in Chapters 24 and 25.

In using natural resources such as land, there is always the issue of individual versus public welfare, for what is best for the individual may not be best for society. Those who exploit a natural resource without regard to long-term effects will find that society will seek to restrict their activities. On the other hand, society may well decide to subsidize and encourage the individual owner to take better care of natural resources in the interest of public good. There arise subsequent issues, however, such as how much encouragement or incentive should be paid, to whom, and for how long. Society does not have unlimited funds for incentive payments to individual owners and therefore, through Congress, must pick and choose the projects that it feels warrant financial support.

Land Development

Historically, clearing forests, draining swamps, and irrigating dry land have been the main techniques for reclaiming land for agricultural uses. The decision whether to develop land was based primarily upon economic considerations—for example, determining the most profitable or productive use for land by evaluating land development and operation costs versus benefits.

However, the decision whether to develop land should depend jointly on economic and environmental considerations. During the export boom of the 1970s and early 1980s, development of new cropland accelerated. Much of the expansion occurred on marginal land, with a direct loss of fragile grassland and wetland ecosystems. There have also been indirect effects, including increased soil erosion and degraded water quality, as a result of farming these vulnerable lands.

Wetlands have been converted into cropland since the early years of European settlement. Draining and developing wetlands caused little concern because of the abundance of natural wetlands. From 1954 to 1975, almost 14 million acres were converted from wetlands to other uses, with 87 percent going to agricultural uses. Decisions to reclaim these lands were made on the basis of relatively short-term economic aspects. However, public awareness of the environmental benefits of wetlands, such as providing wildlife habitat, flood protection, and scenic recreational sites, has been increasing. Thus, environmental aspects need to be considered, as well as the economic aspects, in the decision whether bring new land into agricultural production.

The question of public land development must be analyzed in terms of public needs, goals, and benefits versus cost or expenditure. Environmental aspects must also be considered, not just economic aspects. These include soil erosion potential, water quality, sedimentation, and wildlife habitat.

Topics for Discussion

1. Discuss the several aspects of land as a natural resource.
2. Discuss the future water needs of the United States.

3. Discuss the nature of mineral resources in the United States.
4. Why is air becoming an economic resource?
5. What gives rise to the value of farm land?
6. What are some of the key issues in taxing farm land?
7. Define and discuss zoning.
8. Define and discuss eminent domain.
9. Contrast land conservation with land development.

Problem Assignment

Estimate annual property taxes for a family-sized farm in your home county, and allocate the amount of such property taxes to county government, schools, and other public services.

Recommended Readings

Anderson, Margot (ed.). *Agricultural Resources and Environmental Indicators.* Agricultural Handbook No. 705. Washington, DC: Economic Research Service, USDA, 1994.

Carriker, Roy R. "Wetlands and Environmental Legislation Issues," *Journal of Agricultural and Applied Economics,* 26(1):80–89 (1994).

Chavas, Jean-Paul. "On Sustainability and the Economics of Survival," *American Journal of Agricultural Economics,* 75(1):72–83 (1993).

Ditwiler, C. D. "Environment Perceptions and Policy Misconceptions," *American Journal of Agricultural Economics,* 55(3):477–483 (1973).

Drabenstott, Mark. "Agriculture's Portfolio for an Uncertain Future: Preparing for Global Warming," *Economic Review,* 77(2):5–20 (1992).

Haveman, Robert H. "Efficiency and Equity in Natural Resource and Environmental Policy," *American Journal of Agricultural Economics,* 55(5):868–878 (1973).

Headley, J. Charles. "Agricultural Productivity, Technology and Environmental Quality," *American Journal of Agricultural Economics,* 54(5):749–756 (1972).

Johnson, Larry A. "Sustainability Issues: How Should Government Coordinate Farm Regulations and Policy?" *Journal of Agricultural and Applied Economics,* 26(1):75–79 (1994).

Kircher, Harry B., Donald Wallace, and Dorothy J. Gore. *Our Natural Resources and Their Conservation,* 7th ed. Danville, IL: Interstate Publishers, Inc., 1992.

Porter, Lynn, Jasper S. Lee, Diana L. Turner, and J. Malcolm Hillan. *Interstate's Environmental Science and Technology.* Danville, IL: Interstate Publishers, Inc., 1996.

Smith, Deborah T. (ed.). *Agriculture and the Environment—1991 Yearbook of Agriculture.* Washington, DC: U.S. Department of Agriculture, 1991.

9

Capital Resources

Capital is defined as goods used to produce other goods. There are different kinds of capital goods: (1) natural capital, such as land, (2) human capital, such as managerial and trade skills, and (3) produced capital goods, such as machinery. In this chapter, we will focus our discussion on the last group and on issues pertinent to it. Capital is often thought of as money, but, as we will see, it includes much more than that.

Farm capital resources are intermediate goods, i.e., goods used to produce other goods with the goal of generating income over time. Included in farm capital resources are farm buildings, equipment, houses, machinery, inventories, livestock, motor vehicles, and other movable property.

Farm capital assets consist of both farm capital resources and natural resources such as land and timber. They also include farm financial assets, which consist of deposits in banks, cash on hand, U.S. savings bonds, and other savings and investments. These financial assets are claims on capital and can be converted into real estate, machinery, or other capital goods.

Capital invested per farm in the United States averages over $4.1 million, up dramatically from about $95,000 in 1970. Continued increases in farm capitalization are expected in the future. This is because farming is a very capital-intensive business, requiring almost $18 in total farm assets to generate $1 in net farm income.

Importance of Farm Capital Assets

Comprising almost 74 percent of total farm assets, farm real estate (land, buildings, and fences) is the single most important farm asset (Table 9-1). Real estate has been increasing in importance over time as prices have trended up and farms have enlarged. For the farm sector as a whole, debt owed on farm real estate is quite low, at only 8.6 percent of total asset values. However, this represents an average; some farmers have much higher debt loads, especially for real estate. Rising real estate values do not provide immediate cash benefit to farmers unless they liquidate, although a higher borrowing base is created. Those who wish to buy land are, of course, disadvantaged by higher land prices.

The second most important farm asset is machinery and motor vehicles, accounting for about 10 percent of total assets. With ongoing technological advances, machinery and equipment will remain an important component of total farm assets.

Livestock is the third most important dollar asset, accounting for about 8 percent of the total. The importance of livestock as a component of farm assets has been quite steady over the past several decades.

The importance of other farm assets is shown in Table 9–1. Investments in farmer cooperatives have been increasing steadily since the early 1950s. Financial assets, including investments in cooperatives, represent about 5 percent of total farm assets.

Assets, debt, and equity

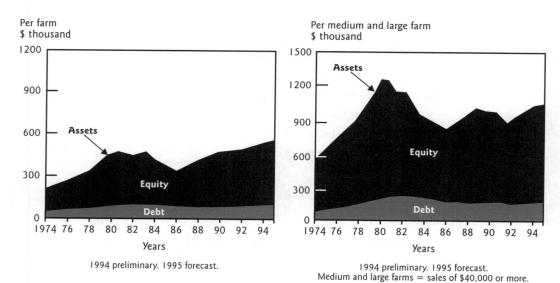

Source: Economic Research Service, USDA

Figure 9–1

Table 9-1

Balance Sheet of the U.S. Farm Sector, 1993

	Percent	
ASSETS		
Physical Assets		94.8
Livestock and Poultry	8.2	
Machinery and Motor Vehicles	9.6	
Crops	2.6	
Purchased Inputs	0.5	
Real Estate	73.9	
Financial Assets		5.2
Investments in Cooperatives	3.5	
Other	1.7	
Total Assets		100.0
DEBT AND EQUITY		
Debt		16.0
Nonreal Estate Debt	7.4	
Real Estate Debt	8.6	
Equity		84.0
Total Debt and Equity		100.0

Source: Economic Research Service, USDA.

Debt Capital and Net Worth

Farmers own almost 84 percent of all their assets (equity capital) and owe for the rest (debt capital) (Table 9-1). This net worth is very high, exceeding that of any other type of U.S. business. Overall, the farm sector is in good shape financially. But as was pointed out earlier, individual farmers may be in financial trouble, even though the farm sector as a whole is financially sound.

The total value of farm real estate increased a dramatic 287 percent during the 1970s. This increase can be attributed in part to the over 15 percent increase in acres of cropland used for crops in the 1970s. But the dominant force driving the value of farm real estate up was land values, which increased 355 percent from 1970 until they peaked in 1982. Many farmers who purchased land and farm equipment during the export boom years of the 1970s assumed large amounts of debt because of the inflated prices for land and equipment. As a result, they found themselves especially vulnerable to the decline in farm land values that occurred in the early 1980s.

About 54 percent of the total debt in the farm sector is real estate debt. The primary credit provider for real estate financing, in terms of outstanding debt, is the Farm Credit System (specifically the Federal Land Bank), followed by commercial banks, life insurance companies, and the Farmers Home Administration (now a part of the Consolidated Farm Service Agency). Individuals and others are also important sources of financing for real estate acquisition.

The remaining debt, the nonreal estate debt, makes up about 46 percent of the total debt of the farm sector. Included in this category are outstanding operating loans used to purchase seed, feed, fertilizer, etc.; machinery and equipment loans; and livestock loans. Commercial banks are the primary source of nonreal estate financing, followed by the Farm Credit System (specifically Production Credit Associations) and the Farmers Home Administration. Again, individuals and others provide much of this type of financing in the farm sector.

Most commercial farmers borrow money in some amount, with much of that debt secured by real estate. Farmers have a broad base of borrowing power: for every $100 of debt-free real estate, they can borrow from $60 to $65. Presently, they have only about $12 borrowed per $100. It may be axi-

Principal farm credit flows

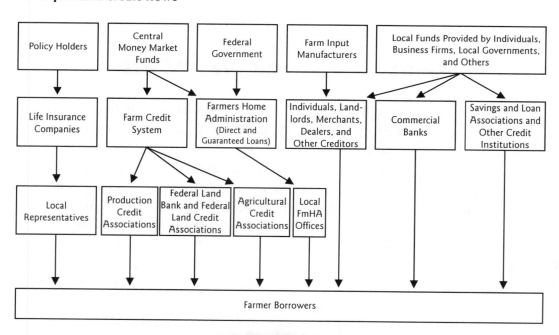

Figure 9–2

omatic to say that farmers can borrow a great deal more than they can pay for out of current farm income. Simply put, many farmers are "land rich" and "money poor."

For most types of farming areas, the greatest increase in demand for loanable funds has occurred for short-term loans to cover operating expenses. Cash operating expenses per farm increased over 700 percent (in nominal dollars) from the early 1960s to the early 1990s.

Agricultural Credit Systems

In agriculture, as in commerce or industry, credit is needed to overcome a shortage of equity capital. Limited capital, fluctuating interest rates, and lack of credit information are major problems facing many farmers.

Credit, in effect, is an economic tool producers can use to increase production, raise the quality of what is produced, or otherwise improve operations to make them more profitable. Its use must at least generate enough additional income to pay for the cost of the borrowed money (the interest) and to assure that the principal is repaid within the specified period of the loan.

A system of credit geared to the requirements of commerce and industry is not, without modification, suited to the needs of financing agriculture. Many farm units are small and operated by the persons who own them, with production normally on a seasonal basis. The primary fact that farming is based on the biological processes of plant and animal life results in a slow rate of turnover of invested capital and a potentially uneven flow of income. The nature of farming is such that risks in agricultural lending are usually greater than in commercial or industrial lending, and cash incomes generated in relation to capital invested are low compared with commerce and industry.

Agricultural credit must necessarily be adapted to the unique aspects of farming. Agricultural loans are typically made for longer periods than loans in commerce or industry. Terms of agricultural loans should be flexible, with repayment schedules that are consistent with income flows.

Farm debt can take different forms. Short-term debt, also called "current debt," includes existing debt obligations that are to be paid within the next year. An example of short-term debt is accounts payable to merchants and other suppliers for seed, feed, fertilizer, fuel, and other operating farm inputs.

Noncurrent debt includes debt obligations that have maturities greater than one year. The noncurrent portion of loans used to purchase land, machinery, equipment, and livestock constitutes noncurrent debt.

Real Estate Debt

Real estate loans generally have terms of from 10 to 40 years and are normally used to purchase farm land or make major capital improvements to farm property. Key sources of long-term or farm real estate financing are discussed in this section. The importance of these sources is reflected as follows:

Type of Lender	Percentage of Total Long-Term Debt Held
Farm Credit System	34
Commercial and savings banks	25
Life insurance companies	12
Farmers Home Administration	8
Individuals and others	21
	100

Farm Credit System

The Farm Credit System (FCS) is a system of federally chartered but privately owned banks, lending associations, and service units organized for the purpose of providing credit and related services to agricultural producers and their cooperatives in the United States. The lending institutions and other entities of the FCS are chartered under federal authority contained in the Farm Credit Act of 1971, as amended. By design, the agricultural borrowers become the system's owners and holders of equity capital.[1]

The Farm Credit System underwent major reorganization during the 1980s and early 1990s. As of early 1994, there were 8 district Farm Credit Banks, 3 Banks for Cooperatives, and 238 local lending associations making up the Farm Credit System. The structure of the Farm Credit System is expected to undergo more changes in the future.

Farm real estate lending is the most important lending program of the Farm Credit System. Long-term real estate loans can be obtained through

---

[1] Peter J. Barry, Paul N. Ellinger, John A. Hopkin, and C. B. Baker, *Financial Management in Agriculture,* 5th ed. (Danville, IL: Interstate Publishers, Inc., 1995), pp. 481–497.

Federal Land Bank Associations (FLBAs), Federal Land Credit Associations (FLCAs), and Agricultural Credit Associations (ACAs).

Loans can be made for maturities ranging from 5 to 40 years, with most maturities falling in the 20- to 30-year range. Up to 85 percent of the appraised value of the land may be borrowed, with the land pledged as security. Repayment programs can be either equally amortized payments or constant payments on the loan principal, plus interest payments on the outstanding principal balance. Loans with variable or adjustable interest rates will have varying payment sizes.

In general, the Farm Credit System is a good source of real estate financing. There are no penalties for early repayment, interest rates are reasonable, and repayment terms accommodate the farmer's ability to repay. In addition, FCS institutions are farmer-owned, with no federal government capital.

Commercial and Savings Banks

Since the mid- to late 1980s, the share of farm real estate debt financed by commercial and savings banks has increased significantly. Through the 1960s and 1970s, commercial and savings banks accounted for about 12 percent of outstanding real estate debt. Reflecting the tightening of agricultural credit in the early 1980s, the commercial and savings banks' share shrank to 7 percent of total real estate debt. Since then, commercial and savings banks have steadily increased their involvement in long-term financing. Much of the growth of commercial bank real estate loans during the late 1980s was due to the use of farm real estate as security for refinancing production and intermediate-term loans.

Legal lending limits to any bank borrower are based on the bank's equity capital. The limits are designed to curtail the concentration of lending risks. For a nationally chartered bank, loans to an individual borrower cannot exceed 15 percent of the bank's unimpaired capital and unimpaired surplus fund, which is a measure of net worth. The limit increases to 25 percent for loans to purchase livestock or for loans fully secured by readily marketable collateral. Loan limits for state banks vary by state but are comparable to those of national banks.

When rural banks cannot meet all of a mortgage request, they rely on correspondent banks to complete the financing. Correspondent banking takes place when a large bank, the correspondent, provides various banking services to a smaller bank in return for compensation provided by the smaller bank, either as payment of fees or as placement of a demand deposit with the corespondent. Correspondent services include check clearing, federal funds transactions, security safekeeping and transactions, leasing services, electronic data processing, investment counseling, and loan participation. Rural banks have long relied on loan participation with correspondent banks to meet farm loan requests that exceed their legal limits and to share lending risks.[2]

Also, many banks enter into contracts with life insurance companies for sharing the farm mortgage business originated by the banks. Such arrangements often enable a bank to provide farm mortgage credit locally in greater volume and on longer terms than would otherwise be possible. Loans guaranteed by the Small Business Association and the Farmers Home Administration may also be available.

In general, commercial banks are a good source of financing because of their knowledge of local real estate and local citizens. The farmer may also obtain a wide range of banking and credit services from local commercial banks.

Life Insurance Companies

Insurance companies have been an important source of real estate credit since the early 1900s. Because of long-term availability of funds from insurance premiums, these companies prefer to invest in offsetting long-term mortgages, which are a sound type of investment for them. The involvement of life insurance companies in real estate lending varies by region and by changing financial market conditions. As lenders, insurance companies have historically preferred lending to larger, low-risk farming operations. They typically make much larger loans, on average, than other lenders.

Laws in some states limit the size of a life insurance company's loan to a designated proportion of the appraised value of the real estate. However, there is considerable flexibility in the choice of appraisal method. While individual loans have no external legal limits, internal policy for given companies may limit loan size.

Life insurance companies have changed their maturities and repayment schedules over time as a way of providing interest rate protection from the effects of high inflation. A typical arrangement has involved a loan maturity of from 10 to 20 years; a longer amortization schedule—30 years, for example; and a provision for an interest rate adjustment every 5 years. Because the loan

---

[2] Ibid., pp. 473–480.

maturity is shorter than the amortization period, a *balloon* payment is due at maturity. The borrower can either meet the balloon payment with cash or refinance it with the life insurance company or another lender.[3]

In general, life insurance companies are a good source of real estate financing, provided the interest rate is reasonable, the period of repayment is satisfactory, and the penalty for early repayment is not excessive.

Farmers Home Administration

The Farmers Home Administration (FmHA), a lending agency of the U.S. Department of Agriculture, was first organized in 1935 as the Farm Resettlement Administration. It was reorganized in 1946 to meet certain credit needs of farmers not being met by other lenders and was renamed the Farmers Home Administration. Currently, the FmHA is undergoing further reorganization. As of this writing, at least part of FmHA operations have been merged with the Agricultural Stabilization and Conservation Service (ASCS) to form the Consolidated Farm Service Agency.

FmHA lending has occurred under four broad loan categories: (1) farm ownership (including expansion), (2) farm operating, (3) emergency, and (4) other, such as soil and water conservation and rural housing. FmHA lending expanded rapidly during the late 1970s and early 1980s, due in large part to the economic emergency loan program. This program was scaled back in 1984. Farm loan assistance can be as direct loans or as loan guarantees. Since the mid-1980s, the FmHA has placed most of its emphasis on loan guarantees rather than on direct lending.

Under the guaranteed loan program, the loans are made and serviced by private lenders, with the FmHA guaranteeing up to 90 percent of each loan if the borrower defaults. The guarantee may result in more favorable loan terms for the borrower, since much of the risk is shifted from the lender to the Farmers Home Administration.

The Farmers Home Administration has played an important role in financing high-risk borrowers. The FmHA has met the needs of many borrowers who have success potential but lack access to adequate financing from commercial sources. As these producers become better established, they usually "graduate" from FmHA borrowing to commercial sources.

---

[3] Ibid., p. 499.

Credit programs in the Farmers Home Administration have specific targeting and accounting requirements. Programs are targeted specifically to beginning farmers and the socially disadvantaged (SDA). Within the beginning farmer and SDA classifications, there are also distinctions made between obligations based on gender and ethnic origin.

Individuals

Individual lending is an important source of long-term farm loans but is less significant in short-term finance. Loans from relatives, neighbors, and wealthy individuals provide considerable funds in some areas. However, the relative importance of this source of funds has lessened over time. As farm credit extension becomes more formalized and institutionalized, individual lending will play a smaller role in real estate financing.

One type of individual loan for financing real estate being increasingly used is the land contract: the owner sells the land but retains title until the buyer has paid either in full or a substantial portion. Land contracts, like mortgages and the other more conventional real estate loans, are simply another way to buy land on a long-term basis. They generally require a lower down payment and carry reasonable interest rates. Less capital is tied up in land, leaving more available to buy machinery and operating inputs. Also, in many situations this type of arrangement can have tax advantages for the seller.

The chief disadvantage for the borrower is that in the event of failure to make payments on schedule, all preceding payments can be treated as rent, with the borrower losing any equity that has been paid.

In general, borrowing from individuals for real estate is a good method of financing, provided the interest rate is reasonable and the length of loan is of sufficient duration to allow a farmer to repay satisfactorily.

Amortized Loans

Most real estate loans are **amortized**—that is, repayments are made periodically until the entire loan is repaid. There are two types of amortized loans: (1) equal-payment and (2) decreasing-payment. In the first type, equal payments are made periodically, with the early payments allocated more to interest and less to principal. In later years, as the outstanding principal balance gets smaller, interest payments decrease, leaving more of each payment to reduce principal. In the second type, equal principal payments are made each period; in addition, interest is paid on the outstanding principal balance. Since the outstanding principal balance decreases over the life of the loan, the corresponding interest payments on that outstanding balance also get smaller, and so do the overall payments. These amortization plans are illustrated in Table 9–2. Most financial management software programs calculate loan amortization schedules.

Table 9–2

Loan Amortization Table: Annual Installments; Principal, $100,000; Interest Rate, 10 Percent; 30 Years

Annual Payment Number	Total Payment	Interest	Principal Payment	Unpaid Balance
equal-payment plan				
1	10,607.92	10,000.00	607.92	99,392.08
2	10,607.92	9,939.21	668.71	98,723.37
3	10,607.92	9,872.34	735.58	97,987.78
10	10,607.92	9,174.48	1,433.44	90,311.32
20	10,607.92	6,889.94	3,717.98	65,181.38
29	10,607.92	1,841.11	8,766.81	9,644.29
30	10,607.92	964.43	9,644.29	0
Totals	318,283.39	218,238.39	100,000.00	—
decreasing-payment plan				
1	13,333.33	10,000.00	3,333.33	96,666.67
2	13,000.00	9,666.67	3,333.33	93,333.33
3	12,666.67	9,333.33	3,333.33	90,000.00
10	10,333.33	7,000.00	3,333.33	66,666.67
20	7,000.00	3,666.67	3,333.33	33,333.33
29	4,000.00	666.67	3,333.33	3,333.33
30	3,666.67	333.33	3,333.33	0
Totals	255,000.00	155,000.00	100,000.00	—

Other Debt

While real estate debt is the largest single type of farm debt, other types of debt are important as well. These include current debt and noncurrent debt for capital assets other than land, such as machinery, equipment, and breeding livestock. Credit used for production purposes—for example, to purchase feed, fuel, seed, fertilizer, and chemicals—is usually borrowed for a period of one year or less and is referred to as "current debt" or "short-term debt." Noncurrent debt for assets other than real estate, which includes money borrowed to purchase durable items, such as tractors, machinery, autos, trucks, and home appliances, is commonly called "intermediate-term credit" because repayment periods may extend from 1 to 10 years. In the following discussion, current and intermediate-term debt are considered together.

Nonreal estate debt credit for farm operations is obtained from the following sources:

Type of Lender	Percentage of Total Nonreal Estate Debt Held
Commercial and savings banks	52
Farm Credit System	16
Farmers Home Administration	11
Individuals and others	21
	100

Trade Creditors

Trade creditors, i.e., merchants and dealers, included in the "individuals and others" category, are an important group of short-term farm lenders. They usually finance producers through a production cycle, for example, lending in the spring and collecting in the fall. Interest is usually charged, except in the case of certain supplies, such as fertilizer, where interest charges are part of the selling price of the item. If, for some reason, a farmer cannot repay at harvest, the loan is usually carried over into the next crop year with interest. Usually, mortgages or crop liens are not taken as collateral—thus, the term "open credit." Sometimes, the price of items acquired and the interest cost on trade credit are higher than if the farmer borrowed money and paid cash.

In recent years, equipment companies have become suppliers of trade credit. Equipment companies have become instrumental in the sale of automated poultry and livestock equipment and certain other farm machinery. As production of certain commodities gets more capital intensive, with more and more specialized equipment, direct equipment company financing will become more significant. While the interest rates charged by equipment companies are usually slightly higher than those charged by other agencies, the repayment terms may be more lenient. An added advantage to financing directly with the seller and installer of equipment is that farmers may get better service and attention to their equipment than if it is financed elsewhere. While most equipment company loans are for five years or less, there are instances where longer-term loans may be made. Also, special financing terms, such as no interest for 12 months, attract the attention of many potential borrowers.

Contract farming is involved in trade or merchant credit. Many farmers, unable or unwilling to obtain short-term credit, have turned to contract farming. The contractors supply the necessary items, and the growers are paid a certain fee for producing. The growers do not obtain credit directly except for the long-term investments they may be required to make, such as for land, buildings, and fixed equipment.

Trade creditors are generally a good source of short-term credit because of the flexibility, simplicity of operation, and numerous services provided. However, wide variation in their credit operations exists, and borrowers should appraise interest rates, repayment terms, prices charged for goods financed, and the reputation of the lender.

Commercial Banks

Commercial banks do considerable farmer financing over the short term. Loans are arranged for from 1 month to 36 months for seasonal crop needs, machinery, and consumer goods for the farm family. Banks are increasingly accommodating intermediate-term financing for from three to five years on tractors, farm machinery, and farm buildings. Interest rates charged by commercial banks are usually somewhat higher than those charged by cooperative credit or government credit sources. However, commercial banks provide a wide range of credit and banking services to the farm family and offer wise counsel in financing operations.

Except for small, unsecured loans, borrowers usually give promissory notes or pledge livestock, crops, and other personal property. Repayments are scheduled to coincide with sales of farm products or receipt of other income. Bank loans for production credit are generally smaller than those made by Production Credit Associations.

Farm Credit System

The Farm Credit System, discussed above in reference to real estate lending, is also a major player in providing nonreal estate credit to the agricultural sector. Production Credit Associations (PCAs) are the primary FCS associations that make direct short- and intermediate-term loans with funds obtained from Farm Credit Banks. Agricultural Credit Associations (ACAs) also make short-, intermediate-, and long-term loans. ACAs were created by the Agricultural Credit Act of 1987 through voluntary mergers of Federal Land Bank Associations and the Production Credit Associations.

Farm Credit Associations make short- and intermediate-term loans to farmers and other eligible borrowers for almost any purpose associated with farming, ranching, farm-related businesses, on-farm services, and rural life. While the maximum loan maturity is 10 years, maturity for most loans for machinery, equipment, and breeding livestock is between 3 and 5 years. The asset being financed is usually used as collateral to secure the loan. Unsecured loans generally involve shorter maturities, high equity positions, and strong credit worthiness by the borrowers.

The terms and repayment patterns of most loans are based on the borrower's projected cash flows. Intermediate-term loans may be of the equally amortized type or the constant-payment-on-principle type, with repayment plans scheduled to coincide with sales of farm products or receipt of other income. Operating loans are usually line-of-credit loans. That is, funds are extended to the farmer as needed and interest is charged only on the time the money is in use. The loans are repaid as income becomes available. This usually produces a lower total interest charge for the farmer. Service fees are usually charged to cover loan origination and servicing costs, and stock or fee requirements of the association's capitalization policy must be met.

In general, the Farm Credit Associations are a good source of short- and intermediate-term credit. Their interest rates and collateral requirements are reasonable; funds are budgeted; payments are geared to farmers' ability to repay; and credit services are both extensive and knowledgeable. Interest rates vary with their cost of money.

Leasing Companies

Leasing companies are playing a more important role in providing farm credit. Farmers can lease equipment from such companies instead of borrowing money to make outright purchases. This frees farmers from making extensive capital investments and provides them with more operating capital. One disadvantage is that lease payments are fixed, are often high, and must be paid out of current revenues.

Commodity Credit Corporation (CCC)

The Commodity Credit Corporation (CCC), established in 1933, functions primarily in connection with commodity price support programs of the U.S. Department of Agriculture to provide a comprehensive financial program that reduces price variability of many commodities. CCC programs also provide inventory management, increased marketing flexibility, and maintenance of farm incomes. The CCC administers longer-term storage and commodity reserve programs that extend the inventory financing and shift some of the marketing control to the government. Loans have also been available for the purchase, construction, and installation of storage and drying equipment on the farm.

A major activity of the CCC is making **nonrecourse** price-support loans on agricultural commodities. The loan amount is equal to the loan rate of the particular commodity multiplied by the quantity of the commodity in storage. The farmer then has two options: (1) if market prices are favorable, sell the stored crop on the open market and pay off the loan plus interest at a rate based on the government's cost of funds, or (2) if market prices remain below the loan rate, at the end of the loan contract, usually 9 months, transfer the commodity pledged as collateral to the CCC. This transfer of the commodity serves as full payment of loan principal and interest, hence the name "nonrecourse loan." The nonrecourse loan assures that the participating farmer receives a minimum price for that commodity. Commodities acquired by the CCC are disposed of through domestic and export sales, transfers to other government agencies, and donations for welfare.

Custom Hiring

While custom hiring is not a direct source of financing, its net effect is to reduce a farmer's need for capital by hiring work on a custom basis. A disadvantage is that custom work may not be available exactly at the time the farmer needs it most.

Banks for Cooperatives

The Banks for Cooperatives are a credit system devoted solely to meeting the needs of farmer cooperatives—i.e., cooperatives involved in processing, handling, or marketing farm or aquatic products; furnishing products or services to farmers; or furnishing services to eligible cooperatives. To be able to borrow from the Banks for Cooperatives, cooperatives must have at least 80 percent of their membership made up of farmers.

Banks for Cooperatives provide a major part of the credit used by farmer cooperatives and thus are an important part of the total credit system available to agriculture. Funds from these banks are generally not loaned directly to farmers. Farmers, however, are indirectly affected through membership in a cooperative. These banks will not be discussed further since they primarily finance business-type cooperatives, not individual farmers.

Credit for Farm Family Needs

In addition to production and real estate credit, the farm family may also need credit for consumer expenditures, such as for household furnishings and appliances, automobiles, medical bills, recreation, remodeling, education, clothing, and a host of other items.

Farm consumer credit may be obtained from some or all of the following sources, some of which have already been discussed: (1) trade credit with retailers, (2) banks, (3) the Farm Credit System, (4) finance companies, (5) credit unions, (6) individuals, and (7) life insurance policy cash values.

Procedures in Borrowing Funds

The more important factors in selecting a creditor are (1) integrity of the lender, (2) collateral and down-payment requirements, (3) length and terms of repayment, (4) foreclosure policies, and (5) interest rates.

Integrity of lenders may be ascertained through Better Business Bureaus and individuals acquainted with lenders and their policies.

Collateral and down-payment requirements can be ascertained by shopping among various lenders. Long-term loans usually require collateral consisting of real estate, cash values in life insurance policies, stocks, and other assets. Shorter-term loans may require collateral such as vehicles, farm machinery, livestock, crop liens, etc. The greater the down payment made on loans, the less total interest to be paid.

Length and terms of repayment will vary with the purpose of the loan and the lender's operating requirements. Commercial banks, for example, are restricted in the number of long-term loans they can make.

Foreclosure policies differ among lenders. The Farm Credit System and the FmHA, for example, may be more lenient in foreclosures than other lenders.

Interest Charges

Interest charged varies among lenders. Some creditors are limited by law as to interest charged. Small loans, unsecured and risky, command a higher interest rate than sound, large, well-secured loans. Interest rates are also affected by the cost of money, operating costs of lending institutions, custom and precedent, and competition between lenders. One way to classify interest rates is as (1) nominal and (2) effective. **Nominal interest rates,** stated annual interest rates compounded periodically, refers to the percentage usually quoted by lenders. **Effective interest rates** refers to actual or true interest, which depends not only upon the nominal rate quoted but also the frequency of repayment required.

The effective interest rate is the rate that compounded only once (one payment per year, for example) produces the same final value as the nominal interest rate. The effective interest rate, EIR, can be calculated as followed:

$$\text{EIR} = (1 + \text{Periodic rate})^{\text{number of periods per year}} - 1$$

$$\text{EIR} = (1 + \frac{i}{m})^m - 1$$

where i is the nominal interest rate and m is the number of payments per year.

Besides the interest charge, many lenders assess notary and recording fees, borrower's life insurance, application fees, and other charges.

Savings and Investments

Farmers, as well as nonfarmers, may use a variety of institutions and instruments for the investment of idle funds, such as (1) commercial banks, (2) savings and loan institutions, (3) credit unions, (4) U.S. savings bonds, (5) U.S. government securities, such as treasury bills, (6) corporate bonds, (7) state and municipal bonds (usually free of income tax), (8) mortgage company obligations, (9) Production Credit Association notes, and (10) preferred stocks. These 10 institutions and instruments usually have low risks and good interest yields but little or no opportunity for capital appreciation. Some of them have their dollar capital insured from loss by the U.S. government or its agencies.

Another group of investment opportunities with greater income and growth potential but with higher risks includes (1) common stocks, (2) mutual funds (closed-end or open-end, load or no-load), and (3) stock warrants (rights to purchase stocks). Mutual funds may provide a good investment vehicle. Some mutual funds are tailored for growth stocks, some for income, and others for both growth and income.

Another group of investment outlets involves considerably greater risk because of variable yield potential, but with the opportunity for capital appreciation. In addition, some may contain income tax sheltering opportunities. Among these are (1) rental income properties, both residential and commercial, (2) speculation in commodity markets (farmers may choose to be hedgers instead of speculators), (3) ownership in nonfarm business ventures, and (4) investments in precious metals, antiques, art objects, etc.

Another type of investment vehicle favored by farmers and others is land, for farming or otherwise. Land on the urban fringe that can be farmed and held for later appreciation in value has typically been an attractive investment opportunity.

Truth in Lending

Federal legislation has been passed that requires merchants and lenders to provide credit borrowers with full disclosure of credit costs, terms, and annual interest rates. The reason for this legislation is to protect consumers and businesses from misleading quotes of a contract interest that could result from different methods of charging interest and noninterest costs. The laws also regulate credit advertising and place certain restrictions on garnishment of wages. Credit borrowers are now afforded better information with which to evaluate interest costs and other credit terms.

Evaluating Credit Investments

The first principle in deciding to borrow money is to determine if the added return from the investment will more than pay the principal and interest charged. For example, if the interest charge is $10 per $100, then the added return from the investment should be more than $110 ($100 principal plus $10 interest). To establish such criteria, budgeting is usually employed.

Second, money borrowed for a long-term investment should not have to be repaid in too short a time. For example, a loan for a new milking parlor or a new tractor should not have to be repaid in one year.

Third, certain ratios should be maintained between current assets and current liabilities and between fixed assets and fixed liabilities. For example, an operation with current liabilities that far exceed its current assets has a poor **liquidity** position. It would have trouble freeing the capital to repay its short-term debt without disrupting normal farm operations.

Fourth, if a choice must be made between borrowing for short-term purposes and borrowing for long-term ones, choose short-term borrowing, particularly for income-generating purposes, such as for better seeds, fertilizer, feed, etc.

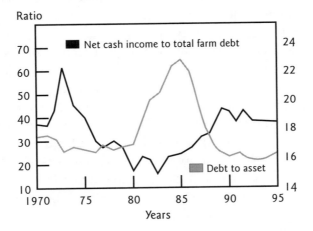

1994 preliminary. 1995 forecast.
Source: Economic Research Service, USDA

Figure 9–3

Fifth, in choosing between productive and consumptive investments, choose productive ones first. They have the potential of "paying for themselves."

Sixth, placing all credit business with one lender is advisable. It will not only be less costly but also will result in better and more comprehensive financial planning and counseling.

Three "R's" of Credit

In borrowing money, a farmer trades for his or her credit worthiness. The farmer should analyze what he or she has to trade on the basis of what a lender looks for in making loans: returns, repayment capacity, and risk—the three "R's" of credit.

The primary reason for using farm credit is to increase net farm returns (the first "R"). The question is, Will the net farm returns be greater with the use of added capital in the form of credit than if it were not used? A farmer must carefully choose among alternatives in using borrowed money. Borrowing should not be based on whims, personal preferences, or quick decisions. It is important to analyze each investment to determine its expected payback, as is discussed later in this chapter.

The second "R" is repayment capacity. Lenders expect their money to be repaid in full plus interest. Farmers must determine their capacity or ability to repay a loan. Lenders will also determine this. Since farmers usually pledge collateral for their loans, inability to repay may lead to foreclosure. The farmer's repayment character is important too, because some who can repay choose not to. Priority should always be given to loans that are self-liquidating, those that have earning capacity. For example, borrowing for laying pullets takes priority over borrowing for an automatic feeder to replace hand feeders. The latter type of borrowing does not have repayment capacity if the farmer cannot use the labor thus saved in a profitable manner. It is apt to become just another expense if no additional earnings result. Repayment capacity is also tied to sacrifices in consumptive living in favor of productive investments. It may be better to continue to drive a good used pickup and apply funds to productive investments than to drive a new pickup.

One must build some degree of owner equity before one can borrow successfully. Borrowing funds only when needed on a budget basis and repaying them promptly when funds become available is very critical in determining repayment capacity. Otherwise, wasteful spending may result when unused funds are left idle before the time they are due to be repaid.

Problems with risk, the third "R," usually arise because of uncertainty in production or technology, health, personal relationships, or prices. Borrowers with a strong "equity position" can usually assume more risk than those in less favored equity positions. Lenders tend to favor those farmers who have the ability to reduce operating and living expenses in adverse periods and who have stability and reliability of income and good equity position initially. Speculative investments must be given low priority because of their large risks. Enterprises with price and income stability are preferred. Farmers can potentially reduce risk by diversifying enterprises, although diversification should not be carried too far lest it reduce overall efficiency. Crop insurance can also lessen risk and is discussed later in this chapter.

Analyzing Capital Investments

Farm managers often are confronted with decisions involving the pay-out or profitability of an investment. Three of the most common methods used to analyze investments are (1) payback period, (2) simple rate of return, and (3) net present value.

Payback Period

The payback period indicates the length of time required for the expected earnings from an investment alternative to pay back the initial capital outlay.

If the projected cash flows for an investment (E) are uniform, the payback period (P), expressed as number of periods, is computed using this equation:

$$P = \frac{I}{E}$$

where I is the initial investment. If the projected cash flows are not uniform, then the payback period is given by the value of n.

$$P = \frac{I}{\sum_{n=0}^{N} E_n} = 1$$

For example, if the initial investment of a new piece of equipment is $10,000 and this equipment will generate additional cash flows of $2,000 per year, the payback period is five years.

$$P = \frac{\$10,000}{\$2,000} = 5 \text{ years}$$

Investments can be ranked based on their payback period, with shorter payback periods preferred. The payback period can also be compared to a standard payback period (set by the decision maker) to determine if the project is acceptable. If the calculated payback period is less than the standard payback period, the investment is acceptable.

The payback period is commonly used because of its simplicity. However, it has some drawbacks. The payback period is not a measure of profitability but simply a measure of how quickly the initial investment will be recovered. It does not consider any earnings after the payback is completed, which penalizes investments with returns patterns that increase over their lifetime. Also, the payback period does not account for any differences in the timing of cash flows while the initial investment is being recovered.

Simple Rate of Return

The simple rate of return expresses the average annual profits generated by an investment as a percentage of either the original investment or the average investment over the investment's expected life. The simple rate of return, SRR, can be calculated as follows:

$$SRR = \frac{Y}{I}$$

where Y is the expected average annual profits with depreciation subtracted and I is the initial investment or the average investment over the investment's life.

Average lifetime investment may be computed by averaging the beginning and ending investment, disregarding the method used for depreciation. Average investment in the preceding example is $5,000, assuming no salvage value. Assuming a life of 10 years, average annual depreciation is $1,000. Using the average investment, the simple rate of return on the investment is 20 percent.

$$SRR = \frac{\$2,000 - \$1,000}{\$5,000} = 20 \text{ percent}$$

Using the initial investment rather than the average investment results in a lower SRR.

$$\text{SRR} = \frac{\$2{,}000 - \$1{,}000}{\$10{,}000} = 10 \text{ percent}$$

Alternative investments can be ranked according to their simple rate of return, with higher simple rates of return preferred. Individual investments can also be judged against a standard or required rate of return set by the decision maker. If the simple rate of return is higher than the required rate of return, then the investment is acceptable. If it is not, the investment is rejected.

The major limitation of the simple rate of return is its failure to take into account the timing of cash flows. This is especially important for investments whose cash flows are variable over time.

Net Present Value

The basis of the net present value method is that cash flows received in future years are not equal to those received currently. Therefore, future cash flows must be discounted using an appropriate discount rate. "Discounting" refers to the process of finding the present value of future cash flows. The future value is discounted to a lower present value to account for the effects of the **time value of money**. The discount occurs because the investor must wait to receive the future payment and cannot invest it at present in an alternative investment opportunity yielding interest rate i.

Net present value, NPV, which incorporates the time value of money concept into the investment analysis process, is calculated using this equation:

$$\text{NPV} = -\text{INV} + \frac{P_1}{(1+i)} + \frac{P_2}{(1+i)^2} + \ldots + \frac{P_N}{(1+i)^N} + \frac{V_N}{(1+i)^N}$$

where INV is the initial investment; $P_1, P_2, \ldots,$ and P_N are the projected cash flows in each period of the investment's life; V_N is the salvage value of the investment; N is the length of the planning horizon; and i is the interest rate or required rate of return, also called the "discount rate" or "cost of capital."

The present value of a dollar received over a period of years at various rates of interest is shown in Table 9–3. The dollar income or cash flow, which is discounted, consists of net income or profit after income taxes, depreciation, and interest on investment funds.

Table 9–3

Present Value of $1 Discounted at Various Rates of Interest and by Year Received

Year	Interest Rate (%)						
	3	5	7	9	11	13	15
	Present value ($)						
1	.97	.95	.93	.92	.90	.89	.87
2	.94	.91	.87	.84	.81	.78	.76
3	.92	.86	.82	.77	.73	.69	.66
4	.89	.82	.76	.71	.66	.61	.57
5	.86	.78	.71	.65	.59	.54	.50
6	.84	.75	.67	.60	.53	.48	.43
7	.81	.71	.62	.55	.48	.43	.38
8	.79	.68	.58	.50	.43	.38	.33
9	.77	.64	.54	.46	.39	.33	.28
10	.74	.61	.51	.42	.35	.29	.25
15	.64	.48	.36	.27	.21	.16	.12
20	.55	.38	.26	.18	.12	.09	.06
25	.48	.30	.18	.12	.07	.05	.03
30	.41	.23	.13	.08	.04	.03	.02
35	.36	.18	.09	.05	.03	.01	.01
40	.31	.14	.07	.03	.02	.01	°
45	.26	.11	.05	.02	.01	°	°
50	.23	.09	.03	.01	.01	°	°

° = less than $.01.

The underlying principle is that, in making the current investment outlay, the investor is actually "buying" a series of future annual incomes. The investor cannot afford to make an outlay larger than the value of the future cash flows discounted to their present value at certain rates of interest.

The sign and size of the NPV of an individual investment are used to determine its ranking and acceptability. If the NPV is greater than zero, given the discount rate used, then the investment is considered acceptable. An investment with a negative NPV is rejected. If several investments under consideration have positive net present values, then the investment with the largest NPV is preferred, with the investment with second highest NPV ranking second best, and so on.

Assume a farmer is considering an investment with an expected useful life of 10 years that will require an initial outlay of $25,000. The farmer expects to generate annual net cash flows of $4,000 over the investment's life and a salvage value of $5,000. Assuming a discount rate of 8 percent, should the farmer make the investment based on its net present value? Using the equation above,

$$\text{NPV} = -\$25,000 + \frac{4,000}{(1.08)} + \frac{4,000}{(1.08)^2} + \ldots + \frac{4,000}{(1.08)^{10}} + \frac{5,000}{(1.08)^{10}}.$$

NPV equals $4,156, which would be acceptable, since the NPV is positive.

Insuring Capital Resources

Capital resources, such as farm homes, buildings, vehicles and equipment, etc., should be adequately insured to protect farmers' investments in them. The types of insurance on capital goods might include fire, windstorm, hail, personal and property damage liability, among others.

Generally, the types of personal and business liability a farmer faces in managing assets are the following:

1. *As an individual,* the operator may be held liable for the consequences of his or her negligible acts.

2. *As a head of a family,* the operator may, under certain circumstances, be held liable for the negligence of family members.

3. *As the owner of land, improvements, and personal property,* the operator may be held liable for negligence in the upkeep of the property or for damage done by the farm's animals.

4. *As an employer,* the operator may be held liable for negligence that results in harm to employees or for injuries to others by employees.

5. *As an owner of machinery, automobiles, and trucks,* the operator may be held responsible for their negligent use in connection with the family or business.

All these hazards can be properly insured by consulting qualified insurance brokers or agents. Generally, a multi-peril or comprehensive coverage policy is cheaper and better suited for the farmer's interests. Deductible clauses, within the farmer's ability to sustain some losses, will reduce the cost of insurance premiums. Also, placing all insurance coverage with one reliable company generally lowers premium cost, provides better service on claims, and affords better insurance planning and coverage.

Crop Insurance

The federal government, through the Federal Crop Insurance Corporation (FCIC), provides insurance on selected crops in certain counties and states. Insurable hazards include drought, excessive moisture, insects, hail, frost, freeze, hurricanes, plant diseases, and certain other hazards. Crop insurance helps farmers not only in reducing financial risks but also in obtaining production credit more readily.

For most crops, the FCIC policies insure the yield and quality required to cover approximate costs of production. Designed only as protection against disaster, FCIC is not allowed to insure profits. For each bushel or pound that the crop falls short of the yield guarantee, the policy pays an indemnity; that is, makes up the loss in cash.

The amount of indemnity payment per bushel or pound is selected by the farmer from a low, medium, or high value at the time the policy is purchased. The premium is figured accordingly and may be paid at time of harvest.

Livestock and Poultry Insurance

For livestock and poultry, private insurance companies offer policies to cover hazards such as death from fire, lightning, and windstorm. Disease insurance, if available, is usually quite costly.

Future Needs for Farm Capital

As farmers become larger and more specialized, they purchase more inputs and require more production credit. As they demand more machinery and equipment, in general, and more specialized machinery and equipment, in particular, they need more intermediate-term credit. For farmers expanding their acreage, more long-term credit is required. Therefore, capital needs of farmers and farm families will likely continue to increase in the years ahead.

However, in relation to increased productivity, all else equal, the relative cost of capital keeps declining. As capital becomes cheaper relative to the cost of labor and land, farmers will continue to use more capital.

Full ownership of resources will not be a necessity for many farmers in the future. More land, machinery, and equipment will be leased instead of purchased, and custom work will also increase. For certain commodities, contract farming will replace production credit to a substantial extent. Young farmers, with low equity, will increasingly use short- and intermediate-term credit instead of acquiring full ownership through long-term credit.

Farm incorporation will provide opportunities for acquiring capital while providing limited liability for stockholders. In many cases, a small corporation, one with 10 or fewer persons (a Subchapter S corporation), is taxed as if it were a partnership. Incorporation likewise provides opportunities for capable young managers to enter farming and gradually acquire an increasing share of the farm business.

The introduction of big-business methods into farming will provide non-farm capital for agriculture and for the hiring of college-trained specialists to manage the vast and complex farming enterprises of the future. A farmer will likely procure most credit from one lending institution.

In the more highly specialized, larger farms of the future, the rewards for success and the penalties for failure will be much greater than in the past. Thus, lending policy for the larger loans will rely more on the operator's qualifications, including the potential for business success, and less on asset security.

Topics for Discussion

1. Define "capital."
2. What is the average U.S. farmer's asset, liability, and net worth positions?
3. Discuss the sources available for long-term and short-term credit.
4. Discuss amortized loans.
5. Discuss the difference between nominal and effective interest rates.
6. Discuss the evaluation of credit investments.
7. Name and discuss the three "R's" of credit.
8. Discuss the three methods for analyzing the feasibility of capital investments.
9. Discuss the insuring of capital resources.
10. Discuss the future capital needs of farmers.

Problem Assignment

Interview a farm operator, and obtain data on his or her enterprise or enterprises' costs and returns situation. Obtain a balance sheet for the end of the last period, and project the next year's credit needs. Take this information to one or more lending agencies, without revealing the farmer's name, and obtain information on how the lender would appraise these credit needs.

Recommended Readings

Ahrendsen, B. L. *Federal Legislation: Agricultural Credit Issues and Institutions.* Special Report No. 171. Fayetteville: Agricultural Experiment Station, University of Arkansas, November 1995.

Barry, Peter J., Paul N. Ellinger, John A. Hopkin, and C. B. Baker. *Financial Management in Agriculture,* 5th ed. Danville, IL: Interstate Publishers, Inc., 1995. Ch. 9, 10, 14, 17.

Bowers, Wendell. *Machinery Replacement Strategies.* Moline, IL: Deere & Company Service Publications, 1994.

Lins, David A. *Time Value of Money and Investment Analysis: Explanation and Applications for Agriculture and Agribusiness.* Report 93-F-5. Urbana: The Center for Farm and Rural Business Finance, Department of Agricultural Economics, University of Illinois, September 1993.

Markley, Deborah M., and Ron Shaffer. "Rural Banks and Their Communities: A Matter of Survival," *Economic Review,* 78(3):73–79 (1993).

Melichar, E. "Financing Agriculture: Demand for and Supply of Farm Capital," *American Journal of Agricultural Economics,* 55(2):313–325 (1973).

Peoples, Kenneth L., et al. *Anatomy of an American Agricultural Credit Crisis: Farm Debt in the 1980s.* Lanham, MD: Rowman & Littlefield Publishers, Inc., 1993.

10

Selected Characteristics of U.S. Farming Operations

Farming operations in the United States are very diverse. There is a wide variety of types of farming enterprises, sizes of farms, incomes, tenure, labor, and other factors. The purpose of this chapter is to briefly explore important aspects of U.S. farming operations. We begin by discussing the varied types of farming areas in the United States.

Types of Farming Areas[1]

The United States has been divided into 10 major farm production regions (Figure 10–1). These regions differ in soils, slope of land, climate, distance to market, and storage and marketing facilities. The diversity of farming operations has some advantages: (1) Crop failures in one area may be overcome by crop successes in other areas. (2) Interregional competition is enhanced as one area tries to produce more efficiently than another. (3) Localized demand for certain farm products is met. (4) Farmers have a wide range of options in selecting farming enterprises.

The Northeast and the Lake States are the major milk-producing areas of the United States. The climate and soils in these regions are well suited for raising grains and forages for cattle and for providing pasture land for grazing. Broiler production is important to the Northeast, specifically Maine, Delaware, and Maryland. Also, fruit and vegetables are important to that region.

---

[1] Adapted from Office of Communications, *Agriculture Fact Book—1994* (Washington, DC: U.S. Department of Agriculture, 1994).

U.S. farm production regions

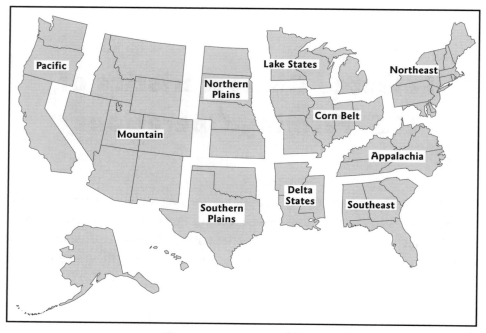

Source: U.S. Department of Agriculture

Figure 10–1

Appalachia is the major tobacco-producing region of the United States. Peanuts, cattle, hogs, and dairy production are also important.

In the Southeast, beef and broilers are the major livestock commodities. Fruit, vegetables, and peanuts are important crops. Agricultural producers in Florida provide much citrus and many winter vegetables.

The principal crops of the Delta States include cotton and soybeans. Rice and sugar cane are also grown. With improved pastures, livestock production has increased in importance. The Delta States are also major broiler and catfish producers.

With its rich soils and good climate, the Corn Belt produces much of the nation's corn and hogs. Beef cattle, dairy products, soybeans, wheat, and other feed grains besides corn are also major farm commodities.

Farm production in the Great Plains region, which extends from Canada to Mexico, is limited by rainfall in the west and short growing seasons in the north. About 60 percent of the nation's winter and spring wheat is produced in the Great Plains. Other small grains, hay, forage crops, and pastures form the basis for raising cattle. Cotton is an important commodity produced in the southern part of the region.

The terrain limits the potential for agricultural production in much of the Mountain States region. Much of the area is suited for cattle and sheep production. Wheat is an important commodity in the northern parts. Irrigation enables the production of hay, sugar beets, potatoes, fruit, and vegetables.

The Pacific region includes the three Pacific Coast states, plus Alaska and Hawaii. Farmers in Washington and Oregon specialize in raising wheat, fruit, and potatoes; vegetables, fruit, and cotton are important in California. Cattle are raised throughout the region. Sugar cane and pineapples are the major crops in Hawaii, while horticultural and dairy products are the major crops in Alaska.

Sizes of Farms

The U.S. Census Bureau defines a farm as "any establishment from which $1,000 or more of agricultural products are sold or would normally be sold during a year." In addition, a distinction is made between commercial and

Number of farms compared with land in farms

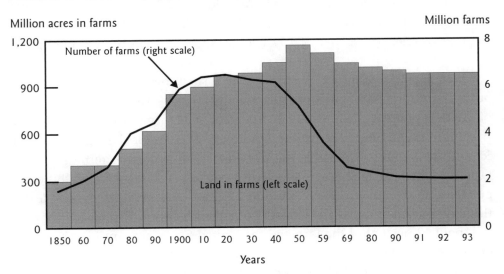

Source: National Agricultural Statistics Service, USDA

Figure 10–2

noncommercial farms. Commercial farms have gross sales of agricultural products of $50,000 or more. Noncommercial farms have gross sales of less than $50,000.

The number of farms has been declining since the 1930s, while the number of acres per farm has been steadily increasing since 1910. Farms below 180 acres account for 59 percent of all farms by number but only 8 percent of the acres in farms. Those from 180 to 999 acres account for 32 percent of all farms and 27 percent of the acreage. Farms with 1,000 to 1,999 acres account for 5 percent of all farms and 15 percent of the acreage. Farms that are 2,000 acres in size or larger account for only 4 percent of all farms but 50 percent of the land in farms.

Economic Classes of Farms

As mentioned previously, there are two types of farms: commercial and noncommercial. Table 10-1 shows that farms are also classified by level of the value of sales. The numbers reported in this section reflect the continuation of trends observed in the agricultural sector for several decades.

Most farms in the United States are small. Almost 50 percent of all farms have gross sales of less than $10,000. An additional 25 percent of U.S. farms have gross sales between $10,000 and $25,000. However, these smaller farms (with gross sales of less than $50,000) account for less than 10 percent of gross sales.

In contrast, large farms constitute only a small percentage of total farms but account for a substantial proportion of gross sales. Farms with gross sales of $100,000 or greater, while only 17.3 percent of all farms, account for over 82 percent of gross sales of agricultural products. The largest class, gross sales of $1 million or more, illustrates the dominance, in economic terms, of large farms in American agriculture. Less than 1 percent of all farms are in this class, but these farms account for 33 percent of gross sales. Average gross sales of this largest class of farms are almost $3.5 million per farm.

About 12 percent of the farms in the United States account for 75 percent of total gross sales. The trend toward larger and larger farming operations is expected to continue. Technical efficiency arising from economies of size is probably the most important factor influencing farm size. The ability to farm very large acreages with sophisticated technology will continue to encourage larger farms.

Table 10-1

U.S. Farms Classified by Value of Sales Class, 1993

Value of Sales Class	All Farms	Cash Receipts
Less than $20,000	53.6	6.2
$ 20,000 – $ 39,999	12.3	4.5
$ 40,000 – $ 99,999	15.2	12.1
$ 100,000 – $249,999	12.5	20.7
$ 250,000 – $499,999	3.9	14.6
$ 500,000 – $999,999	1.7	12.6
$1,000,000 or more	0.8	29.3
Total	100.0	100.0

Source: Economic Research Service, USDA.

Farm Income

Farm income can be measured in a variety of ways. Conventionally, cash receipts from farm product sales are added to direct government payments and other farm-related income to yield total cash farm income. Net cash farm income equals gross cash farm receipts less cash expenses. Subtracting depreciation, operator wages, and rent from net cash farm income yields self-employment farm business income.

The household share of self-employment farm business income is calculated by multiplying self-employment farm business income by the share the household receives. To this, operator wages and rent are added to yield the share of self-employment income from farming. Farm operator household income is found by adding total off-farm income to the household share of self-employment income from farming. Total off-farm income includes wages, salaries, and off-farm business income received, along with interest, dividends, and transfer payments received.[2]

In 1992, the average household share of farm income was just over $4,330 and the average off-farm income was $35,731, yielding an average household income of $40,061.

---

[2] Economic Research Service, *Economic Indicators of the Farm Sector: National Financial Summary, 1993*, ECIFS 13-1 (Washington, DC: U.S. Department of Agriculture, 1994).

Table 10-2

Gross Income, Production Expenses, and Net Income per Farm, by Value of Sales Class, 1993

Value of Sales Class	Per Farm ($)			Percentage of Gross Income		
	Gross Income	Production Expenses	Net Income	Gross Income	Production Expenses	Net Income
Less that $20,000	11,084	9,274	1,810	100.0	83.7	16.3
$ 20,000 – $ 39,000	35,727	26,813	8,915	100.0	75.0	25.0
$ 40,000 – $ 99,999	81,374	58,954	22,419	100.0	72.4	27.6
$ 100,000 – $249,999	167,827	120,714	47,113	100.0	71.9	28.1
$ 250,000 – $499,999	363,495	262,773	100,722	100.0	72.3	27.7
$ 500,000 – $999,999	688,699	511,800	176,899	100.0	74.3	25.7
$1,000,000 or more	3,085,234	1,887,585	1,197,649	100.0	61.2	38.8

Persistence of Small Farms

A logical question at this point is, Why do small farms persist? Many of the small farm operators are older and are either unable or unwilling to adjust to other activities. Since they usually own their small farms, operators have low overhead costs. Also, family demands on income have decreased as the children have grown and moved away. These operators also lack the operating capital to expand and would be reluctant to borrow capital even if it were available.

The viability of small farms is dependent on the nonagricultural economy in rural areas, as well as the agricultural economy. Most small farm operators are quite dependent on off-farm employment to supplement their farm income. The number of farms with less than 50 acres fell by 7 percent between 1987 and 1992. This decrease is a result, in large part, of a decrease in manufacturing and other jobs in rural areas.

Importance of Nonfarm Income

Off-farm income consists of wages, rent, interest, dividends, social security, pensions, welfare payments, and income from nonfarm businesses and professions. Off-farm income is becoming relatively more important to total farm family income each year. For the agricultural sector as a whole, off-farm income accounts for almost 90 percent of average total farm family income.

In 1992, over 66 percent of farm operator households reported receiving off-farm wages, salaries, and nonfarm business income, and about 68 percent reported receiving interest and dividend income. Just under 30 percent of farm households had household incomes of less than $15,000, while 37 percent had household incomes between $15,000 and $37,000. The remaining 34 percent had household incomes greater than $37,000. The reliance on off-farm income is expected to continue to increase in the future.

Farm Tenure

Farmers may be divided into three basic tenure groups: (1) full owners, (2) part owners, and (3) tenants.

Full owners of farm land account for about 58 percent of all farm operators but operate only 31 percent of the total farm acreage in the United States. Part owners comprise about 31 percent of all farm operators and farm 56 percent of the land in farms. Tenants account for 11 percent of all farm operators and farm only 13 percent of the land.

Farms and land in farms by tenure of farm operator, United States, 1992

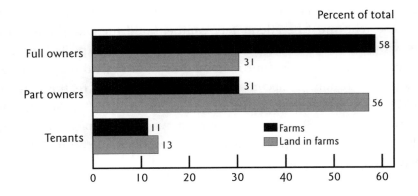

Source: U.S. Census of Agriculture, 1992

Figure 10–3

The percentage of farm land operated by full owners has been declining, while the portion operated by part owners and tenants has been increasing. Many full owners are operators of small farms, some too small to provide an adequate living.

Leasing has evolved from the entry rung on the tenure ladder to merely another device for gaining access to land. In 1940, part owners (those who rent part and lease part of the land they farm) leased 33 percent of all land leased. In 1987, part owners leased 68 percent of all land leased. Part owners have become the dominant tenure group in terms of production agriculture.[3]

Types of Leases

There are two basic types of farm renters: (1) cash tenants and (2) share tenants. A cash tenant pays a set fee per acre of land farmed and has about the same managerial responsibility as an owner. A share tenant pays rent by sharing a portion of production or returns with the landowner. The specific share depends on the amount of resources and managerial responsibility furnished by each of the two parties. In general, landlord participation in day-to-day farm operations and seasonal production decisions is somewhat limited. Landlords select renters and decide on rental acreage, buildings and improvements, environmental effects of land use, and conservation.

About 65 percent of farm land leased is under cash rentals, 30 percent under share rentals, and the remainder under combination or some other arrangement. Because cash rents are less likely to vary by crop or market conditions than are share rents, cash rents are favored by risk-averse landlords (those who must be adequately compensated for taking risks) and by renters who prefer to absorb risk for a somewhat higher expected return.[4]

Leasing as a way to gain control of additional farm land may have several adverse effects. It may (1) become exploitative in nature, creating a tendency to neglect soil conservation; (2) be unfair to one or both parties in lease or rental arrangements; (3) favor row-cropping over other types of farming; and (4) fail to compensate tenants for any improvements made on rental land and property.

---

[3] Margot Anderson (ed.). *Agricultural Resources and Environmental Indicators*, Agricultural Handbook No. 705 (Washington, DC: Economic Research Service, USDA, 1994), p. 20.

[4] Ibid.

Percentage of cash and share leases by region, 1988

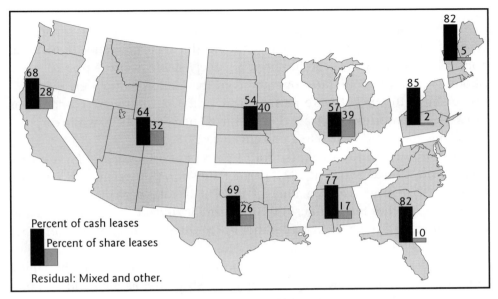

Source: Economic Research Service, USDA

Figure 10-4

These disadvantages may be overcome by (1) agreeing on compensatory payments to tenants who improve land and property, (2) negotiating lease terms longer than for one crop year, and (3) negotiating leases in writing and recording them.

Other Types of Tenure

There are two other types of farm operators: (1) hired managers and (2) contract farmers. Hired farm managers are important in the operation of very large farms and multiple farm units. Farmers who operate on contract with processors or integrators are most frequently engaged in production of broilers, table eggs, turkeys, processed vegetables, sugar beets, and other specialized crops. Contract hog production has increased considerably in some parts of the country. Contract farmers usually own their farms and provide permanent capital improvements (broiler houses, for example), while integrators provide capital, production supplies, and marketing assistance.

Farm Business Organization

Farm operations in the United States are organized into one of three basic types of business organization: sole proprietorships, partnerships, or corporations. The percentage of each has been relatively stable over the past 20 years. About 86 percent of the farm operations in the United States were sole proprietorships, 10 percent were partnerships, and 4 percent were corporations in 1992. Sole proprietorships are down from 87 percent in 1978, while corporations have increased from 3 to 4 percent since 1978.

Sole proprietorships accounted for 54 percent of total sales by farm operations in the United States in 1992, partnerships for 18 percent, and corporations for 27 percent. The remaining 1 percent of total agricultural sales was from firms of other types of business organization, including cooperatives and estates or trusts. Family-held corporations accounted for 21 percent, and other corporations for 6 percent of total farm sales.

The most common type of corporation involved in farming is the family farm that has adopted the corporate business structure. Family-held corporations constitute 89 percent of all farm corporations. Of these, 97 percent have less than 10 stockholders.

A second type, making up 11 percent of farm corporations, is one formed by a group of unrelated individuals for the purpose of owning or operating a farm. Of corporations of this type, 86 percent have less than 10 stockholders.

The family farm corporation, as discussed earlier, is actually a desirable alternative, since it provides the farm family with the advantages of a corporation without many of its disadvantages. By incorporating, the farm family can borrow money more readily, will retain the farm within the family when the operator dies, and may pay less income taxes than it might otherwise.

Labor Inputs Used in Agriculture[5]

Labor use on U.S. farms has changed dramatically over the last several decades. Average farm employment dropped from 9.9 million in 1950 to 3.1 million in 1992. This decrease was primarily the result of the trend toward fewer and larger farms, increased farm mechanization and other technological innovations, and higher off-farm wages. However, farm employment appears

---

[5] *Agriculture Fact Book—1994*, p. 24.

to have stabilized somewhat over recent years as increases in mechanization and labor-saving technology have leveled off and the downward trend in farm numbers has slowed.

Family workers, including farm operators and unpaid workers, accounted for 64 percent of farm labor in 1992, while hired workers accounted for 28 percent. A recent change in farm labor use patterns has been the increased use of service workers, including crew leaders and custom crews, who accounted for 8 percent of all workers on farms in 1992 compared to less than 2 percent in 1980.

A significant portion of total farm production expenses is spent on farm labor. According to the 1992 Census of Agriculture, hired and contract labor totaled almost $15.3 billion in 1992, nearly 12 percent of total farm production expenses. About 36 percent of all farms had hired labor expenses, and 12.5 percent had contract labor expenses.

The importance of labor varies considerably by type and size of farm. The proportion of total farm production expenses attributed to hired and contract labor was greatest on horticultural specialty farms (45.4 percent), fruit and tree-nut farms (40.1 percent), and vegetable and melon farms (36.7 percent). These types of farms are least mechanized, with many of the commodities they produce still harvested by hand. In contrast, labor expenses comprised less than 6 percent of total farm expenses on cash grain, livestock, and poultry farms.

Larger farms are more likely to have labor needs that exceed what can be provided by the farm family. Farms of less than 260 acres (68 percent of all farms) paid 30 percent of total expenses for hired and contract labor. About 53 percent and 56 percent of the farms that hired labor and contracted labor, respectively, were in this group of smaller farms. In contrast, farms of over 1,000 acres in size accounted for 17 percent of the farms that hired labor and contracted labor but accounted for 41.5 percent of total farm expenditures for hired and contract labor. In fact, only 26 percent of all farms paid 75 percent of hired farm labor expenses in 1992.

Of the farms that hired labor, 89 percent had workers who worked less than 150 days in 1992. In contrast, about 43.5 percent of the farms that hired labor had workers employed for 150 days or more. These statistics point to the seasonal or temporary nature of much of the employment in production agriculture. People employed at nonfarm jobs who do seasonal farm work to supplement their nonfarm earnings make up an important part of hired farm workers.

The average wage rate for hired farm workers in the United States in 1992 was $6.06 per hour. Wages varied by type of worker: livestock workers averaged $5.55, field workers averaged $5.69, and supervisors averaged $9.58.

Farm wage rates per hour are highest in the Pacific region, New England, and California and lowest in the Southeast and the Delta States. Minimum wage legislation applicable to workers on large farms has tended to narrow somewhat the differentials in hourly wages paid in various regions of the country. It has also caused widespread unemployment for workers not skilled enough to earn the minimum wages. While farm wage rates paid per hour are lower than nonfarm wage rates, farm workers probably have lower living costs in terms of housing and have other benefits in terms of gardens, poultry, etc.

Some of the steps necessary to provide hired farm workers a better economic and social environment include:

1. Developing scales of compensation more in line with those of other types of employment.

2. Providing more continuous employment for seasonal farm workers.

3. Training supervisors to respect their workers as individuals and cooperators in the farm production process.

4. Upgrading housing and giving greater consideration to family integrity and community status of workers.

Topics for Discussion

1. List and discuss the types of farming in your area, state, or country. Why are these enterprises typical of your location?

2. Distinguish between commercial and noncommercial farms.

3. What are some reasons for the persistence of small farms? Why has the number of small farms declined over recent years?

4. Discuss the importance of nonfarm sources of income.

5. Discuss the various types of farm tenure in the United States.

6. Discuss the role of hired farm workers in the agricultural economy.

Problem Assignment

Select a specific type of farm tenure and present its advantages and disadvantages relative to those of other types of tenure.

Recommended Readings

Anderson, Margot (ed.). *Agricultural Resources and Environmental Indicators.* Agricultural Handbook No. 705. Washington, DC: Economic Research Service, USDA, 1994.

Barkema, Alan, and Michael L. Cook. "The Changing U.S. Pork Industry: A Dilemma for Public Policy," *Economic Review,* 78(2):49–65 (1993).

Barkema, Alan, and Mark Drabenstott. "Agriculture Rides Out the Storm," *Economic Review,* 79(1):29–43 (1994).

Economic Research Service. *Economic Indicators of the Farm Sector: National Financial Summary.* Washington, DC: Economic Research Service, USDA. Annual issues.

Office of Communications. *Agriculture Fact Book—1994.* Washington, DC: U.S. Department of Agriculture, 1994.

Smith, Deborah T. (ed.). *Americans in Agriculture: Portraits of Diversity—1990 Yearbook of Agriculture.* Washington, DC: U.S. Department of Agriculture, 1990.

Smith, Stew. "`Farming'—It's Declining in the U.S.," *Choices,* 7(1):8–10 (1992).

Stanley, Julie A. "Agricultural Biotechnology: Dividends and Drawbacks," *Economic Review,* 76(3):43–55 (1991).

Tweeten, Luther, and Carl Zulauf. "Is Farm Operator Succession a Problem?," *Choices,* 9(2):33–35 (1994).

Urban, Thomas. "Agricultural Industrialization: It's Inevitable," *Choices,* 6(4):4–6 (1991).

PART THREE

11 Specialization and Comparative Advantage

12 Physical Production Relationships

13 Costs and Revenue

14 Optimum Levels of Output

15 The Supply Concept

16 The Demand Concept

17 The Interaction of Supply and Demand: Prices

18 Market Structures and Competition

PART THREE

Specialization and Comparative Advantage

Enterprise Relationships

Most businesses are operated with the objective of earning a profit. Economists usually assume that entrepreneurs maximize profits. However, in reality, other objectives also influence decision making. Businesses are interested in increasing profits and will attempt to do so unless this activity interferes with something valued more highly, such as leisure time.

The choices students face are analogous to those faced by businesses. If we assume grades are to a student what profits are to a business, we can see the motive to make good grades. But do students really want to maximize their grade point average? Only if to do so does not interfere with something valued more highly—perhaps organizational involvement, social activities, sporting events, etc.

Businesses have identified acceptable methods that increase profits by increasing returns or reducing costs. One of the most common methods is to combine several enterprises into one business or to vary the combination of enterprises. These enterprises cannot be randomly chosen, however. They must fit together in a special way if increased profits are to result.

Supplementary Enterprises

Some businesses, or at least portions of those businesses, are idle during parts of the year. In such a business there may be opportunities to add enterprises that will "fill in" or utilize the firm's idle resources without interfering with the primary enterprise. For this reason, sporting facilities house not only diverse sports events but also concerts, conferences, etc. Cotton gin operators sell feed, seed, and fertilizer in the off-season, and many grain farmers grow cattle. The newly adopted activity that utilizes otherwise idle resources is a **supplementary enterprise.** It supplements income received from the primary activity without interfering with the business activity of that enterprise.

Complementary Enterprises

The fact that some organisms get along better in the presence of other organisms is common knowledge. It is not so widely known that some business enterprises also exhibit this relationship. Merchants know that sales are better when credit is offered at the same place where the goods are sold and that credit agencies thrive best when they operate in connection with retail establishments. Farmers discover that rice yields are increased when soybeans or legume-type forages are grown in rotation with the rice crop. The academic success of students is usually enhanced when they do internships related to their major. When one activity contributes to the success of another, the two are said to be **complementary.** Businesses search for complementary enterprises as a means of expanding profits.

Competitive Enterprises

An activity that was begun as supplementary or complementary to the primary activity may be expanded to the point that one detracts from the successful performance of the other. Then, the enterprises are said to be **competitive.** When students spend so much time working or participating in extracurricular activities that they do not have time to study, their work or other activities become competitive with good grades. Retailers who overextend credit must use employees to collect bad debts and maintain credit accounts when they might otherwise use them to expand sales. A rice farmer who increases soybean or forage acreage begins to lose money because of the associated reductions in rice acreage.

Entrepreneurs hope to avoid competitive enterprises as much as possible and confine their business organizations to enterprises that are supplementary, complementary, or at least independent. If two enterprises are **independent,** then one neither adds to nor detracts from the other. If resources are adequate, then enterprises that might ordinarily be competitive may be independent enterprises. For example, a farmer with access to sufficient land may grow both rice and cotton without experiencing either a mutual benefit or a disadvantage.

Specialization Versus Diversification

We live in an age of **specialization**. Individuals, farmers, and businesses find that they can be more successful if they engage in only one activity and reach perfection in their mastery of that one task or one type of business. For that reason different persons may specialize in welding, brain surgery, teaching, computer programming, cotton farming, dairy production, and so on. With specialization, the total output of all is greater than what would occur if each person attempted to carry on some of all these activities.

But there are disadvantages to specialization. What happens to a heart surgeon whose hands become affected by arthritis, to a cotton farmer when synthetic fibers capture a large share of the textile market, or a strawberry farmer whose crop is destroyed by a freeze? Without supplementary, complementary, or joint enterprises to fall back on, specialized businesses or individuals are exposed to adverse movements in their specialties. For that reason there is incentive for **diversification**, engaging in several enterprises or activities to provide protection against the risk of adversity. When one is diversified, seldom would a single calamity affect all activities to the same degree. For example, should changes in taste and preferences lower demand for fresh butter, the sales of cottage cheese, skim milk, and frozen yogurt may not be similarly affected; thus, a dairy marketing firm handling diverse products would not lose all its business. A hail storm may destroy a farmer's cotton crop, with little or no effect on income from small grains and livestock enterprises.

Despite the advantages of diversification, the efficiencies gained from specialization would seem to be more than sufficient to offset them. Specialization in all areas is occurring at an ever-increasing rate. The higher standard of living in specialized economies is evidence of the benefits to be gained from specialization.

The Principle of Comparative Advantage

It is interesting to note that the form of specialization varies from one region of the country to another, just as it does from one individual to another. Not all regions attempt to specialize in cotton, electronics, or furniture production. Neither do all individuals attempt to become eye specialists or construction workers. It is obvious that neither individuals nor areas are equal with respect to resources and capabilities of performing or supporting certain functions. Each person is able to do some activities better than others, just as some areas can produce some products much better than others. Thus, each producing unit tends to specialize in the activity to which it is best suited and then sells its surplus production to others engaged in different specialties. This phenomenon is known as the **principle of specialization.** It is a universal economic law that applies wherever free economic choices are made. Barriers erected by governments sometimes interfere with the free practice of specialization by preventing the supply of surplus goods from one area to flow into another. The standard of living of the population as a whole always is lower than it could be when this is the case.

Beginning economics students may have difficulty understanding why it is necessarily true that specialization of production and trade among areas results in a greater quantity of things available for consumption than would otherwise be the case. But, if economic life is reduced to the simplest of circumstances, the gains from specialization and free trade become more apparent.

Suppose that two people, Smith and Jones, live on a primitive island and have no contact with the outside world. Their needs for survival are very basic, namely, food and clothing. Both commodities must be produced with the labor that is available—Smith's and Jones' efforts. Since each needs both food and clothing to survive, they must each produce some of both products unless some trade arrangement can be worked out between them.

Assume that each individual requires 30 units of food and 10 units of clothing for survival. Assume also that by working full time, both Smith and Jones are just able to meet their requirements.

Situation I—Before Specialization

	Smith	Jones
Food (units)	30	30
Clothing (units)	10	10

At some point, it may occur to one of these individuals that he could become much more skilled at producing either food or clothing if he worked at that job continuously rather than attempting to do both tasks. In other words, if Smith were to specialize in food production, he could produce, say, 70 units, 10 more than the quantity required by both individuals. Jones, on the other hand, may discover that by specializing in clothing production, he could also become more skilled at that task, able to produce 25 units, 5 more than the sum of what he and Smith could produce before specializing. Thus, total production after specializing would increase by 10 units of food and 5 units of clothing, yet with Smith and Jones working the same amount of time.

Situation II—After Specialization

	Production		Trade and Consumption		Extra Output
	Smith	Jones	Smith	Jones	
Food	70	0	30	30	10
Clothing	0	25	10	10	5

As a result of specialization, Smith's and Jones' standards of living may rise if they can work out a fair trading arrangement. Both can consume more food and wear more clothing, or they may choose to work less and enjoy more leisure time rather than produce surplus commodities. Whatever the decision, it is incontestable that specialization and trade can improve the lot of both people.

Though the advantages of specialization are easily seen in this example, when extending the possibilities to large sections of a country or to whole countries, it is easy to lose sight of the advantages of specialization. One region may insist on producing every product that is consumed within its boundaries and erect complicated legal and economic barriers to keep out the production from competing areas. In so doing, it fails to realize that the advantages of specialization are sacrificed.

We assumed that Smith and Jones were equal in their respective abilities to produce food and clothing. Such an assumption is normally unrealistic. No two individuals are exactly equal in their abilities or desire to produce various commodities. Neither are geographic areas identical with respect to their ability to produce different products. For example, one person may have the manual dexterity to perform the most intricate wiring of electronic computers. Yet that same person may be unable or unwilling to manage human resources. Someone else, an expert at verbal communication, for example, may be inept at performing mechanical tasks. Likewise, one region of the country, due to attributes of natural resources, climate, and topography, may be able to grow excellent sugar cane but be unable to produce peanuts. Another region may be unexcelled in producing peanuts but be unable to compete successfully with other regions in corn or lettuce production. In such cases, it is to the advantage of an individual or region to specialize in the activity in which its advantage is greatest, compared to other individuals or regions.

Consider, for example, two regions of a country, Region I and Region II, that can each produce cotton and rice. However, due to differences in soil and topography, yields vary greatly by region.

Situation III—Before Specialization

	Region I	Region II
	—(Yield per acre)—	
Cotton (lbs. of lint)	600	300
Rice (cwt.)	28	38

If producers in these regions insist on producing both products, the total output from two acres of cotton (one in each region) would be 900 pounds of lint and from two acres of rice would be 66 hundredweight (cwt.). But, with each region producing the crop in which its advantage is greatest, total output from two acres of cotton in Region I would be 1,200 pounds—a 300-pound gain over the former arrangement. Total output from two acres of rice in Region II would be 76 cwt.—a gain of 10 cwt. Assuming that equal amounts of resources are required for crop production in each region, a substantial gain in output occurs with no additional resources employed.

Since Region I can produce more cotton (600 pounds) than Region II (300 pounds) on one acre, Region I is said to have an **absolute advantage** in cotton production. Likewise, Region II has an absolute advantage in rice production.

What about regions that do not have an absolute advantage? For some individuals, no matter what activity is undertaken, someone else is better at the job. In such a case, is there any justification for engaging in any activity or producing any product at all? Of course, there is. Many regions and countries fit this situation, and they do not cease their activities. The example shown by Situation IV illustrates why.

Situation IV—Absolute Advantage

	Region I	Region II	Total Product Output
	—(Yield per acre)—		
Cotton (lbs. of lint)	600	300	900
Rice (cwt.)	40	30	70

In this case, Region I has the absolute advantage in both cotton and rice production, since it can produce more per acre of both commodities. However, there are still advantages to specialization. Region I can produce twice (600/300) as much cotton as Region II, but only $1\frac{1}{3}$ (40/30) times as much rice. Therefore, compared with Region II, Region I has a greater absolute advantage in producing cotton. Region I will tend to specialize in cotton production, even though it can also produce more rice per acre than Region II.

Region II is at a disadvantage in the production of both commodities, but the disadvantage is not equal. Although Region II produces only 1/2 (300/600) as much cotton per acre as Region I, it grows 3/4 (30/40) as much rice. Therefore, compared with Region I, the disadvantage of Region II is less in rice production. Region II will tend to specialize in rice production. Total production as a result of specialization would be 1,200 pounds of cotton lint and 60 cwt. of rice. While rice output is 10 cwt. less than before, cotton output is one-third larger after specialization. Overall net gain would result from specialization.

This example illustrates the principle of **comparative advantage**, which states that regions tend to specialize in the production of commodities in which their comparative advantage is greater or their comparative disadvantage is lesser. The regions then purchase other commodities needed with their surplus production, rather than attempt to produce everything locally.

Government intervention sometimes prevents free operation of this principle, however. Producers in Region II may seek protection and support so that they can grow cotton, regardless of better alternatives awaiting them in other lines of endeavor. They may secure such a high level of protection that profits for producing the disadvantaged commodity are artificially high, but this always occurs at the expense of the consumers within the affected region. Producers in artificial markets must always be concerned that eventually consumers may begin to consume lower-priced substitute products. Taxpayers may consider the costs they pay to provide such protection. If the costs exceed the benefits, they will stop supporting legislation that creates artificially high prices for commodities.

History supports the view that short-run gains are possible through protection. But, in the long run, the very existence of the protected enterprise is seriously jeopardized.

Topics for Discussion

1. Name and discuss the three types of enterprise relationships.

2. Contrast "specialization" with "diversification."
3. Discuss the principle of "comparative advantage."
4. Contrast "absolute advantage" with "comparative advantage."

Problem Assignment

Examine the different agricultural enterprises conducted in your county, state, region, or country, and discuss those that have a comparative advantage and those that have an absolute advantage, if any.

Recommended Readings

Colander, David C. *Economics,* 2nd ed. Chicago: Richard D. Irwin, Inc., 1995. Ch. 1 and 37.

Gwartney, James D., and Richard L. Stroup. *Economics: Private and Public Choices,* 7th ed. Orlando, FL: The Dryden Press, 1995. Ch. 2.

Kay, Ronald D., and William M. Edwards. *Farm Management,* 3rd ed. New York: McGraw-Hall, Inc., 1994. Ch. 6.

Seitz, Wesley D., Gerald C. Nelson, and Harold G. Halcrow. *Economics of Resources, Agriculture, and Food.* New York: McGraw-Hill, Inc., 1994. Ch. 9.

12

Physical Production Relationships

How do entrepreneurs decide how much of a product to produce? Do feed manufacturers mix every ton of feed that can possibly be processed in their feed mills in a given period? Do textile manufacturers weave all the cloth that it is physically possible for their plants to produce? Do feeders put every pound of gain on their cattle that is possible before selling them for slaughter? The answer, of course, is usually no. The purpose of this chapter is to begin exploring how entrepreneurs decide how much to produce. The physical relationship and the price relationship between the inputs and the output provide necessary information for that decision. The focus of this chapter is on the physical relationship between the inputs and the output.

Production Functions

A **production function** describes the technical relationship that transforms inputs (resources) into outputs (commodities) within a given period of time. It represents an input-output relationship. Production functions are often referred to as "input-output functions" by economists and as "response functions" by biological scientists.

The Factor-Product Model

The **factor-product model** is the starting point in a discussion of production functions. The factor-product model uses a single-input production function in which one variable input is used to produce one output. A single-input production function can be expressed as

$$y = f(x_1 / x_2, \ldots, x_n)$$

The classical production function

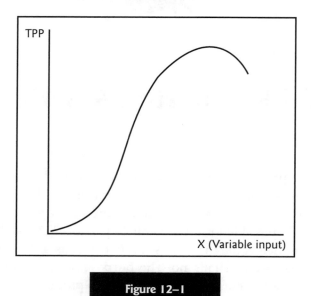

Figure 12-1

where y is the output, also called **total physical product (TPP)**, and $x_1, \ldots, x_n$ are inputs, the resources used to produce y. The production function indicates that output is a function of the **variable input**, x_1, given the level of the **fixed inputs**, $x_2, \ldots, x_n$. Variable inputs are those that the producer can vary or change in the production process, such as fertilizer. Fixed inputs are inputs of which the producer has a set amount, such as land or machinery.

The number of inputs that are variable or fixed in a production function depends on the length of the planning period. For example, at the beginning of a planting season, farmers can select what to plant from among a number of crops. Seed, then, is a variable input. However, once that decision is made and the seed drilled, the seed obviously becomes a fixed input.

Marginal physical productivity (MPP) refers to the change in output that results from an incremental change in the variable input. MPP indicates how output changes as a result of adding the last unit of input. The relationship between the amount of a single variable input and the amount of output that results can take on three general forms: increasing marginal physical productivity, constant marginal physical productivity, and diminishing marginal physical productivity. These relate to the slope (MPP) and curvature of the production function.

The three types of marginal productivity (increasing, constant, and diminishing) can be combined into one graphic presentation called the "classical production function" (Figure 12–1). It has an area of increasing marginal productivity, an area of constant marginal productivity, and an area of diminishing marginal productivity.

Increasing Marginal Productivity

An unlikely condition under which a manager might never stop expanding production is one in which each additional unit of input produces a greater output than did the previous unit of input. A production function with such an input-output relationship exhibits *increasing marginal productivity.*

Suppose a cattle feeder finds that each time the animals are fed another ton of feed, they gain more weight than they ever did before. When would the feeder stop feeding and begin to market the cattle? Theoretically, the feeder never would stop, because gains would be greater from continuing to feed these cattle than from starting over with a new group of animals.

This kind of relationship is shown graphically in Figure 12–2. With each additional unit of input (shown on the bottom or x axis), the units of output

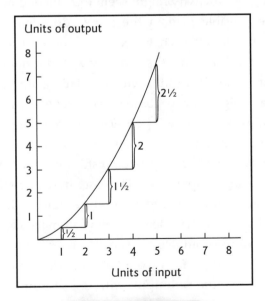

Figure 12–2

(represented on the vertical or y axis) increase at an increasing rate. The first unit of input produced 1/2 unit of output, the second unit produced 1 unit, the third produced 1½ units, etc. This relationship between input and output creates a curve that moves upward to the right at an ever-increasing slope. Because each additional unit of input produces more output than the preceding one, the relationship is called "increasing marginal productivity." The production function is increasing at an increasing rate.

Although most agricultural producers would be happy with such a relationship, it is encountered only rarely, usually under extreme conditions. For example, when an animal approaches starvation, its weight response to added units of feed exhibits increasing productivity initially, but only for a short period. Plants originating in media completely devoid of plant food (such as pure sand) respond with increasing productivity to the first plant nutrients received, but they, too, depart from this pattern quickly. Because most soils have some plant food in them initially, such a relationship is rarely, if ever, seen under natural conditions.

Constant Marginal Productivity

While increasing productivity might seem too much to hope for, you might think it would be possible to gain the same additional output from each added unit of input. After all, each added pound of fertilizer or added unit of feed is chemically identical to the first, so why would it not generate the same response as the first one? If it did, then the relationship between input and output would be a straight line, or *constant marginal productivity* (Figure 12–3).

Such relationships are sometimes observed in natural occurrences, over a limited range of input levels. Laying hens convert feed into eggs in accordance with the constant productivity relationship, within limits. That is, hens produce eggs in constant proportion to increases in feed intake, up to the limit of feed capacity. Dairy cattle are said to produce milk in direct proportion to the amount of feed consumed, up to the limit of stomach capacity. Under such circumstances, if it were profitable to feed a hen or a cow 1 pound of feed, then all succeeding pounds of feed would also be profitable, since they would produce an identical amount of output, with units of input costing the same. It would be a happy circumstance for producers if such a relationship were infinite; unfortunately, there always seem to be some limiting factors.

A production function exhibiting constant marginal productivity

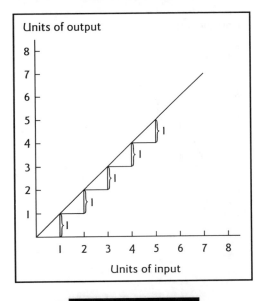

Figure 12–3

Diminishing Marginal Productivity

Except for limiting factors, the world's food supply might be produced in a flower pot. As plant nutrients are added, yields usually do increase but by less as each additional unit is added. A point is reached where there is not adequate space for further production in a given area or where there are not enough other necessary elements to combine with plant nutrients to obtain increasing or even constant additions to output as the variable input is increased. Reaching this point does not occur suddenly, but rather somewhat gradually, as shown in Figure 12–4.

For example, when the first unit of fertilizer is applied to a given area of land, the crop yield may increase by three units. The second unit of fertilizer, although just like the first, may add only two units to total yield. The third unit may increase yield by one unit, the fourth may add one-half unit of output, and the fifth unit might fail to increase yield at all. If further fertilizer applications were made, a toxic level of some elements in the soil might even cause a reduction in total yield.

A production function exhibiting diminishing marginal productivity

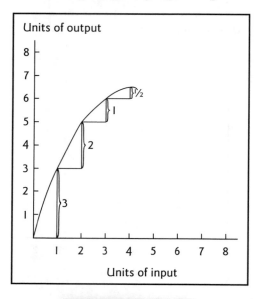

Figure 12-4

This most common type of input-output relationship is called *diminishing marginal productivity*, in which the function is increasing at a decreasing rate. It is the basis of a fundamental law of production economics called the **law of diminishing marginal returns**, which states that as more and more units of a variable input (e.g., fertilizer) are added to units of one or more fixed inputs (e.g., land), after a point, each additional unit of input produces less and less additional output. Figure 12–4 shows that total output is increasing with each additional input but that each successive increase in output is smaller than the previous one. Additions to total output diminish in size as additional units of input are applied.

The law of diminishing marginal returns has traditionally been called the "law of diminishing returns." However, this name is technically incorrect because the law applies to the incremental or marginal changes in output as units of the variable input are added.

The curve showing diminishing marginal productivity in Figure 12–5 may be extended for some production functions to show the area of negative marginal productivity, in which additional units of input actually decrease total output. Figure 12–5 shows that when inputs are increased beyond the fifth unit, output not only does not increase but it actually begins to decline.

A production function exhibiting decreasing productivity

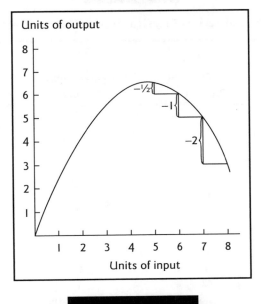

Figure 12–5

Some practical examples can be cited to illustrate this case. When too many sales personnel are employed in a department store, they tend to get in each other's way, causing total sales to fall. When animals are given too much feed, they can founder and begin to actually lose weight. When fertilizer is applied to a crop, a point is reached where further additions become toxic and yields are reduced. It is clear that regardless of the cost of the inputs, it is unwise to continue applications when total output begins to decrease. From an economic standpoint, there is no interest in the area of decreasing productivity beyond that of knowing where it begins.

Average and Marginal Physical Product

A hypothetical production function is represented by the data in Table 12–1. One variable input, such as fertilizer, is related to the output produced. The first column shows the input level ranging from 0 to 10 units, and the second column shows the total physical product (TPP) at each level of input. The TPP curve in Figure 12–6a shows graphically the data in the first and second columns.

Table 12-1

Hypothetical Physical Input–Output Data

(1) Input	(2) Total Physical Product	(3) Marginal Physical Product[1]	(4) Average Physical Product[2]
\- \- \- \- \- \- \- \- \- \- \-	\- \- \- \- \- \- \- \- \- \- Units \- \- \- \- \- \- \- \- \- \-		\- \- \- \- \- \- \- \- \- \- \-
0	0		—
		5	
1	5		5.0
		11	
2	16		8.0
		10	
3	26		8.7
		9	
4	35		8.8
		7	
5	42		8.4
		5	
6	47		7.8
		4	
7	51		7.3
		3	
8	54		6.8
		2	
9	56		6.2
		−1	
10	55		5.5

[1] Differences between successive total physical product (Δy) divided by differences between successive total input units (Δx). Marginal data displayed between successive units provide a closer approximation of exact marginal values computed as derivatives.

[2] APP equals TPP (Col. 2) divided by the input level (Col. 1).

The rate of change, or slope, of the TPP curve at any given point determines whether production is occurring in the area of increasing, constant, or diminishing marginal productivity. This rate of change, marginal physical product (MPP), is given in the third column of Table 12–1. MPP is the amount of output that is added to TPP (Δy, which is the change in y) resulting from the addition of each successive unit of input (Δx, which is the change in x).

$$MPP = \frac{\Delta y}{\Delta x}$$

TPP, with MPP and APP curves, defining the stages of production, for the classical production function

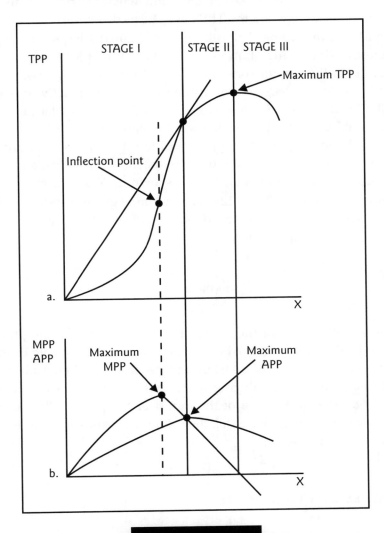

Figure 12–6

It is called "marginal" because it refers to the last increment of output resulting from the last increment of input.

Figure 12–6b shows that, for the classical production function, MPP increases at first but that it reaches a peak and then begins to decline. Maximum MPP corresponds to the *inflection point* on the production function, where the function goes from increasing at an increasing rate to increasing at a decreasing rate. It is at the inflection point that the law of diminishing marginal returns begins. MPP equals zero at the point where TPP is at a maximum, then becomes negative when the TPP curve turns downward.

When MPP is increasing, TPP is in the area of increasing marginal productivity. Output is increasing at an increasing rate. When MPP reaches a peak, production is in the area of constant marginal productivity, however briefly. When MPP begins to decline, diminishing marginal productivity sets in. Although TPP is increasing in this area, it increases by a smaller amount with each additional unit of input. When MPP equals zero, TPP is at its highest point. Any additional units of input will not result in increases in production. Total output may, in fact, actually decrease beyond this point, and marginal physical product will become negative, as shown in Figure 12–6b.

The fourth column of Table 12–1 shows the **average physical product** at each input level. It is obtained by dividing TPP by the number of units of input at each level of application.

$$APP = \frac{y}{x}$$

APP is useful because of the information it provides regarding the logical area of production. As long as APP is rising, entrepreneurs have incentive to add inputs, because each additional unit places them in a better position than previous ones; they are producing more units of output per unit of input used. So, regardless of the cost of the inputs or the price of the output produced, if any inputs are added at all, they should continue to be added at least until APP reaches its peak.

The Stages of Production

The classical production function can be divided into three regions, or **stages of production.** Stage I is defined as the area in which MPP is greater than APP. TPP increases throughout Stage I, as does APP. MPP increases, reaches its maximum, and then begins to fall in Stage I. Assuming unconstrained access to additional inputs, Stage I is an irrational region of production. If production at any level is profitable, it is unreasonable or illogical to stop adding inputs to the production process (regardless of their cost) anywhere within this area, because each added unit results in a higher average physical product or output per unit. Note in Figure 12–6b that the MPP curve intersects the APP curve at the highest point of the APP curve. This point of intersection defines the boundary between Stage I and Stage II.

Stage II is the area in which APP is greater than MPP and MPP is positive. TPP increases throughout Stage II and reaches a maximum at the boundary between Stage II and Stage III. This is the logical area of production, because it is not until this area or stage is reached that there is any question as to whether additional inputs should be utilized. Both MPP and APP are decreasing, while TPP is increasing at a decreasing rate. It is somewhere within Stage II that an entrepreneur will reach the decision that the added output from the last unit of input is not sufficient to cover the added cost. If inputs are free, they will continue to be used as long as they add anything at all to total physical product (until MPP becomes zero). If units of input are very costly in relation to the value of the product, production may cease very early in Stage II (shortly after APP begins to decline).

Stage III occurs when MPP becomes negative. Adding more units of the input actually causes TPP to decline. No one would logically want to continue adding an input if each unit resulted in a decrease in total output (unless he or she was being *paid* enough to use the input). Stage III, then, is also an irrational region of production. If production occurs in this area, total output may always be increased by reducing inputs until Stage II or the rational area of producing is again reached.

In summary, businesses with access to sufficient additional inputs are interested only in Stage II of the production function with diminishing marginal productivity. Stage I is rarely seen in actual practice, and if it is experienced, production is increased to the Stage II area, because it is increasingly profitable to do so if production is profitable at all. When production occurs in Stage III of the production function, inputs are rapidly reduced to the Stage II area, because to do so increases total product. Therefore, Stage II is the rational region of production.

Topics for Discussion

1. Explain what is happening when yield response to additional units of fertilizer is characterized by increasing marginal productivity.

2. Explain what is happening when yield response to additional units of fertilizer is characterized by constant marginal productivity.

3. Explain what is happening when yield response to additional units of fertilizer is characterized by diminishing marginal productivity.

4. Using the law of diminishing marginal returns, explain why we cannot produce the world's food supply from a flower pot.
5. What is a "production function"?
6. Discuss the three stages of production, and indicate which stage is the most logical one for production.

Problem Assignment

Establish a production function in your major field of study, and graph the respective total, average, and marginal physical products.

Recommended Readings

Boehlje, Michael D., and Vernon R. Eidman. *Farm Management.* New York: John Wiley & Sons, Inc., 1984. Ch. 3.

Colander, David C. *Economics,* 2nd ed. Chicago: Richard D. Irwin, Inc., 1995. Ch. 23.

Debertin, David L. *Agricultural Production Economics.* Macmillan Publishing Co., 1986. Ch. 2.

Doll, John P., and Frank Orazem. *Production Economics: Theory with Applications,* 2nd ed. Melbourne, FL: Krieger Publishing Co., 1992 reprint of 1984 ed. Ch. 2.

Luening, Robert A., Richard M. Klemme, and William P. Mortenson. *The Farm Management Handbook,* 7th ed. Danville, IL: Interstate Publishers, Inc., 1991.

Schlegel, A. J. and J. L. Havlin. "Corn Response to Long-Term Nitrogen and Phosphorus Fertilization," *Journal of Production Agriculture,* 8(2):181–185 (1995).

Steward, Jim, Raleigh Jobes, James E. Casey, and Wayne D. Purcell. *Farm and Ranch Business Management,* 3rd ed. Moline, IL: Deere and Company, 1992. Ch. 5.

13

Costs and Revenue

We determined in the last chapter that the rational region of production is within Stage II of the production function. But the question of how much to produce within Stage II still remains. The purpose of this chapter is to discuss the economic information, referred to as **costs** and **revenue,** needed to answer to that question.

Costs

Costs are usually thought of as the expenditures incurred in the production of products or services. Actually, costs are not all alike, and sometimes they become quite difficult to determine. Because of the many differences in costs, some are often overlooked in efforts to determine the total cost of producing a particular good or service. For example, when asked the cost of producing a crop, farmers often think in terms of seed, fertilizer, fuel, and labor used. Many neglect to include costs reflecting the use of machinery and land in the production process. The latter items may not necessarily represent monetary outlays made in a particular growing season. However, they are expenditures necessary for production and must be considered in any accurate computation of total production costs.

Variable Costs

The costs that usually first come to mind when considering production expenses are the immediate out-of-pocket expenditures, referred to as **variable costs.** Total variable cost (TVC) reflects the cost of purchasing variable inputs, those that vary directly with the level of production. For the factor-product model, TVC is given by

$$TVC = r * x$$

where r is the price per unit of the variable input and x is the number of units of the variable input used. If there is no production, variable costs equal zero; but as production increases, variable costs also increase. To increase output of rice, a farmer must also increase expenditures for such inputs as seed, fertilizer, labor, fuel, etc. If a clothing factory increases the output of shirts, then the expenditure for raw materials, labor, and fuel will also increase. Variable costs are so named because they change, or vary, with the level of production.

Column 5 of Table 13–1 illustrates how units of input are related to total variable cost (TVC). In this example, as inputs increase from 0 to 10, total variable cost increases from 0 to $40. Each additional input adds $4 to total variable cost or total out-of-pocket expenditures.

Table 13–1

Basic Input–Output Data and Total Costs, Variable Costs, and Fixed Costs

Physical Input–Output Data		Total Costs		
(1)	(2)	(5)	(6)	(7)
Input[1]	Total Physical Product[2]	Total Variable Cost (Col. 1 × $4)[3]	Total Fixed Cost[4]	Total Cost (Col. 5 + Col. 6)
---------- Bushels ----------		---------------------- Dollars ----------------------		
0	0	0	20	20
1	5	4	20	24
2	16	8	20	28
3	26	12	20	32
4	35	16	20	36
5	42	20	20	40
6	47	24	20	44
7	51	28	20	48
8	54	32	20	52
9	56	36	20	56
10	55	40	20	60

[1] Variable inputs represent all variable items. Examples commonly include fertilizer, feed, seed, labor, etc.
[2] Same physical input–output relationships as Table 12–1. For columns 3 and 4, which are not shown here, refer to Table 12–1.
[3] Each unit of variable input costs $4.
[4] Fixed costs per acre are assumed to be $20.

Fixed Costs

Costs that businesses, particularly farm businesses, are likely to overlook are those that do not change as production and inputs are varied. These are called **fixed costs.** Fixed costs are also important expenditure items, but they may not be incurred on a regular basis or be associated with the production of any particular crop or commodity.

Farmers typically buy land once and may produce many crops from it in the years that follow. Because they do not repurchase the land each year, it is easy to see why they may forget to include an allowance for it in their production costs. A machine, such as a tractor or combine, may last for several years, yet it is purchased once and its cost is not directly related to production in any given year. It is also true that such costs are incurred whether anything is produced in a given year or not. Taxes, insurance premiums, depreciation from obsolescence, and interest on investments are examples of costs that farmers incur even if they stop producing.

Column 6 of Table 13–1 shows the total fixed cost (TFC) related to each level of input in the production example. Note that total fixed cost remained at $20 regardless of the level of inputs or output. Thus, total fixed cost does not change or vary with changes in output.

Total Cost

The total cost of production consists of both fixed and variable costs. Neither can be ignored in obtaining an accurate statement of total cost.

Column 7 of Table 13–1 shows that **total cost (TC)** is obtained by adding total variable cost (Column 5) to total fixed cost (Column 6). Total cost equals $20 at the zero level of output, because fixed costs exist whether anything is produced or not. Total cost increases as variable costs increase. In this example, and in agriculture generally, fixed costs are relatively more important at low levels of production than are variable costs. Variable costs become more important at higher levels of production.

Average Costs

Up to this point, costs have been considered on the basis of the total cost of the units of input employed. It is also useful to consider costs on the basis of the units of output produced. For example, an auto's gasoline cost may be obtained in terms of the total cost of filling the tank (the cost of input). It may be more meaningful, however, to have gasoline cost on the basis of miles traveled (output). The latter cost can be obtained by simply dividing the cost of gasoline consumed by the total miles traveled or by computing the average gasoline cost per mile. For example, $2.50 worth of gasoline to travel 50 miles would give a cost of $.05 per mile.

Average Variable Cost

Average out-of-pocket expenditures, or **average variable cost (AVC),** can be obtained by dividing the total variable cost by the total units of output (y) at any given output level. AVC is the variable cost per unit of output.

$$AVC = \frac{TVC}{y}$$

Column 8 of Table 13–2 shows the average variable cost at each level of output, which is obtained by dividing total variable cost (Column 5) by total output (Column 2).

When five units of output are produced, the average variable cost equals 80¢. As output increases to 35 units, AVC declines to its low point, 44¢ per unit. AVC rises again to 73¢ per unit when 10 units of input are used to produce 55 units of output.

Average Fixed Cost

Average fixed cost (AFC) equals total fixed cost divided by total units of output.

$$AFC = \frac{TFC}{y}$$

Remember that total fixed cost does not change regardless of the level of output. Average fixed cost would then be expected to decline as output increases, because larger and larger quantities of output are divided into a constant amount of total fixed cost.

Column 9 of Table 13–2 shows that average fixed cost in our example is obtained by dividing total fixed cost, $20 (Column 6), by total output (Column 2) at each respective level. Average fixed cost thus begins at the relatively high level of $4 per unit but declines rapidly to $.36 per unit at the 55-unit-of-output level.

Table 13-2

Basic Input–Output Data with Total and Average Costs

	Physical Input–Output Data			Economic Data					
				Total Costs			Costs per Unit of Output		
(1)	(2)	(3)	(4)	(5)	(6)	(7)	(8)	(9)	(10)
Input	Total Physical Product	Marginal Physical Product[1]	Average Physical Product (Col. 2 ÷ Col. 1)	Total Variable Cost (Col. 1 × $4)[2]	Total Fixed Cost[3]	Total Cost (Col. 5 + Col. 6)	Average Variable Cost (Col. 5 ÷ Col. 2)	Average Fixed Cost (Col. 6 ÷ Col. 2)	Average Total Cost (Col. 8 + Col. 9)
Units	--------- Bushels ---------			------------------------- Dollars -------------------------					
0	0		0.0	0	20	20	—	—	—
		5							
1	5		5.0	4	20	24	0.80	4.00	4.80
		11							
2	16		8.0	8	20	28	0.50	1.25	1.75
		10							
3	26		8.7	12	20	32	0.46	0.77	1.23
		9							
4	35		8.8	16	20	36	0.44	0.57	1.01
		7							
5	42		8.4	20	20	40	0.48	0.48	0.96
		5							
6	47		7.8	24	20	44	0.51	0.43	0.94
		4							
7	51		7.3	28	20	48	0.55	0.38	0.93
		3							
8	54		6.8	32	20	52	0.59	0.37	0.96
		2							
9	56		6.2	36	20	56	0.65	0.36	1.00
		−1							
10	55		5.5	40	20	60	0.73	0.36	1.09

[1] Difference between successive total products divided by differences between successive total input units. Marginal data displayed between successive units provide a closer approximation of exact marginal values computed as derivatives.
[2] Each unit of variable input costs $4.
[3] Fixed costs per acre are assumed to be $20.

Average Total Cost

Average total cost (ATC) shows the average cost of producing a unit of output at each level of output. Like average fixed and average variable costs, average total cost can be obtained by dividing total cost by total output at any level. However, since its components have already been determined, ATC (Column 10) is most easily obtained by adding average variable cost (Column 8) to average fixed cost (Column 9).

$$ATC = \frac{TC}{y} = \frac{TVC + TFC}{y} = \frac{TVC}{y} + \frac{TFC}{y} = AVC + AFC$$

Average total cost is influenced mostly by fixed costs initially, but as output increases, average fixed cost declines and average variable cost makes up the larger proportion.

Note that in Column 10 of Table 13–2, average total cost per unit of production is reduced sharply at first by increasing output. This should provide a clue to the reason many businesses seek to expand output and grow larger. Farmers particularly have been increasing output per acre at a tremendous rate in recent years.

Marginal Input Costs and Marginal Costs

Another helpful way to view costs when analyzing the optimum level of production is to look at the additional cost incurred from adding another unit of input or the additional cost incurred from producing another unit of output.

The additional cost incurred from adding the last unit of input is referred to as **marginal input cost (MIC).** "Marginal input cost" is defined as the change in total variable cost given a change in the input level.

$$MIC = \frac{\Delta TVC}{\Delta x} = r$$

MIC for the example is given in Column 11 of Table 13–3. Each additional unit of input costs $4, so the marginal input cost equals $4 at any input level.

The additional cost incurred when changing the production level can also be expressed in terms of changes in output. **Marginal costs** reflect the additional costs incurred from producing the last unit of output.

$$MC = \frac{\Delta TC}{\Delta y} = \frac{\Delta TVC}{\Delta y}$$

Table 13–3

Basic Input–Output Data and Related Costs

	Physical Input–Output Data			Total Costs		Costs/Output	Marginal Costs	
(1)	(2)	(3)	(4)	(5)	(7)	(10)	(11)	(12)
Input	Total Physical Product	Marginal Physical Product[1]	Average Physical Product (Col. 2 ÷ Col. 1)	Total Variable Cost (Col. 1 × $4)[2]	Total Cost (Col. 5 + $20)[3]	Average Total Cost (Col. 7 ÷ Col. 2)	Per Unit Additional Input (Col. 5 ÷ Col. 1)	Per Unit Additional Output (Col. 11 ÷ Col. 3)
Units	---------- Bushels ----------			------------------------- Dollars -------------------------				
0	0		0.0	0	20	—	—	—
		5						
1	5		5.0	4	24	4.80	4	0.80
		11						
2	16		8.0	8	28	1.75	4	0.36
		10						
3	26		8.7	12	32	1.23	4	0.40
		9						
4	35		8.8	16	36	1.01	4	0.44
		7						
5	42		8.4	20	40	0.96	4	0.57
		5						
6	47		7.8	24	44	0.94	4	0.80
		4						
7	51		7.3	28	48	0.93	4	1.00
		3						
8	54		6.8	32	52	0.96	4	1.33
		2						
9	56		6.2	36	56	1.00	4	2.00
		−1						
10	55		5.5	40	60	1.09	4	—

[1] Difference between successive total products divided by differences between successive total input units. Marginal data displayed between successive units provide a closer approximation of exact marginal values computed as derivatives.
[2] Each unit of variable input costs $4.
[3] Fixed costs per acre are assumed to be $20.

This equation states that marginal costs are equal to the change in total cost (or total variable cost) divided by the change in output. Column 12 of Table 13–3 is obtained by dividing the marginal cost per unit of input (Column 11) by the marginal product (Column 3). The cost of the last unit of output varies a great deal from the early output levels to the final level of production. A point is reached where eventually the cost of obtaining an additional unit of output becomes prohibitively high.

Cost Curves

The average and marginal costs discussed above are often plotted as curves, as shown in Figure 13–1. The relationship between the various costs is more readily seen in graphic form. Both the average total cost and average fixed cost curves are high at the outset but drop rapidly as output increases. In fact, the average fixed cost curve continues to fall as long as output increases. Thus, the greater the output, the lower the fixed cost per unit of output. Fixed cost is spread over more units of output by producing higher levels of output.

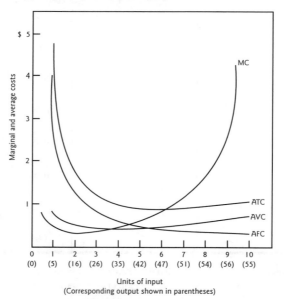

Figure 13–1

The average variable cost curve falls initially but soon begins to rise as output increases. It eventually overpowers the effect of the declining average fixed cost curve on the average total cost curve, and the latter also begins to rise.

The marginal cost curve reflects the addition to total cost of producing an additional unit of output at all points along the scale. At first, through the area of increasing and constant returns, additional units of output are even less costly, or at least no more costly, than the preceding units. But as production moves into the area of diminishing marginal return, the cost of producing each additional unit of output rises rapidly. The marginal cost curve falls initially but soon begins to rise at an increasing rate.

Note that the marginal cost curve intersects both the average variable and the average total cost curves at their lowest points. Both curves begin to rise immediately thereafter.

Near the point of maximum output, returns diminish to such an extent that almost no additional production is gained from further expenditures for inputs. The marginal cost curve becomes nearly vertical at this point, indicating that further additions to output at that level are extremely costly and finally become prohibitively expensive.

Revenue

Cost information is indispensable in determining the optimum or most profitable level of production. Equally important are the returns generated, or the dollar value of the output. Obviously, unless monetary returns exceed the production costs, there can be no profit. So, before determining where in Stage II of the production function to produce, a firm must know its returns as well as its costs.

Total Revenue

Once total physical output is known, it is relatively easy to derive total monetary return for agricultural enterprises, since the output usually is sold at a constant price per unit. When the output price is multiplied by TPP at any level, the result is **total revenue (TR).**

$$TR = P_y * y = P_y * f(x)$$

Returning to our production example, Table 13–4 shows that when total physical product (Column 2) is multiplied by a price of $2 per unit, the total return at each level of input is as shown in Column 13. Total revenue reaches its highest level of $112 at nine units of input.

Table 13-4

Basic Input–Output Data and Related Costs and Revenue

	Physical Input–Output Data			Total Costs			Revenue			Marginal Revenue	
(1)	(2)	(3)	(4)	(5)	(6)	(7)	(13)	(14)	(15)	(16)	(17)
Input	Total Physical Product	Marginal Physical Product[1]	Average Physical Product (Col. 2 ÷ Col. 1)	Total Variable Cost (Col. 1 × $4)[2]	Total Fixed Cost	Total Cost (Col. 5 + Col. 6)	Total Revenue (Col. 2 × $2)	Profit (Col. 13 − Col. 7)	Average Revenue (Col. 13 ÷ Col. 2)	Per Unit Input (Col. 3 × $2)	Per Unit Output (Col. 16 ÷ Col. 3)
Units	---------- Bushels ----------			---------------------------- Dollars ----------------------------							
0	0		0.0	0	20	20	0	−20	—		2
		5								10	
1	5		5.0	4	20	24	10	−14	2		2
		11								22	
2	16		8.0	8	20	28	32	4	2		2
		10								20	
3	26		8.7	12	20	32	52	20	2		2
		9								18	
4	35		8.8	16	20	36	70	34	2		2
		7								14	
5	42		8.4	20	20	40	84	44	2		2
		5								10	
6	47		7.8	24	20	44	94	50	2		2
		4								8	
7	51		7.3	28	20	48	102	54	2		2
		3								6	
8	54		6.8	32	20	52	108	56	2		2
		2								4	
9	56		6.2	36	20	56	112	56	2		2
		−1								−2	
10	55		5.5	40	20	60	110	50	2		2

[1]Difference between successive total products divided by differences between successive total input units. Marginal data displayed between successive units provide a closer approximation of exact marginal values computed as derivatives.
[2]Each unit of variable input costs $4.
[3]Fixed costs per acre are assumed to be $20.

Profit

When total revenue and total cost are known, the next logical step is to determine **profit (π)**, or net return. Profit is the difference between total cost and total return at any production level.

$$\pi = TR - TC = TR - TVC - TFC$$

It indicates whether an operation is making or losing money.

Column 14 of Table 13–4 shows profit, the difference between total return (Column 13) and total cost (Column 7). Profit ranges, in our example, from a net loss of $20 at first to a net gain of $56 when either eight or nine units of input are applied. It is interesting to note that no profit was realized until the second unit of input was applied. Profit actually decreased when 10 units of input were applied.

Average Revenue

Just as it was possible to determine the average cost of each unit of output, so is it possible to determine the **average revenue (AR)** at any level of production. "Average revenue" refers merely to the revenue produced by each unit of output.

$$AR = \frac{TR}{y} = \frac{P_y * y}{y} = P_y$$

It is, therefore, the total revenue at any level divided by the total output at the corresponding level, as shown in Column 15 of Table 13–4. Average revenue equals the price per unit of output in our production example. If each unit that is produced is sold for $2, then the average price of, or average revenue from, all units would have to be $2. Thus, under conditions where one producer is too small to affect the price of the product, as is often the case in production agriculture, price and average revenue are equal.

Marginal Value Product and Marginal Revenue

The marginal concept has already been introduced in the discussion of marginal cost. The marginal cost concept is the same with regard to marginal revenue, except that here we refer to the additional revenue generated from the last unit of output.

The value of additional output units may be regarded in two ways. One is the value of additional output that results from using the last unit of input. This is referred to as **marginal value product (MVP).**

$$\text{MVP} = \frac{\Delta \text{TVP}}{\Delta x} = \frac{\Delta(P_y * y)}{\Delta x} = P_y * \frac{\Delta y}{\Delta x} = P_y * \text{MPP}$$

The MVP for our production example is shown in Column 16 of Table 13–4. It is computed by multiplying marginal physical product (Column 3) by $2 (the price per unit of output). MVP first increases rapidly and then declines until it finally disappears at the tenth unit of input. Clearly, the latter units of input did not produce as much return as did earlier ones.

The second method of looking at the value of output is to consider the value of each additional unit of output produced, referred to as **marginal revenue (MR).** Marginal revenue reflects the change in total revenue that results from producing the last unit of output.

$$\text{MR} = \frac{\Delta \text{TR}}{\Delta y} = \frac{\Delta(P_y * y)}{\Delta y} = P_y * \frac{\Delta y}{\Delta y} = P_y$$

Marginal revenue for the example is shown in Column 17 of Table 13–4.

Note that MR in our example equals AR and the output price. If each unit produced is sold for a price of $2, then the marginal revenue, the return for each additional unit of output, must also be $2. So, for an individual producer under market conditions observed for many commonly produced agricultural commodities, marginal revenue equals average revenue, which in turn equals the price per unit of output.

Topics for Discussion

1. Compare and contrast "variable costs" and "fixed costs."

2. Discuss the concepts of "total cost" and "average total cost."

3. Define "marginal input costs" and "marginal costs."

4. Discuss the concepts of "total revenue," "profit," "average revenue," and "marginal revenue."

Problem Assignment

Using the data of the production function in the problem assignment for Chapter 12, apply the appropriate costs and revenue estimates, and calculate average variable cost, average fixed cost, average total cost, marginal cost, average revenue, total revenue, and marginal revenue.

Recommended Readings

Boehlje, Michael D., and Vernon R. Eidman. *Farm Management.* New York: John Wiley & Sons, Inc., 1984. Ch. 3.

Colander, David C. *Economics,* 2nd ed. Chicago: Richard D. Irwin, Inc., 1995. Ch. 25.

Debertin, David L. *Agricultural Production Economics.* Macmillan Publishing Co., 1986. Ch. 3 and 4.

Doll, John P., and Frank Orazem. *Production Economics: Theory with Applications,* 2nd ed. Melbourne, FL: Krieger Publishing Co., 1992 reprint of 1984 ed. Ch. 3.

Fales, S. L., et al. "Stocking Rate Affects Production and Profitability in a Rotationally Grazed Pasture System," *Journal of Production Agriculture,* 8(1):88–96 (1995).

Gwartney, James D., and Richard L. Stroup. *Economics: Private and Public Choices,* 7th ed. Orlando, FL: The Dryden Press, 1995. Ch. 18.

Luening, Robert A., Richard M. Klemme, and William P. Mortenson. *The Farm Management Handbook,* 7th ed. Danville, IL: Interstate Publishers, Inc., 1991.

Seitz, Wesley D., Gerald C. Nelson, and Harold G. Halcrow. *Economics of Resources, Agriculture, and Food.* New York: McGraw-Hill, Inc., 1994. Ch. 4.

Steward, Jim, Raleigh Jobes, James E. Casey, and Wayne D. Purcell. *Farm and Ranch Business Management,* 3rd ed. Moline, IL: Deere and Company, 1992. Ch. 5.

14

Optimum Levels of Output

We introduced several methods of determining costs and revenue in the last chapter. These computations are a necessary part of arriving at our ultimate objective—the best, or *optimum*, level of production or output. The purpose of this chapter is to discuss how to determine the optimum level of output in an economic sense. This is the level where profits are maximized or, if profits cannot be made, where losses are minimized.

Maximum Profit

A fundamental assumption of basic production economics is that producers will use their resources to maximize profit. This does not mean that they will produce all they can until they attain maximum output. It was explained previously that maximum output is the most profitable production point only if inputs are free (or if the output price is infinitely high). In other words, a rational corn producer will apply nitrogen fertilizer to the point where corn yields are maximized only if the fertilizer is free—an unlikely prospect.

How, then, is the optimum input level to employ in production derived? Inputs should be applied until the difference between total cost and total revenue, or profit, is at a maximum. The production example illustrated in Table 14–1 shows that the maximum profit is obtained at the eighth and ninth units of input. To increase output further actually decreases profit.

The same conclusion can more easily be reached in another way, although it may require more thought by the reader at the first encounter with the method. It has already been suggested that in no case would a producer spend more money for added inputs than he or she gets back in added revenue. By the same token, money can be profitably expended as long as the producer gets back more money from each added input than its cost, or as long as the added revenue exceeds the added cost.

Table 14-1

Basic Input–Output Data and Related Costs and Revenue

(1)	(2)	(3)	(4)	(5)	(6)	(7)	(8)	(9)	(10)	(11)	(12)	(13)
	Physical Input–Output Data			Total Costs			Marginal Costs				Revenue	
Input[1]	Total Physical Product	Marginal Physical Product[2]	Average Physical Product (Col. 2 ÷ Col. 1)	Total Variable Cost (Col. 1 × $4)[3]	Total Fixed Cost[4]	Total Cost (Col. 5 + Col. 6)	Per Unit Additional Input (Col. 5 ÷ Col. 1)	Per Unit Additional Output (Col. 8 ÷ Col. 3)	Total Revenue (Col. 2 × $2)	Profit (Col. 10 − Col. 7)	Marginal Revenue per Unit of Input (Col. 3 × $2)	Marginal Revenue per Unit of Output (Col. 12 ÷ Col. 3)
Units	Bushels			Dollars								
0	0		0.0	0	20	20			0	−20		
1	5	5	5.0	4	20	24	4	0.80	10	−14	10	2
2	16	11	8.0	8	20	28	4	0.36	32	4	22	2
3	26	10	8.7	12	20	32	4	0.40	52	20	20	2
4	35	9	8.8	16	20	36	4	0.44	70	34	18	2
5	42	7	8.4	20	20	40	4	0.57	84	44	14	2
		5					4	0.80			10	2

(Continued)

Table 14–1 (Continued)

(1)	Physical Input–Output Data				Economic Data							
	(2)	(3)	(4)	Total Costs			Marginal Costs		Revenue			
				(5)	(6)	(7)	(8)	(9)	(10)	(11)	(12)	(13)
Input[1]	Total Physical Product	Marginal Physical Product[2]	Average Physical Product (Col. 2 ÷ Col. 1)	Total Variable Cost (Col. 1 × $4)[3]	Total Fixed Cost[4]	Total Cost (Col. 5 + Col. 6)	Per Unit Additional Input (Col. 5 ÷ Col. 1)	Per Unit Additional Output (Col. 8 ÷ Col. 3)	Total Revenue (Col. 2 × $2)	Profit (Col. 10 − Col. 7)	Marginal Revenue per Unit of Input (Col. 3 × $2)	Marginal Revenue per Unit of Output (C. 3)
Units	------ *Bushels* ------			---------------------------------- *Dollars* ----------------------------------								
6	47		7.8	24	20	44			94	50		
		4					4	1.00			8	2
7	51		7.3	28	20	48			102	54		
		3					4	1.33			6	2
8	54		6.8	32	20	52			108	56		
		2					4	2.00			4	2
9	56		6.2	36	20	56			112	56		
		−1					4	—			−2	2
10	55		5.5	40	20	60			110	50		

[1] Variable inputs represent all variable items. Examples commonly include fertilizer, feed, seed, labor, etc.
[2] Difference between successive total products divided by differences between successive total input units. Marginal data displayed between successive units provide a closer approximation of exact marginal values computed as derivatives.
[3] Each unit of variable input costs $4.
[4] Fixed costs per acre are assumed to be $20.

We can use this idea also to determine the most profitable level of output in the production example shown in Table 14–1. In this case we compare the added or marginal cost (Column 9) with the added or marginal revenue (Column 13). Note that at the second input level, marginal cost, the cost of producing the last unit of output, is $.36. The value of each unit of output at this level is $2. In effect, then, the producer receives $2 for a unit of output that cost $.36 to produce, expended when the second unit of input was applied. Was the added production profitable? If we became aware of an opportunity to exchange $.36 for $2, it is likely that we would trample each other trying to get to the place of exchange as long as we have an additional $.36. So, the added production was clearly profitable.

The producer cannot expect to receive the same rate of return for additional expenditures indefinitely, however. Remember the principle of diminishing marginal returns. You can see that it is working in this example, because when seven units of input are added, the marginal cost per unit of output rises to $1. As production increases it becomes more costly to get additional output. Each added unit of output is still worth $2, however, and the rate of exchange ($1 cost for $2 return) is still favorable.

In our example, profit can be further increased by adding more inputs until the ninth unit is added. At this level the marginal cost of the added unit of output is $2, the same as the value of that unit of output. Obviously, no further gain is realized if the cost of an added unit exactly equals the return realized from it. This is the maximum profit level of production. It can be confirmed by looking again at the profit column, which is the difference between total cost and total revenue. At no level are profits higher than at the ninth unit of input.

Minimum Net Losses

It sometimes happens that producers are unable to make a profit at any level of output. In this case one might first wonder, Why produce at all? But it is often true that continuing in production in the short run can reduce the losses that would occur if production completely stopped. The economic problem then becomes one of determining the level of production that minimizes losses.

In Table 14–1 the fixed cost (the cost that does not vary with production) is $20. Thus, even when no production occurs, there is a cost incurred of $20. It may be possible to reduce this loss by producing, even if total cost always exceeds total revenue. The key to whether this is possible is the size of variable costs in relation to total revenue. If total revenue exceeds variable or out-of-pocket expenses at any level of production, then it is possible, in the short run, to reduce losses by producing at that level. In such a case additional out-of-pocket expenses are covered, with something left to apply toward fixed costs. Otherwise, the fixed cost would remain completely uncovered.

Table 14–2 shows our production example with a few changes in prices. The price per unit of input is now $5 instead of $4, causing an increase in variable and total costs. The price per unit of output is now $1 instead of $2, causing total revenue and marginal revenue to be reduced by half. Under these conditions, Column 11 shows that profit is negative; thus, there is a net loss at all levels of output. However, the net loss is not the same at all levels. At the zero level of output, the net loss is $20, the full amount of the fixed cost. But as output is increased, Column 10 shows that total revenue exceeds total variable cost (Column 5) by a sufficient amount to offset some of the fixed costs, reducing the net loss. Thus, by producing, net losses are reduced from $20 when no production occurs to a minimum of $3 when six units of input are applied. From that point net losses begin to increase again, indicating that production has been carried beyond the minimum loss level.

As with the point of maximum profit, the point of minimum net loss can also be determined more readily by comparing marginal cost and marginal revenue. The criteria are exactly the same. Inputs are added as long as the marginal value product (Columns 12 and 13) exceeds the marginal input cost (Columns 8 and 9). After the first unit of input, marginal input cost was less than the marginal value product until the sixth unit was applied. There MIC and MVP were equal, indicating that no further gain was possible by exchanging inputs for outputs. The data in Column 11 confirm that net losses were minimized at the sixth unit of input.

In agriculture, particularly, fixed cost as a proportion of total cost is high. High investment requirements for land, machinery, and buildings leave farmers with many costs that cannot be reduced even if production becomes unprofitable. For this reason farmers usually continue to produce in the short run, even when losing money, because the losses would be even greater if they did not produce. It is important to know, then, how to determine the level of production then minimizes losses.

Table 14-2

Basic Input–Output Data and Related Costs and Revenue

Physical Input–Output Data				Economic Data								
				Total Costs			Marginal Costs		Revenue			
(1)	(2)	(3)	(4)	(5)	(6)	(7)	(8)	(9)	(10)	(11)	(12)	(13)
Input[1]	Total Physical Product	Marginal Physical Product[2]	Average Physical Product (Col. 2 ÷ Col. 1)	Total Variable Cost (Col. 1 × $5)[3]	Total Fixed Cost[4]	Total Cost (Col. 5 + Col. 6)	Per Unit Additional Input (Col. 5 ÷ Col. 1)	Per Unit Additional Output (Col. 8 ÷ Col. 3)	Total Revenue (Col. 2 × $1)	Profit (Col. 10 − Col. 7)	Marginal Revenue per Unit of Input (Col. 3 × $1)	Marginal Revenue per Unit of Output (Col. 12 ÷ Col. 3)
Units	---- Bushels ----			---------- Dollars ----------								
0	0		0.0	0	20	20			0	−20		
		5					5	1.00			5	1
1	5		5.0	5	20	25			5	−20		
		11					5	0.45			11	1
2	16		8.0	10	20	30			16	−14		
		10					5	0.50			10	1
3	26		8.7	15	20	35			26	−9		
		9					5	0.56			9	1
4	35		8.8	20	20	40			35	−5		
		7					5	0.71			7	1
5	42		8.4	25	20	45			42	−3		
		5					5	1.00			5	1

(Continued)

Table 14–2 (Continued)

	Physical Input-Output Data				Total Costs			Marginal Costs		Economic Data		Revenue	
(1)	(2)	(3)	(4)		(5)	(6)	(7)	(8)	(9)	(10)	(11)	(12)	(13)
Input[1]	Total Physical Product	Marginal Physical Product[2]	Average Physical Product (Col. 2 ÷ Col. 1)		Total Variable Cost (Col. 1 × $4)[3]	Total Fixed Cost[4]	Total Cost (Col. 5 + Col. 6)	Per Unit Additional Input (Col. 5 ÷ Col. 1)	Per Unit Additional Output (Col. 8 ÷ Col. 3)	Total Revenue (Col. 2 × $2)	Profit (Col. 10 − Col. 7)	Marginal Revenue per Unit of Input (Col. 3 × $2)	Marginal Revenue per Unit of Output (C. 3)
Units	------- Bushels -------				--- Dollars ---								
6	47		7.8		30	20	50			47	−3		
		4						5	1.25			4	1
7	51		7.3		35	20	55			51	−4		
		3						5	1.67			3	1
8	54		6.8		40	20	60			54	−6		
		2						5	2.50			2	1
9	56		6.2		45	20	65			56	−9		
		−1						5	—			−1	1
10	55		5.5		50	20	70			55	−15		

[1] Variable inputs represent all variable items. Examples commonly include fertilizer, feed, seed, labor, etc.
[2] Difference between successive total products divided by differences between successive total input units. Marginal data displayed between successive units provide a closer approximation of exact marginal values computed as derivatives.
[3] Each unit of variable input costs $5.
[4] Fixed costs per acre are assumed to be $20.

The Principle of Profit Maximization and Loss Minimization

As you may have already surmised, there is a general economic principle that states the conditions for the optimum economic level of production. When inputs are added in the production process to the point where the added cost of obtaining the last unit of output is exactly equal to its value, there profit is maximized or loss is minimized. In other words, maximum profit (or minimum loss) occurs at the point where *marginal cost equals marginal revenue*. The rule always applies and is a relatively easy method of determining the optimum production level (Figure 14–1).

Optimum level of output under perfect competition

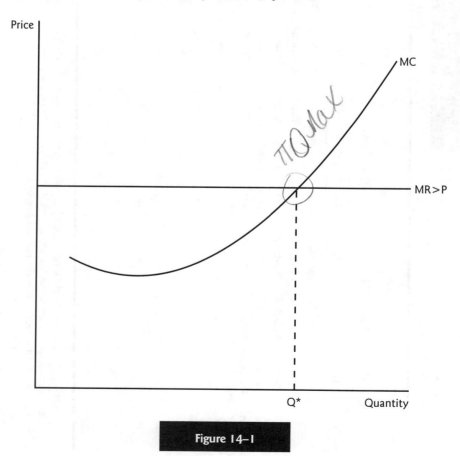

Figure 14–1

Allocating Limited Resources

Entrepreneurs do not always have all the resources they can profitably utilize. In fact, rarely does production occur with resources in unlimited supply. The problem for the producer then becomes one of making the most profitable use of resources that are available. A given amount of capital may be used in a number of ways. Which use or combination of uses will result in the greatest possible revenue?

Equimarginal Returns

The key to determining the best allocation of scarce resources among all possible uses is an economic principle known as the **principle of equimarginal returns.** It states that resources should be allocated among alternative uses in such a way that returns from the last unit employed in each use equal returns from the last unit employed in all other uses.

For example, a producer may have $1,000 of capital to allocate to all possible uses. If this capital is divided into $100 increments, then the best allocation will be one in which each enterprise is producing approximately equal returns from the last $100 invested in it. If Enterprise A returns $300 to the last $100 invested while all other alternatives are returning only $200, then a greater total return can be attained by reducing the capital allocated to other enterprises and increasing the amount allocated to Enterprise A. It is only when the returns to the last units of capital expended are equal in all enterprises that total revenue has been maximized.

A farmer who has $1,000 of capital allocated among various enterprises would probably not spend it all for livestock feed, fertilizer, insecticides, improved breeding stock, or better machinery. To maximize profit the producer must allocate available capital among all these alternatives in accordance with the principle of equimarginal returns.

Opportunity Cost

Closely associated with the equimarginal returns principle is the concept of **opportunity cost,** which states that the cost of using a resource in one way is the return that could be obtained from using that resource in its best alternative use. In our example, opportunity cost refers to the returns foregone because resources were used for Enterprise A rather than Enterprise B. A farmer who grows corn on a tract of land cannot grow cotton there. If corn returns $1,000 and cotton $1,200, the opportunity cost of growing $1,000 worth of corn is $1,200 in cotton.

It is usually true that no matter how a particular resource is used, it could also have been used in another manner to produce some return. In other words, to use a resource in a particular way always requires giving up the opportunity to gain returns that the resource could have earned in an alternative use. The return given up, or foregone, is the cost of the opportunity passed by and hence is referred to by economists as the "opportunity cost." It is hoped that the return of a resource in its present use is always greater than what could have been earned elsewhere. Opportunity cost is then said to be the return foregone from the next best alternative use of a resource. If the opportunity cost is greater than the return produced in the present use, the resource is not being used most efficiently.

A practical example of opportunity cost in agriculture is provided by the use of land. An acre may be used to grow cotton, soybeans, or pasture for livestock. Assume that the net return in each use would be $150, $80, or $15 per acre, respectively. The opportunity cost of using land to grow cotton is $80 per acre, the net return from soybeans, which would be the next best alternative. Assume that because of acreage restrictions, the farmer is limited in the acres of cotton that can be planted. Soybeans may be produced on the remaining land at an opportunity cost of $15 per acre, which is the net return from producing livestock. Or, assuming an average land value of $500 per acre and an annual interest rate on savings of 4 percent, the opportunity cost would be the $20 of interest on savings ($500 × .04 = $20) the farmer forgoes by investing in land rather than in a savings account.

Can you think of an application of opportunity cost to your personal expenditure of time as a student?

Optimum Allocation of Capital Among Enterprises

Assume the following relationships:

Cost of $100 of Aditional Capital[1]	Returns from the Use of $100 of Additional Borrowed Capital, by Enterprise				
	A	B	C	D	E
	($)				
9	15	12	25	9	15
9	15	11	20	8	9
9	14	10	15	7	5
9	12	9	9	6	0
9	9	8	5	5	–5

[1]Annual interest rate of 9.0 percent, not adjusted for income tax purposes.

We will operate on the principle that as long as returns equal or exceed the cost of borrowed capital, allocation of borrowed capital is a wise management choice. Next, we will assume that the firm is operating in each of two possible situations: (1) with unlimited capital or no constraints on borrowing except for paying the required costs of borrowing and (2) with limited available capital of $1,000.

Assuming unlimited capital, the capital allocation would be as follows:

Enterprise A	$ 500
Enterprise B	400
Enterprise C	400
Enterprise D	100
Enterprise E	200
Total	$1,600

Assuming capital is limited to $1,000, the following allocation would logically result:

Enterprise C	$ 200
Enterprise A	200
Enterprise E	100
Enterprise C	100
Enterprise A	100
Enterprise B	100
Enterprise A	100
Enterprise B	100
Total	$1,000

To summarize by enterprise:

Enterprise A	$ 400
Enterprise B	200
Enterprise C	300
Enterprise D	0
Enterprise E	100
Total	$1,000

With limited capital, all five enterprises would be restricted, with Enterprise D eliminated entirely. Under the limited capital scenario, Enterprises A and C would receive 70 percent of the available capital.

Other Input-Output Functions

Thus far, we have discussed the factor-product model, using only one variable input to produce only one output. In Chapter 27, we will extend this discussion to situations somewhat closer to reality when we cover functions consisting of more than one input and one output.

Topics for Discussion

1. Discuss the methods used to arrive at maximum profit.

2. Discuss the principle of minimizing net losses.

3. Discuss the principle of equimarginal returns.

4. Discuss the principle of opportunity cost.

5. Discuss the principles that apply in situations where there are more than one input to produce a given output.

Problem Assignment

Using the data prepared for the assignment given in Chapter 13, determine the optimum level of output.

Recommended Readings

Boehlje, Michael D., and Vernon R. Eidman. *Farm Management.* New York: John Wiley & Sons, Inc., 1984. Ch. 3.

Colander, David C. *Economics,* 2nd ed. Chicago: Richard D. Irwin, Inc., 1995. Ch. 25.

Debertin, David L. *Agricultural Production Economics.* Macmillan Publishing Co., 1986. Ch. 3 and 4.

Doll, John P., and Frank Orazem. *Production Economics: Theory with Applications,* 2nd ed. Melbourne, FL: Krieger Publishing Co., 1992 reprint of 1984 ed. Ch. 3.

Heady, E. O., and John L. Dillon. *Agricultural Production Functions.* Ames: Iowa State University Press, 1961.

Luening, Robert A., Richard M. Klemme, and William P. Mortenson. *The Farm Management Handbook,* 7th ed. Danville, IL: Interstate Publishers, Inc., 1991.

Mjelde, J. W., et al. "Integrating Data from Various Field Experiments: The Case of Corn in Texas," *Journal of Production Agriculture,* 4(1):139–147 (1991).

Seitz, Wesley D., Gerald C. Nelson, and Harold G. Halcrow. *Economics of Resources, Agriculture, and Food.* New York: McGraw-Hill, Inc., 1994. Ch. 4.

Vaughn, B., D. G. Westfall, and K. A. Barbarick. "Nitrogen Rate and Timing Effects on Winter Wheat Grain Yield, Grain Protein, and Economics," *Journal of Production Agriculture,* 3(3):324–328 (1990).

15

The Supply Concept

The quantity of a particular good or service available for purchase is generally referred to as the **supply** of that item. We know that the quantity of goods and services tends to vary with price. For example, if the price of a given commodity increases *ceteris paribus* (while other things remain unchanged), the quantity supplied usually will sooner or later also increase. On the other hand, if the price of a commodity falls *ceteris paribus*, the quantity supplied will tend to decrease. This is true because price is related to profitability when other factors do not change. An increase in price means that it is more profitable to produce a given product, and producers generally respond by increasing output. A reduction in price, with other things unchanged, means lower profits and eventually less production of a commodity.

This general relationship can be shown graphically by constructing a diagram with price on the vertical axis and the quantity produced per unit of time on the horizontal axis (Figure 15–1). The usual relationship between price and quantity is shown by a line or curve sloping upward to the right. This is known as a **supply curve** or **supply schedule.** For example, the quantity produced at a price of $2 is found by extending a horizontal line from that price level until it reaches the supply schedule. Then, a vertical line dropped from that point on the supply schedule to the horizontal axis shows a quantity of 2 units. Likewise, when price increases to $5, quantity supplied is similarly found to increase to 5 units. Thus, the supply schedule shows that as price increases, the quantity offered increases, and as price decreases, the quantity offered decreases.

Where Does a Supply Curve Come From?

The supply schedule is closely related to the costs of production discussed in Chapter 13. A marginal cost curve, like the one from our example (Column

A supply schedule

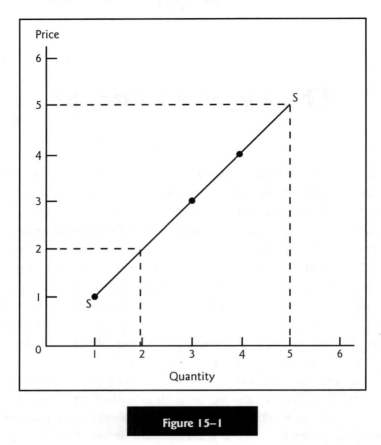

Figure 15-1

12 of Table 13–3), graphed with cost on the vertical axis and quantity of output per unit of time on the horizontal axis, resembles the supply schedule in Figure 15–1. The portion of the marginal cost curve above minimum average variable costs (see Figure 13–1) is, in fact, a supply schedule for a firm. It shows the cost of producing the last unit of output at any given level of production. Firms will not produce beyond the point where marginal cost equals marginal revenue. This level of production is at the price level where the cost of producing the last unit of output just equals the additional revenue from that unit of output. Thus, the marginal cost curve indicates the level of output at varying price levels. That is, of course, exactly what is shown by a firm's supply schedule. The supply curve for an industry, which indicates the same price-output relationships, is simply the aggregation (adding together) of individual firms' supply schedules.

Changes in Supply

A bit of reflection may cause you to question the validity of the supply schedule as presented thus far. There are many commodities for which the price has decreased in recent years and yet the quantity produced has increased tremendously. Personal computers are an obvious example. On the other hand, it is also possible for the price of a commodity to rise while the quantity produced decreases.

The validity of the supply schedule is not negated by these seemingly irregular occurrences. It is just that the supply schedule has the capability of *changing* or *shifting*. The prices that result after a shift in a supply schedule simply reflect the effects of the supply shift.

Figure 15–2 shows how it is possible for output to increase even though prices are lower. The supply schedule simply changes or shifts. For example, on the original supply schedule, SS, at a price of $3, the quantity offered for sale is 3 units. After the supply curve has shifted to the right, S_1S_1, the price may fall to $2, but the quantity offered for sale would be 4 units. Thus, a greater quantity is offered for sale even at a lower price because of the shift in the supply curve. The supply schedule still retains the same shape. It merely shifted to the right; total supply has increased.

But what if there is less of the good offered on the market, yet its price is higher? Schedule S_2S_2 shows the original schedule shifted to the left. It can be seen that even though price increases to $4 on the new schedule, the quantity offered for sale is only 2 units, less than the 3 units offered at a lower price of $3 on the original schedule. In this case, the total supply has decreased although the shape of the supply schedule is still unchanged.

Thus, the complete supply schedule has the capability of changing or shifting. The criteria for determining when supply has shifted or changed are as follows: (1) Supply has increased (shifted to the right) when either a larger quantity is offered for sale at the same price or the same quantity is offered at a lower price. (2) Supply has decreased (shifted to the left) when either the same quantity is offered at a higher price or a lesser quantity is offered at the same price. It will be helpful to graph these conditions for a change in supply or closely examine Figure 15–2 to prove that these conditions exist when supply changes.

Shifts in the supply schedule

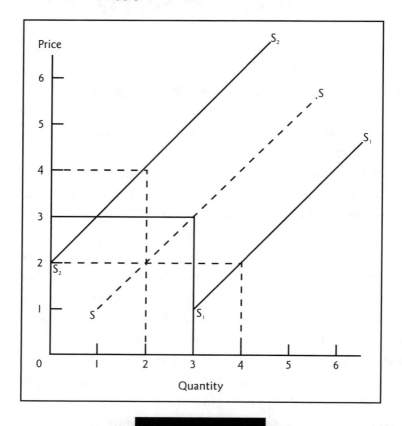

Figure 15-2

Factors That Cause a Supply Function to Shift

A number of factors can cause the supply curve to shift. These include:

1. *Production technology.* New and better technology enables producers to increase their production without affecting total production costs. Thus, average costs (costs per unit of output) decrease. This increased productivity is reflected in a shift to the right of the adopting firm's cost curves, including the marginal cost curve, which reflects the firm's supply function. With a shift to the right, the firm will produce more at a given price level than before adopting the new technology.

2. *Input price changes.* Changes in the prices of inputs used in the production process also affect the firm's, and consequently the industry's, supply curves. A decrease in input prices shifts the cost curves to the right, resulting in the same effects as adopting new technology discussed above. Conversely, an increase in input prices shifts the cost curves to the left, with the firm producing less at a given price level than before.

3. *Taxes or subsidies on inputs.* Taxes or subsidies placed on inputs used to produce a particular product will influence the supply of that product in the same manner as price changes, because they influence what the firm must pay to use those inputs. A tax, then, is effectively equivalent to an input price increase, which results in less of the input being used and consequently, less output. Conversely, an input subsidy effectively lowers the input price, allowing the producer to purchase more of the input and produce more output *ceteris paribus*.

4. *Weather.* For many industries weather generally has little or no influence on supply. However, in the agricultural sector, weather clearly is an important factor determining supply, especially in dryland agriculture.

5. *Prices of other products.* The prices of other outputs that can be produced using the same resources also influence supply of a given product. As we have discussed, most farmers must choose from a number of enterprises when they decide what to produce. If the price of one commodity rises *ceteris paribus*, the farmer is likely to allocate more resources to that commodity. The quantity supplied of that commodity thus increases, while the supplies of other commodities decrease.

6. *Expected prices.* Closely related to No. 5 is the effect of anticipated or expected prices. If farmers expect the price of a commodity—corn, for example—to increase, then they will plant more corn, thereby increasing the supply.

7. *Number of sellers.* One assumption underlying perfect competition is the free mobility of resources. Firms can enter and exit an industry easily. If existing firms are generating profits, new firms have incentive to enter that industry to capture those profits, resulting in increased supplies. Conversely, firms suffering continued losses will exit the industry, decreasing supplies.

8. *Government intervention.* Government policy can also influence supply. The government can provide incentive to increase or curtail production

using a variety of methods, some of which will be discussed in Chapter 25.

A Change in Quantity Supplied

A change in quantity supplied and a change in supply, although sounding the same at first, actually refer to quite different phenomena. A **change in the quantity supplied** refers to what happens to the quantity offered for sale when the price of a product varies. In this case we are speaking of a movement *along a supply schedule,* not a movement *to a new schedule.* In Figure 15–3 a change in quantity (Q) resulting from a variation in price (P) is demonstrated by a movement from Q_2 at P_2 to Q_4 at P_4. A **change in supply,** by contrast, refers to the movement or shift of the whole supply schedule, such as the shift from SS to S_1S_1 in Figure 15–3. Remember the criteria for determining when

A change in the quantity supplied versus a change in supply

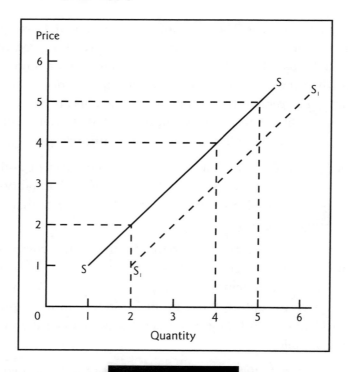

Figure 15–3

there has been a change in supply, and you will not be confused by these similarly sounding economic terms.

The Shape of the Supply Schedule

The normal shape of the supply schedule is a line sloping upward to the right. The slope of this line may vary widely from one product to another or even from one season to the next for a particular product. Understanding the shape of the supply schedule is essential to anyone who wants to be informed about possible supply responses to changes in market conditions for a given good or service.

Figure 15–4 shows some of the variations that may occur between supply schedules. Some may be nearly horizontal as is shown by schedule S_1S_1. With such a schedule, the variation in quantity accompanying even a small change in price would be relatively great. If you were producing a product that had such a supply curve, you could expect a substantial increase in price to result in a tremendous flood of increased production.

Other supply curves are shaped similarly to schedule S_2S_2, where a change in price produces a nearly proportional change in the quantity offered for sale. This schedule slopes upward at approximately a 45° angle.

Some supply schedules may be almost vertical, such as schedule S_3S_3 in Figure 15–4. In this case, a change in price produces hardly any change in the quantity of a product supplied. If you were producing a product with a supply curve of this nature, you would expect a relatively stable quantity of the goods to be produced even though price fluctuated a great deal.

The change of quantity supplied in response to price changes is referred to in economics as the **price elasticity of supply.** To understand the concept of a supply elasticity, think of supply as a piece of elastic. A "tug" or change in price may produce a great "stretch" or change in the quantity supplied. In such cases, supply is said to be *highly elastic,* as shown by schedule S_1S_1 in Figure 15–4. In other cases, a "tug" in price may produce very little "stretch" or change in the quantity supplied, as shown by schedule S_3S_3 in Figure 15–4, and supply is said to be *highly inelastic.*

In theory, supply curves may vary all the way from vertical lines, called "perfectly inelastic supply schedules," to completely horizontal lines, called "perfectly elastic supply schedules." In actual practice, however, such extreme curves are not often encountered. Many agricultural supply curves tend to be relatively inelastic, meaning that price changes are accompanied by less than proportionate changes in quantity supplied. Due to a high level of fixed costs,

Supply curves of varying elasticity

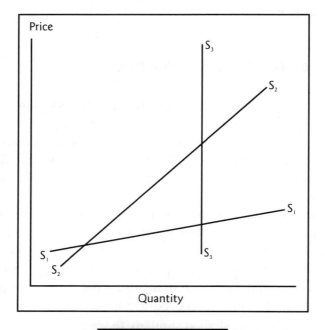

Figure 15-4

farmers are not as quick to respond to price changes (particularly price reductions) as are other types of businesses. Therefore, in the short run, production tends to remain relatively stable even though product prices change.

Although the slope of a supply curve may provide a general indication of the price elasticity of supply, a much more precise method of determining elasticity is available. Actually, the price elasticity of supply may vary between different points along the curve. The *arc* elasticity, E_s, between any two points on a supply schedule may be determined using this formula:

$$E_s = \frac{\text{Percentage change in quanitity supplied}}{\text{Percentage change in price}}$$

which can also be expressed as:

$$E_s = \frac{\frac{Q_1 - Q_2}{Q_1 + Q_2}}{\frac{P_1 - P_2}{P_1 + P_2}}$$

where

Q_1 = the quantity offered for sale before a price change
Q_2 = the quantity offered for sale after a price change
P_1 = the original price, or the price accompanying Q_1
P_2 = the new price, or the price accompanying Q_2

The price elasticity of supply, E_s, is a comparison of the percentage change in quantity to a percentage change in price. The coefficient, E_s, which ranges from 0 (perfect inelasticity) to infinity (perfect elasticity), is an expression of the percentage change in quantity resulting from a 1 percent change in price. If the coefficient is less than 1, say, 0.65, that segment of the supply schedule is said to be *relatively inelastic* because the percentage change in quantity is relatively less than the percentage change in price. If the coefficient is greater than 1, say, 2.3, that segment of the supply schedule is said to be *relatively elastic* because the percentage change in quantity is greater than the percentage change in price. An elasticity coefficient of 1 indicates that the percentage change in quantity supplied exactly equals the percentage change in price, and the supply curve displays *unitary elasticity*.

The elasticity between two points on curve SS in Figure 15–3 can be computed to provide an example of the use of the elasticity formula. When price increased from \$2 to \$4, the quantity supplied increased from 2 to 4 units. Substituting these values into the formula, the price elasticity of supply can be computed as follows:

$$E_s = \frac{\frac{2-4}{2+4}}{\frac{2-4}{2+4}} = \frac{-2}{6} * \frac{6}{-2} = 1$$

Supply curve SS in Figure 15–3 thus exhibits unitary elasticity because the change in quantity supplied was proportionate to the change in price. What would the elasticity coefficient be if Q_2 were 3 units instead of 4?

Cross Elasticity of Supply

The **cross elasticity of supply** refers to the changes in quantity supplied of one good or service in response to changes in the price of another good or service. Commodities that are complementary in supply, like soybean oil and soybean meal, tend to have a positive cross price elasticity of supply. Competitive commodities, those that compete for the same resources, have a negative cross price elasticity of supply. For example, if the price of cotton declined, the quantity of soybeans produced in the South would be expected to increase *ceteris paribus*. Normally, the higher the coefficient of cross elasticity, the greater the competition between two products. Vegetable crops, for example, have a high cross elasticity of supply with other vegetable crops, since the quantity produced of one vegetable depends not only on the price of that vegetable but also on the price of closely related vegetables.

One formula for calculating the cross elasticity of supply is y_1 as follows:

$$E_{sy_1y_2} = \frac{\dfrac{Q^0_{y_1} - Q^1_{y_1}}{Q^0_{y_1} + Q^1_{y_1}}}{\dfrac{P^0_{y_2} - P^1_{y_2}}{P^0_{y_2} + P^1_{y_2}}}$$

where y_1 and y_2 are two goods, $Q^0_{y_1}$ and $P^0_{y_2}$ are the quantity of y_1 supplied and the price of y_2 before a change in the price of y_2, and $Q^1_{y_1}$ and $P^1_{y_2}$ are the quantity of y_1 and price of y_2 after the change.

Short Run Versus Long Run

The *short run* includes that period of time when entrepreneurs cannot vary the fixed capacity of plants (i.e., firm size) and equipment. They may vary units of variable inputs applied to the fixed factors of production and logically should operate where short-run marginal revenue equals short-run marginal costs.

In the short run, a farmer's costs are both fixed and variable. In addition to the fixed costs mentioned earlier, the farmer may decide upon several variable costs, such as how many seeds to plant per acre, how much fertilizer to apply, how much pre-emergence herbicide to use, etc.

The *long run* is that period of time when the entrepreneur can vary or change the fixed capacity of plants and equipment. In the long run, expanding or contracting plant capacity is an option. The goal is to allocate resources so that the most efficient plant size is adopted and managerial capability is best utilized.

In the long run, all costs are variable, because the producer has an opportunity to sell land or rent or purchase additional land, discharge labor or hire more, sell or buy machinery, disperse or acquire more livestock, etc. Thus, the farmer has the opportunity to redirect enterprise selection, whether agricultural or not, in an attempt to choose those endeavors offering the best economic prospects.

In the short run, a firm will produce as long as its marginal revenue (the price per unit of output) exceeds average variable costs so that some contribution can be made toward fixed costs. In the long run, a firm cannot continue to do this, since the firm must meet its average total costs to stay in business. If prices received do not cover long-run average total costs, a firm will probably choose to exit at the first available opportunity. If prices are above average total costs in the long run, a firm may choose to expand, and/or new firms will want to enter the business. This process will continue until it is no longer profitable for an existing firm to expand or for new firms to move in—that is, until profits in excess of normal returns are eliminated in the industry. This will occur when each firm, old or new, is operating at the minimum point of its long-run average total cost curve. In other words, if at the short-run equilibrium level of output, the short-run average total cost curve is above the long-run average total cost curve, a profit-maximizing firm will expand its scale of operation (plant size) until its short-run average total cost curve is tangent at its minimum point to the long-run average total cost curve.

As firms move in or expand, they may (1) bid up the prices of resources, (2) not affect them at all, or (3) cause them to decline. In No. 1, the long-run average total cost or supply curve will move upward to the right; in No. 2, the long-run supply curve will be perfectly elastic or horizontal; and in No. 3, the long-run supply curve will move downward to the right.

Increasing, Constant, or Decreasing Costs

In a long-run framework of analysis, certain industries exhibit certain cost tendencies. Some industries exhibit *increasing cost* tendencies. In agriculture, as more and more land is brought into production, it is usually less and less productive. An example could be bringing pasture land into cultivation for soybean production. This increases the average total cost per unit of production. While it is true that fixed costs per unit decrease as output expands, it is also true that average variable cost per unit increases faster than average fixed cost per unit declines. Mining for coal, for example, exhibits increasing cost tendencies.

Constant cost industries are typified by garment and apparel firms. Their fixed costs are relatively small, and as production expands, total unit costs stay about the same. As production expands, more sewing machines and more labor are added, proportional to the extra output. Since fixed costs are small, total cost per unit of output tends to stay constant.

On the other hand, public utilities often exhibit *decreasing cost* tendencies. As output expands, the average unit fixed costs of utilities decrease faster than their average unit variable costs increase; thus, average total unit costs decrease as output is increased. Utilities are known for their high fixed costs.

Topics for Discussion

1. What is the usual relationship between the quantity supplied and price?
2. Discuss the factors related to a change or shift in supply.
3. Contrast a "change in supply" with a "change in the quantity supplied."
4. Define "elasticity of supply."
5. Name and discuss the three "elasticities of supply."
6. Define and discuss "cross elasticity of supply."

Problem Assignment

Using the concept of supply elasticity, locate examples of crop and/or livestock enterprises that demonstrate elastic, inelastic, and unitary supply responses to price changes.

Recommended Readings

Bowles, Samuel, and Richard Edwards. *Understanding Capitalism: Competition, Command, and Change in the U.S. Economy*, 2nd ed. New York: HarperCollins College, 1993. Ch. 4 and 5.

Colander, David C. *Economics*, 2nd ed. Chicago: Richard D. Irwin, Inc., 1995. Ch. 8, 23, 24.

Gwartney, James D., and Richard L. Stroup. *Economics: Private and Public Choices*, 7th ed. Orlando, FL: The Dryden Press, 1995. Ch. 3.

Seitz, Wesley D., Gerald C. Nelson, and Harold G. Halcrow. *Economics of Resources, Agriculture, and Food*. New York: McGraw-Hill, Inc., 1994. Ch. 5 and 8.

16

The Demand Concept

Each individual consumer requires certain types of goods. The most pressing requirements are those necessary to sustain life—food, clothing, and shelter. But after these basic needs have been met, there is virtually an endless number of additional commodities for which consumers express varying degrees of want. The purpose of this chapter is to discuss how we decide what needs and wants we meet, given the limited amount of money we have available to purchase goods and services.

The strength of the desire to have a particular product or commodity is basically reflected in the effort one is willing to expend (or the leisure one is willing to sacrifice) in order to attain it. If you were starving, for example, you would be willing to sacrifice most, if not all, your leisure time to get food to survive. However, your desire for food might be quickly satisfied. After you were satisfied, you would quickly come to the point where you would not be willing to give up any additional leisure for more food for immediate consumption.

In the sophisticated economies of much of the world today, money has become a common denominator, a standard of value indicating the relative strength of desire for particular commodities. We measure the strength of desire for a given commodity by the amount of money one is willing to give up or the price one is willing to pay to attain the commodity.

It is well understood that for most items, rational people are willing to pay higher prices when a commodity is scarce than when it is in plentiful supply. Or, assuming that substitute products are available, when the price of a commodity is high, consumers will purchase less of the high-priced commodity and switch to a lower-priced substitute. As price rises, less product or service will be consumed; or conversely, as price declines, more will be purchased. This relationship is expressed in the **law of demand**, which states that the quantity of a good or service purchased is inversely related to the price of that good or service, *ceteris paribus*. Simply put, consumers buy more at lower prices.

Consumers gain **utility**, or satisfaction, from the products and services they consume. However, as a consumer uses additional amounts of a commodity, the degree of satisfaction gained from the last units consumed decreases. This, in economic terminology, is referred to as the **law of diminishing marginal utility**.

When successive units of a particular product or service are consumed, with the consumption of all other products and services held constant, a point is reached where the additional utility of a product or service associated with each successive unit consumed begins to diminish. Two logical reasons marginal utility diminishes are that (1) commodities are not perfect substitutes one for another and (2) no particular want is insatiable in a given period of time.

When the prices of commodities are applied to the diminishing marginal utility function, then the consuming unit has a basis for making economical choices between commodities. In most cases the concept of weighted marginal utility is used. Thus, when prices are applied to the commodities in question, the consumer should equate the marginal utility of Product A divided by its price to the marginal utility of Product B divided by its price to the marginal utility of Product C divided by its price, etc.

$$\frac{MU_a}{P_a} = \frac{MU_b}{P_b} = \frac{MU_c}{P_c} = \ldots$$

To express the concept of diminishing marginal utility in another way, assume that owning one automobile provides you with great utility. Depending on your situation, a second may also provide utility, but somewhat less than the first. A third car may provide no additional utility at all, while a fourth car may be such a nuisance and expense that it provides negative utility. Therefore, because of diminishing marginal utility from each successive car, you will not be willing to pay as much for the second car as you did the first, and you will not be willing to pay nearly as much for the third as for the second. Not only will you not be willing to pay for the fourth car, but you will want to be paid before you will take it.

Therefore, the law of diminishing marginal utility, which parallels the law of diminishing marginal returns, specifies that as more and more of a product or service is consumed, there comes a time when the utility of consuming that product or service starts to decline. For example, to induce customers to buy additional cars, the price must come down. Demand curves, therefore, slope downward, from left to right, as opposed to supply curves, which normally slope upward, from left to right.

The relationship between the quantity of a product purchased and its price level can be shown graphically (Figure 16–1). The curve labeled "DD" in Figure 16–1 is called a **demand curve** or **demand schedule.** It shows the quantity of a product that consumers are willing to purchase at varying price levels. For example, if the price, shown on the vertical axis, were $5, the quantity purchased in a given period of time, shown on the horizontal axis, would be 1 unit. If the price should fall to $3, the quantity purchased would increase to 3 units. For consumers to purchase 5 units of the commodity, the price would have to fall to $1.

A demand schedule

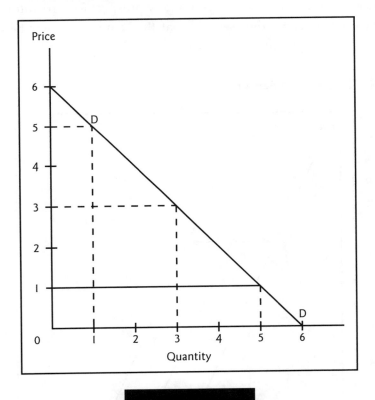

Figure 16–1

The demand schedule is a representation of the consumer's demand or strength of desire for a given commodity in a given period of time. It shows that, as more of the item is obtained, the price one is willing to pay for additional units decreases.

A Change in Demand

The demand schedule for any given commodity can and usually does change as time passes. Like the supply curve discussed in the previous chapter, the demand curve can shift either to the right or to the left. A shift to the right as shown by curve D_1D_1 in Figure 16–2 indicates that demand has increased. Thus, at any price level, a larger quantity will be taken than would be true under the original schedule DD, *ceteris paribus*. Or, any given quantity that would have been purchased under schedule DD will be bought at a higher price under schedule D_1D_1.

An increase in demand is always a happy prospect for agricultural producers. It means they can either sell more at the same price or obtain a higher price for the same quantity of production.

A change in demand

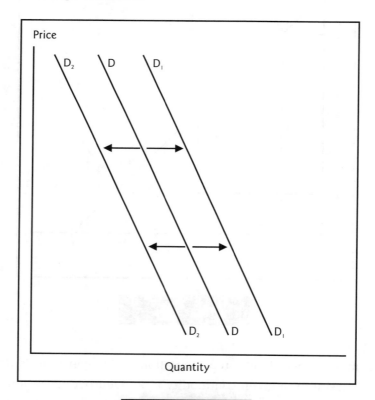

Figure 16–2

The demand schedule may also shift to the left, however, indicating a decrease in demand. This unhappy occurrence for producers is shown by schedule D_2D_2 (Figure 16–2). In this case consumers will take the same quantity of goods only at a lower price. If price remains the same, they will buy less than they would have purchased under the original demand schedule DD.

Several factors cause demand curves to shift. First, it is obvious that consumers must have income to enable them to purchase goods or have an *effective demand*. A change in income will cause demand to change. The usual condition is that demand shifts to the right over time because of rising incomes. Incomes may fall during recessions, however, causing demand to shift to the left.

A second factor influencing demand is a change in the tastes and preferences of consumers. As individuals grow from children to adults, the kinds and amounts of products they demand change. For example, a community of young families may experience a heavy demand for baby food. As the children of these families grow older, the demand for baby food will decline, while the demand for meats, fruits, and vegetables (and pizza) will increase, or shift to the right.

The preferences of individuals change over time regardless of age. An example of this occurrence from the more recent past has been the decline in demand for fat pork and the rapid increase in demand for the leaner cuts of meat.

Of course, we are all aware of the effects of changing preferences in our demand for clothing. Consider how clothing styles change over time, and you will see how clothing preferences influence demand. This is especially evident with trendy lines of apparel that are connected to popular cartoon or movie characters. Once the popularity of the cartoon series or movie wanes or is replaced by a next release, demand for that apparel quickly falls.

Most products available in today's markets compete with goods that are fairly close substitutes. A change in the prices of substitute products is a third factor that may cause demand to shift. For example, when margarine became available at a price much below that of butter, the demand for butter shifted drastically to the left because of the availability of this lower-priced substitute.

A rise in beef prices typically causes an increase in the demand for pork and chicken. In such a case the demand curves for pork and chicken shift to the right because consumers substitute pork and chicken for beef. If pork and chicken prices increase while beef prices remain stable, however, the demand for beef shifts to the right. Consumers substitute beef purchases for purchases of chicken and pork.

A fourth major demand shifter is a change in population. As the population grows, the demand for most products increases because more people have wants and needs to satisfy.

There may be population losses. Over time, many rural areas of the United States have experienced massive population losses as residents have emigrated to more attractive employment opportunities in urban areas. The demand for goods and services in these rural areas has shifted drastically to the left, forcing many businesses to close down as the population has declined.

A Change in the Quantity Demanded

The quantity of a good or service purchased by consumers may change simply because of changes in the price of that good or service. Figure 16–3 shows that when price decreases from $4 to $2, the quantity of the product purchased changes from 2 to 3 units. This is not a change in demand but a

A change in the quantity demanded

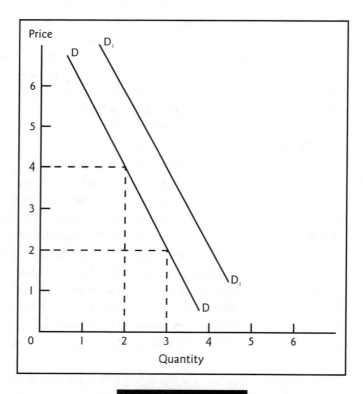

Figure 16–3

change in the *quantity demanded*. For demand to have changed (increased), a larger quantity of goods would have had to be purchased at the same price or the same quantity would have had to be bought at a higher price. A new demand curve, such as D_1D_1, would have been formed to the right of DD if demand had increased. Instead, a change in the quantity demanded is shown by movements along the same demand curve.

The Shape of the Demand Curve

Most demand curves slope downward from left to right, indicating that larger quantities are only purchased at lower prices. Not all demand curves have the same degree of slope, however. They may vary from nearly vertical lines to nearly horizontal lines, depending upon consumers' reactions to price changes for a given good or service.

If consumers buy essentially the same amount of a commodity regardless of price variations, then a graphic presentation of the demand schedule would look like curve DD in Figure 16–4. A demand curve that is completely vertical is said to represent a perfectly *inelastic* demand. Consumers make no changes in the quantity purchased of such a commodity even though price may vary greatly. Salt is usually cited as an example of a product for which consumers have highly inelastic demand. A consumer would use only a fixed supply of salt even if it were totally free. The amount used would not change appreciably if the price of salt were to triple or quadruple.

While there are very few goods and services with perfectly inelastic demand, a large number have relatively inelastic demand curves, or curves that slope sharply downward from left to right, such as curve D_1D_1 in Figure 16–4. Under such a curve, a relatively large change in the price of the good or service results in only a small change in the amount purchased.

The aggregate or total demand curve for food and the total demand curve for most individual food items fit the relatively inelastic model. These demand curves tend to be steep because of the somewhat fixed capacity of the human stomach. Regardless of how high food prices go, consumers still require a certain quantity to support life. But, once that basic requirement has been met, the demand for additional food is very slight. Thus, even if food is available as free goods, the limit is quickly reached, beyond which no additional food can be consumed and no more is demanded.

The nature of the demand curve for food, and for most agricultural products, offers an explanation for why farmers tend to suffer perpetually from

Demand curves showing varying degrees of elasticity

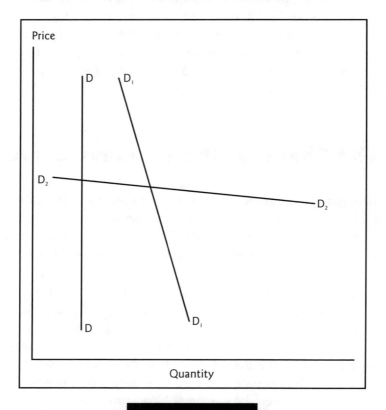

Figure 16–4

overproduction. The hope for increasing sales of agricultural products comes primarily from shifts in the demand curve to the right due to population growth. Offering farm products for sale at lower prices does little to increase total purchases of agricultural products. Technological advances in agriculture have enabled producers to expand output at a much faster rate than that at which demand for farm products has increased.

Curve D_2D_2 in Figure 16–4 is a relatively *elastic* demand curve. With highly elastic demand, small changes in price are accompanied by relatively large changes in quantity demanded. Our agricultural problem of overproduction would probably be nonexistent if this were the shape of the demand curve for most agricultural products. At any given time, farmers would be able to sell all of their output by accepting a slightly lower price.

A perfectly elastic demand curve is a horizontal line. Such a curve suggests that any change in price results in an infinitely large change in quantity purchased. Such a relationship exists only in the theoretical sense for the aggregate demand curve. As will be shown later, an individual producer of an agricultural commodity may face a horizontal demand curve within limits, but even in this case the curve must eventually turn downward if the producer grows a large enough amount of produce.

Price Elasticity of Demand

The **price elasticity of demand** is defined as the percentage change in the quantity of a good or service demanded, given a percentage change in price *ceteris paribus*.

$$E_d = \frac{\text{Percentage change in quantity demanded}}{\text{Percentage change in price}}$$

Two types of measures reflect the price elasticity of demand, *own-price elasticity of demand* and *cross-price elasticity of demand*.

The Own-Price Elasticity of Demand

The **own-price elasticity**, given by

$$E_d = \frac{\frac{Q_1 - Q_2}{Q_1 + Q_2}}{\frac{P_1 - P_2}{P_1 + P_2}}$$

where Q_1 and Q_2 are the quantities demanded at prices P_1 and P_2, respectively, measures the elasticity within a segment, or arc, of the demand curve. It reflects the effects of changes in the price of a given product on the quantity demanded of that product. The own-price elasticity is negative, indicating that price and quantity move in opposite directions in demand relationships. An increase in price is accompanied by a decrease in the quantity demanded. In contrast, the quantity demanded increases when price decreases.

Based on the example illustrated in Figure 16–3, when price decreases from $4 to $2 and quantity increases from 2 to 3 units, the own-price elasticity of that segment of the demand curve is –0.6.

$$E_d = \frac{\frac{2-3}{2+3}}{\frac{4-2}{4+2}} = \frac{-\frac{1}{5}}{\frac{2}{6}} = -\frac{1}{5} * \frac{6}{2} = -\frac{6}{10} = -0.6$$

This can be interpreted to mean that for a 1 percent change in price, quantity demanded will change by 0.6 percent. So if price decreases by 1 percent, the quantity demanded will increase by 0.6 percent.

The own-price elasticity reflects the degree to which quantity demanded changes, or how responsive quantity demanded is to changes in price. The demand curve in the example is relatively *inelastic*, because the absolute value of the elasticity is less than 1. Had it been greater than 1, that segment of the curve would have been relatively *elastic*.

A reliable test of the elasticity of a demand curve is always possible by determining what happens to total revenue (price * quantity) when price falls. If a demand curve is relatively inelastic, total revenue will decline when price falls. Quantity does not increase in proportion to the drop in price under an inelastic curve. Total revenue will increase after a price drop with a relatively elastic demand curve. In this case the increase in quantity as price falls will be proportionately greater than the decrease in price. If total revenue remains the same after a price change, then the demand curve is said to be of *unitary elasticity*; the own-price elasticity equals –1. In this case, the percentage change in quantity purchased is exactly equal to the percentage change in price.

The Cross-Price Elasticity of Demand

Cross-price elasticity of demand refers to the effect that a change in price of one good or service has on the quantity demanded of another good or service. The cross-price elasticity, $E_{d\text{-}xy}$, is the percentage change in the quantity demanded of one good, good x, for example, given a change in the price of another good, good y.

$$E_{d\text{-}xy} = \frac{\text{Percentage change in quantity of x demanded}}{\text{Prcentage change in price of y}}$$

If two products are *substitutes,* the cross-price elasticity will be positive, indicating that the quantity demanded of good x will increase when the price of good y increases. If the two products are *complements,* the cross-price elasticity will be negative. The quantity demanded of good x will fall when the price of good y increases.

Determinants of Price Elasticity

A number of factors influence the price elasticity of demand, including the availability of substitutes, the importance to a consumer's budget, how often the good is purchased, and time.

If there are substitute products available in the marketplace that can easily replace a given product, the elasticity of demand will be greater than if no or few substitutes are available. Consumers are more sensitive to price increases if they can easily switch to another product.

The relative importance of a good or service in a consumer's budget also influences price elasticity of demand. Demand tends to be more elastic for products that represent a large portion of a consumer's budget. Price changes for "big-ticket" items, such as automobiles, household appliances, furniture, etc., have a greater impact on the consumer's real income than do comparable relative price changes for less expensive items, like salt or toothbrushes.

The price elasticity of demand is affected by how often consumers purchase a good or service. The demand for products that are purchased more frequently tends to be more elastic than that for products purchased only occasionally. Consumers have the opportunity and incentive to obtain more information about frequently purchased products, such as prices and availability of substitutes. Consumers will likely know more about soft drink prices at competing grocery stores, for example, than, say, shoe polish prices.

The length of time over which the demand function is specified also influences price elasticity of demand. Normally, as consumers have a longer period of time to adjust to a change in price, the more elastic demand will be, *ceteris paribus*. With more time, consumers can gather more information—about prices and substitutes, for example—and they can rearrange their expenditures, based on modifying their behavior, to a change in price.

The shape of the demand curve for the product one produces may be vitally important in determining the success of the operation. For products with relatively inelastic demand curves, the possibilities for profitable expansion of the industry may be severely limited. In fact, it would usually be to the advantage of the producers in such an industry to collectively restrict production and discourage new entrants rather than to expand production. What have we been doing in this regard with respect to agriculture?

Income Elasticity of Demand

Income elasticity refers to the percentage change in the quantity demanded given a percentage change in income, *ceteris paribus*.

$$E_{d\text{-}y} = \frac{\text{Percentage change in quantity demanded}}{\text{Precentage change in income}} = \frac{\dfrac{Q_2 - Q_1}{Q_2 + Q_1}}{\dfrac{Y_2 - Y_1}{Y_2 + Y_1}}$$

The income elasticity can be either positive or negative. If $E_{d\text{-}y}$ is positive, as is most common, the quantity demanded of a good increases when income increases. Goods of this kind are called *normal goods*. Goods whose income elasticities are positive and large, $E_{d\text{-}y} > 1$, are called *superior goods*.

If $E_{d\text{-}y}$ is negative, the quantity demanded of a good decreases when income increases. Goods of this kind are called *inferior goods*. For example, assuming a poor family's income has increased sufficiently, they will decrease, say, rice and bean consumption and instead purchase more costly foods.

Indifference Curve Analysis

Almost without exception our utility gained from consuming a product drops the larger the number of units of that product we consume. This is referred to as **diminishing marginal utility** and is the basis for the demand explanation using the *cardinal utility approach*. Cardinal utility requires that utility be measured by actual numbers. No economist today believes the numbers have any meaning. However, by using another approach to explain demand theory, the *ordinal utility approach*, we can understand more fully how the factors influencing our demand for a product cause us to act as we do in the marketplace. Ordinal measures simply require a ranking of goods that people reveal when they choose one good over another.

The ordinal utility analysis of demand is usually referred to as the **indifference curve** analysis because it uses what are called "indifference curves" to show preference or indifference between two bundles of goods. Indifference curves are merely graphical displays of indifference schedules. An indifference schedule is a list of combinations of two products that is arranged in such a way that a consumer is indifferent to the combinations, preferring no one combination over another. The following are two indifference schedules:

Schedule A Product			
X	ΔX	Y	ΔY
10		0	
	3		1
7		1	
	2		1
5		2	
	1		1
4		3	
	1		1
3		4	

Schedule B Product			
X	ΔX	Y	ΔY
14		0	
	4		1
10		1	
	3		1
7		2	
	2		1
5		3	
	1		1
4		4	

Under Schedule A, the consumer can have 10 units of X and 0 of Y, or 7 units of X and 1 of Y, or one of the other combinations. Obviously, the consumer desires both products, but the question is the relationship between consumer desires and the quantities of the products. Under Schedule A, the quantities are arranged in such a way that the consumer is indifferent between the five combinations. All combinations are equally desirable; the consumer cannot distinguish between them.

Similarly, all combinations under Schedule B are equally desirable. But, notice that all combinations under Schedule B have more units of X than found in the combinations under Schedule A although they have the same number of units of Y. Under the assumption that more is preferred to less, each combination under B is preferable to any combination under A. Of course, the schedules could be extended to include more units of Y and fewer of X, and additional schedules could be devised offering more X or more Y. These two indifference schedules are illustrated in Figure 16–5.

Now, assume that X costs $.50 per unit and Y costs $1 per unit. How much of each will the consumer purchase? The optimum purchase pattern for Schedule A is 6 units of X and 1½ units of Y because of the following:

$$\frac{\Delta Y}{\Delta X} = \frac{P_x}{P_y} = \frac{1}{2} = \frac{\$.50}{\$1.00}$$

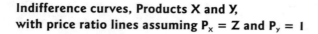

Indifference curves, Products X and Y, with price ratio lines assuming $P_x = Z$ and $P_y = 1$

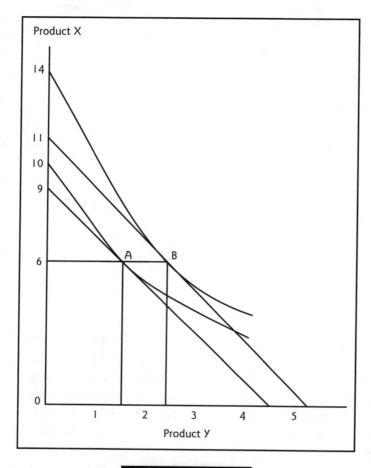

Figure 16–5

Minimum purchase cost is $4.50.

For Schedule B, the optimum purchase pattern is 6 X and 2½ Y because

$$\frac{\Delta Y}{\Delta X} = \frac{P_x}{P_y} = \frac{1}{2} = \frac{\$.50}{\$1.00}$$

Minimum purchase cost is $5.50.

If the price of X and/or Y changes, different purchases may result. As incomes rise and expenditures increase, a higher indifference curve could be attained.

Topics for Discussion

1. Define a "demand."
2. What is meant by "change in demand"?
3. Describe some of the factors that cause a demand curve to shift.
4. What is meant by a "change in the quantity demanded"?
5. Discuss the concept of "own-price elasticity of demand."
6. Discuss the concept of "cross-price elasticity of demand."
7. Discuss "income elasticity."

Problem Assignment

Locate data illustrating the three types of price elasticity of demand (inelastic, unitary, and elastic) for agricultural products.

Recommended Readings

Beierlein, James G., Kenneth C. Schneeberger, and Donald D. Osburn. *Principles of Agribusiness Management,* 2nd ed. Prospect Heights, IL: Waveland Press, Inc., 1995. Ch. 2, 4, 10, 11, 12, 17.

Buse, Rueben C. (ed.). *The Economics of Meat Demand.* Proceedings of the Conference on The Economics of Meat Demand, Charleston, SC. Department of Agricultural Economics, University of Wisconsin, 1986.

Colander, David C. *Economics,* 2nd ed. Chicago: Richard D. Irwin, Inc., 1995. Ch. 8 and 22.

Seitz, Wesley D., Gerald C. Nelson, and Harold G. Halcrow. *Economics of Resources, Agriculture, and Food.* New York: McGraw-Hill, Inc., 1994. Ch. 2 and 8.

Shugart, William F., II, William F. Chappell, and Rex L. Cottle. *Modern Managerial Economics: Economic Theory for Business Decisions.* Cincinnati, OH: South-Western Publishing Co., 1994. Ch. 2.

17

The Interaction of Supply and Demand: Prices

When you are at a supermarket or at a gas station, do you ever wonder why the prices of the products you are buying are set where they are? Your impulse may be that the prices are dependent entirely upon what the store manager decides to charge. At least then you have someone handy to blame when prices go up. But are you justified in doing so? How are prices set, anyway? The purpose of this chapter is to explore how prices are determined in a free market economy.

Saying that the price level for goods and services is determined by the interaction of supply and demand may have little meaning at the moment. To help visualize the process, consider, instead of a supermarket, a traditional farmers' market, where farmers bring their products to a market square for display and consumers come to make their purchases. Here, the transaction price becomes a matter of individual bargaining between the farmer and the consumer. If the farmer has a large supply of a particular item and that item is moving rather slowly, the farmer is likely to accept a relatively low price from the shopper. He or she will be reluctant to let a potential customer pass by without purchasing, because the chances of making sales to others later are somewhat limited. For this particular commodity it may be said that supply is heavy and demand is relatively weak.

On the other hand, if the farmer has a product that is selling so briskly that stocks are rapidly depleting, the bargaining shopper is likely to have little success in haggling for a lower price. In fact, the farmer may even raise the price as the supply becomes depleted, should demand remain strong. Or, it may happen that several customers are on hand wanting to buy the remaining stock of the commodity and they begin to bid against each other for the product, thereby raising the price. In this case, the demand could be described as heavy or strong, while the supply is light or short—a combination tending to produce a high price.

The interaction of supply and demand and equilibrium price and quantity

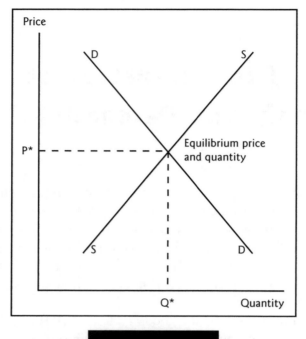

Figure 17–1

This scenario is a practical example of the interaction of supply and demand in the marketplace. Although a shopper in a supermarket may be unaware of its occurrence, at some prior level in the marketing process there has probably already been a confrontation between suppliers of commodities and some intermediary in the marketing system (processor, wholesale distributor, jobber, etc.) that has established the price of the commodity offered for sale. The price at the retail level may simply reflect the retailer's added costs of handling and a normal profit margin.

In free markets, those that operate without restraint from outside forces, the price for a given good or service is determined at the point where the total or aggregate demand and the total or aggregate supply are equal. Figure 17–1 shows that this is the point where the demand curve and the supply curve intersect. At this point demand and supply are said to be in **equilibrium.** The price established by the interaction of the two curves is the **equilibrium price.** At this price, the quantity offered for sale by producers exactly equals the quantity buyers are willing to purchase. At a higher price, the quantity buyers would be willing to purchase would be less than the quantity producers would wish to supply, creating a surplus of the product. The efforts of producers to market the surplus would rapidly drive the price back down to the equilibrium level.

At prices below the equilibrium level, producers would offer less quantity for sale than buyers would be willing to purchase, creating a shortage. In bidding against each other for the short supply of the commodity, buyers would rapidly push the price back up to the equilibrium level.

It may at first appear that a market in equilibrium has reached a happy state of affairs from which there is never any reason to depart. All buyers and all sellers are satisfied with the price that exists, and the quantity that has been produced is moving into consumption leaving no surplus or shortages. This utopia is short lived, however, because changes in supply and demand continuously occur. It will be recalled that demand shifts with changes in income, tastes and preferences, size and composition of the population, and prices of competing products. Supply also shifts, mainly because of changes in input

The effects of demand shifts with supply held constant

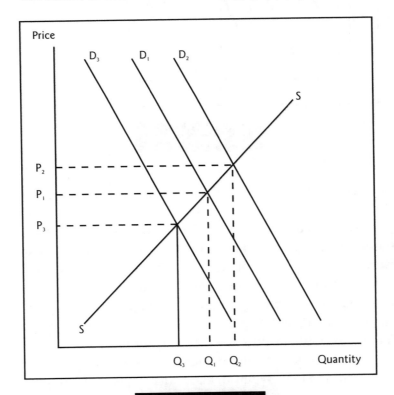

Figure 17–2

The effects of supply shifts with demand held constant

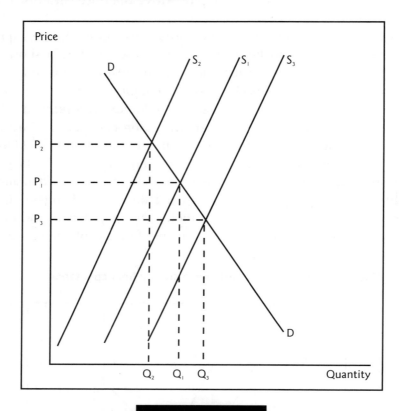

Figure 17-3

prices and changes in technology, which may allow more production at lower costs per unit.

Figure 17-2 illustrates the effect of a shift in demand on price when supply remains unchanged. Curve D_1D_1 represents the original demand curve, which forms the equilibrium price, P_1, at the intersection with the supply curve. If demand increases to D_2D_2, there is a new intersection between the supply and demand curves, higher and to the right. The new equilibrium is formed at price level P_2 and quantity Q_2. As we would expect with an increase in demand, more would be purchased at a higher price.

Should demand decrease, the demand curve would shift to the left, as shown by schedule D_3D_3. When demand decreases with supply remaining unchanged, the new equilibrium price and quantity are lower than before, as indicated by P_3 and Q_3.

Supply may also shift while demand remains unchanged, as illustrated in Figure 17–3. Should the supply curve shift to the left, as shown by curve S_2S_2, price will rise to P_2, because a lower quantity is now offered for sale and consumers will bid up the price of the reduced supply. If supply increases, common for most agricultural commodities, the supply schedule will shift to the right, as shown by S_3S_3. An increase in supply with demand unchanged always results in a lower price (P_3) for the commodity (ruling out the possibility of a perfectly elastic aggregate demand curve).

The more inelastic the demand curve, the more severe will be the drop in price from a supply increase.

Perhaps the more usual situation for demand and supply curves is that both change simultaneously. Demand may increase at the same time that supply decreases, producing a more extreme increase in the equilibrium price. Or, demand may decline while supply increases, a combination of changes that will produce drastic price declines unless some price protection program is initiated from outside sources.

For most agricultural products, it appears that both aggregate supply and demand have been increasing, but changes in supply have been proportionately greater than changes in demand. The result has generally been a steadily declining price level for agricultural commodities. To counter this trend, many proposals have been advanced for increasing demand for agricultural products while limiting supply. We will discuss this in Chapters 24 and 25, which deal with agricultural policy.

Cobweb Theorem

Since changes in both supply and demand continuously occur, the market is typically always in search of, but never quite reaches, equilibrium. No sooner are adjustments to the new location of equilibrium made than it has moved again, requiring further adjustment. Most markets are in a state of *disequilibrium* rather than equilibrium.

An explanation advanced by Mordecai Ezekiel regarding the continual fluctuation of price and quantity of a product in search of equilibrium is shown in Figure 17–4. This convergent cobweb model shows that if price were at an initial level above equilibrium, consumers would demand an amount indicated at point a. However, at that price producers would offer for sale an amount shown at point b. There would be surplus production since consumers would be willing to take only that quantity at a price indicated by point c on the demand curve.

252 / ECONOMICS: APPLICATIONS TO AGRICULTURE AND AGRIBUSINESS

In response to the lower than anticipated price, producers, in the following season, would produce only the quantity shown at point d. Consumers would bid up the price of that short supply, however, to point e. The following season, producers, in response to the higher price, would produce at point f, again creating a surplus. The adjustments would continue in a lagging fashion until finally the point of equilibrium was reached, where the amount supplied is precisely equal to the quantity demanded. Thus, price and production variations tend to converge on equilibrium in this model where the demand curve is relatively more elastic than the supply curve.

The opposite case, the divergent or explosive model, is shown in Figure 17–5. Here a price near equilibrium at the outset will stimulate production adjustments that diverge farther and farther from equilibrium. At the initial price level, the quantity demanded at point a is less than the amount supplied at point b. The resulting price fall to point c produces a production cut to point d in the following season. The resulting shortage stimulates an even

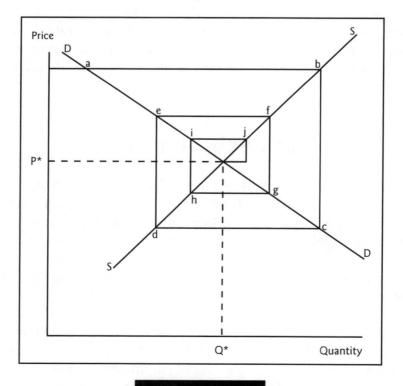

Figure 17–4

Cobweb theorem: divergent model

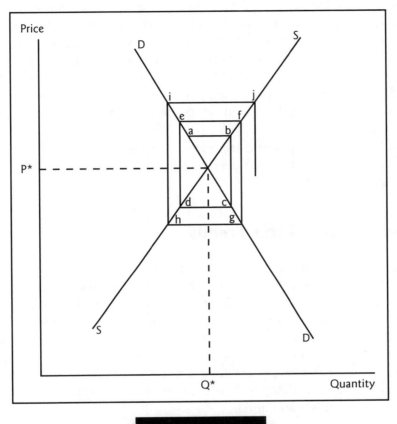

Figure 17-5

higher price than before. The ever-growing production adjustments finally become immeasurable. The "cobweb" grows in ever-widening circles. In this case, the elasticity of supply is relatively greater than the elasticity of demand, creating the exploding effects in price and production variation.

Price Trends and Cycles

Prices of farm products tend to follow the same broad trends as prices of other commodities and services over the long run. However, prices at the farm level fluctuate much more than prices at wholesale and retail levels.

When the price of a product is relatively high, farmers tend to produce more as soon as they can boost production. If the price is relatively low, they are likely to decrease production. In the short run, roughly within a crop year, there is not much opportunity for farmers to modify production in response to the price level. The average farmer has so many fixed, or committed, resources that he or she cannot readily adjust to price. Occasionally, a shift can be made from corn to hogs or from cotton to soybeans, but here again it is difficult to change techniques, introduce new machines, etc., within a short time period.

There are several types of price cycles for agricultural products: (1) secular or long-term trends, (2) year-to-year cycles, (3) seasonal cycles, and (4) intraseasonal changes.

Secular or Long-Term Trends

Secular price trends relate to the behavior of commodity prices over a period of years. For example, broiler chicken prices have exhibited a downward secular price trend since 1945. The main reason for this downward trend has been cost-reducing technology that has afforded growers the opportunity to produce more broilers with the same amount of resources or the same number of broilers with fewer resources. This has lowered per unit production costs, shifted the supply curve of broilers to the right, and, hence, lowered prices. As cost-reducing technology is adopted over time, the downward secular price trend is maintained.

Year-to-Year Cycles

Year-to-year price cycles characterize many agricultural products. The source of the cyclical price movements is typically producer response to higher commodity prices. Because of the time lags required for biological production, producer response is somewhat slow, compared to an industrial response, for example. For many commodities, like most annual crops, producer response is observed in the next production period. For example, if corn prices remain strong relative to the prices of other commodities, producers will have incentive to plant more corn. As they do, corn stocks increase, driving prices down. With lower corn prices, producers have less incentive to plant corn, which lowers corn stocks. This in turn causes prices to increase, and the cycle repeats.

Beef cattle prices also exhibit a cyclical pattern, but unlike the hog price cycle, which averages about 4 years, the cattle price cycle averages 10 years and is smoother.

Beef production and steer prices

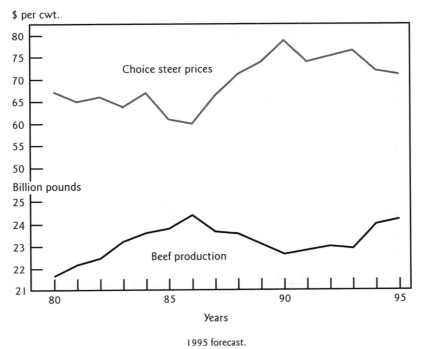

1995 forecast.
Source: Economic Research Service, USDA

Figure 17-6

Fluctuations in feed grain supplies, along with the attempts of farmers to respond to price changes, generate the cycles just mentioned. The length of the cycle for cattle is much longer than for hogs because it requires more time to breed and prepare cattle for market. Furthermore, cattle are not as prolific as hogs, and more breeding stock is required to produce a given number of animals.

The response of farmers to price changes also tends to aggravate the cyclical movements. After livestock producers decide to increase production, their first step is to increase the size of their breeding herds. As more replacement heifers or gilts are held back, fewer animals are marketed, reducing supply and forcing prices even higher. This, in turn, encourages still more production. Eventually, increased supplies will reach the market, causing prices to drop. With lower prices, producers decide to cut production, which involves liquidating some breeding stock. This boosts supply even more, forcing prices downward still further.

Pork production and hog prices

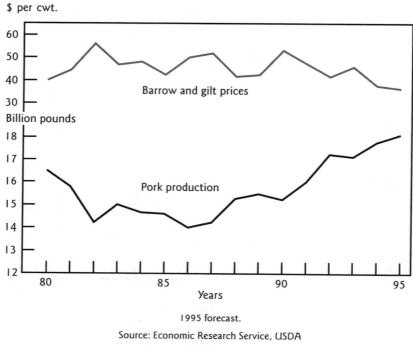

1995 forecast.
Source: Economic Research Service, USDA

Figure 17–7

Seasonal Cycles

In agriculture, there are also seasonal price cycles, those that occur within the year. For eggs, prices are traditionally low during the spring laying flush, when practically all hens are laying. Prices rise in the summer, however, when egg production declines while layers molt and hot weather prevails.

Most row crops have a fall harvesting period, which results in lower prices at harvest time than at other periods during the year. By storing products in the fall for sale during other times of the year, seasonal price cycles are smoothed (Figure 17–8).

Feeder cattle prices display similar patterns throughout the year. They decrease as larger supplies of feeder cattle are marketed in the late fall and winter months (Figure 17–9).

Corn—monthly average U.S. price received by farmers, crop year

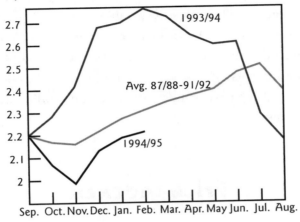

Source: Livestock Marketing Information Center, Cooperative Extension Service

Figure 17–8

Med. frame #1 steer calves, 500-600 lbs., weekly average, Southern Plains

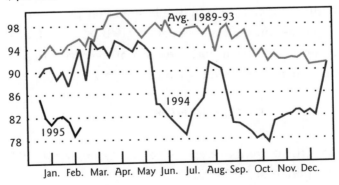

Source: Livestock Marketing Information Center, Cooperative Extension Service

Figure 17–9

Intraseasonal Changes

Daily or weekly (intraseasonal) price movements are characteristic of such commodities as eggs, broilers, fresh vegetables, and cash grain markets. It is not unusual for their prices to vary hourly as prices are most sensitive to supply and demand conditions across the nation. For example, on Mondays, livestock auction markets often experience traditionally heavier receipts since farmers often spend their weekends gathering and hauling livestock to market for Monday sales.

Price Indexes

An index number is a relative figure that aids in making comparisons of changes among diverse items. It indicates the percentage that one figure is of a chosen base figure. The base value may be derived from a single value—say, the average price in a given year—or more frequently an average for what is considered a normal period—several years, for example. Although indexes may be computed to show changes in quantities, qualities, or other values, they are used here to show changes in commodity prices over a period of time.

A *simple index* number, sometimes referred to as a "price relative," is computed by dividing a given price by the base period price and multiplying by 100. For example, assuming the base period (1987–89 = 100) average price for eggs was 73 cents per dozen and the current average price is 79 cents per dozen, then the current price index would be 108.22:

$$(79 \div 73) * 100 = 1.0822 * 100 = 108.22$$

This means that the current average price per dozen eggs is 108.22 percent of that in the base period of 1987–89.

A *weighted index* means that the computed average value takes into account the relative importance of its various parts. Referring again to the average price of eggs, this could be a simple average or a weighted average. For example, assume the following data:

Type	Market Price ($)		Market Price (%)		
Large eggs	0.80	*	50	=	0.4000
Medium eggs	0.75	*	35	=	0.2625
Small eggs	0.69	*	15	=	0.1035
Total			100		
Weighted average					0.7660
Simple average					0.7467

Note that the simple average [(80 + 75 + 69)/3] is 74.67 cents per dozen, while the weighted average is 76.60 cents per dozen.

Indexes are used frequently to compare changes in commodity values that may be measured in different units. Changes in prices of a ton of hay, a dozen eggs, a cord of wood, a bushel of corn, and a hundredweight of rice can all be compared if converted to index numbers. This is the technique used to develop a commodity price index that represents a composite of price indexes for a large number of commodities. A number of price index series are published by the U.S. Department of Commerce, the U.S. Department of Agriculture, and other federal and state government agencies.

Parity Prices and Parity Ratios

The U.S. Department of Agriculture has developed an index comparing prices received by farmers with prices paid by farmers to determine *parity ratios*.

Indexes of prices paid by farmers consider prices of goods and services farmers use, as well as interest, taxes, and farm wage rates. Indexes of prices received by farmers take into account the prices of all farm products sold by farmers (Figure 17–10). By dividing the index of prices received by the index of prices paid, a parity ratio is derived.

If the parity ratio equals 100, then prices of the commodities farmers sell have retained the same relationship to prices of goods and services farmers buy as existed during the base period. If the parity ratio is below 100, then prices of goods and services farmers buy have increased more relative to prices that farmers receive for their farm products. An example of the computation follows:

Hypothetical Parity Ratio Example
(Assume a 5-year base period = 100.)

Index of Prices Received by Farmers	Index of Prices Paid by Farmers	Parity Ratio
207	190	109
219	214	102
210	221	95
203	229	89
215	252	85

Indexes of prices paid and received by U.S. Farmers, 1977=100

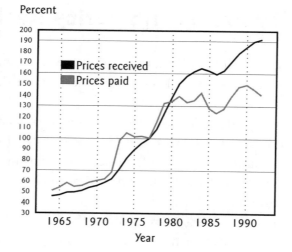

Source: U.S. Department of Agriculture

Figure 17-10

Price Controls

Probably the most futile thing a government can do is to attempt price controls, unless a national emergency clearly warrants such. The main reason for this futility is the distortions that price controls create. Also, an economy usually gets two sets of prices in a price control scheme: (1) the official control price and (2) the "black market" price. For No. 1, the price may be controlled, but there is relatively little supply. For No. 2, the price may be much higher, but there is plenty of supply for those who can afford it.

Businesses circumvent price controls in a variety of ways. They may reduce contents within packages, reduce package sizes, lower quality of contents while keeping price constant, simply not handle or inventory the product, or introduce "new products" to which the price controls do not apply.

Several pitfalls of price controls exist: (1) Price controls have never worked for an extended period of time. (2) Price controls impose a heavy burden on the taxpayer. They necessitate a diversion of human and other resources from productive channels. (3) Price controls impose heavy burdens on business enterprises. Compliance diverts regular employees to handle the added workload and wastes the time of management. Small businesses are hit particularly hard. (4) Price controls delude the consumer. They do not guarantee sufficient products at the controlled prices. They result in quality deterioration; the need for "knowing the right people"; and, if continued long enough, standing in line for "a share" of scarce items. (5) Price controls distort and curtail production. They distort productive effort from what would be produced as a result of "competitive market prices" and substitute something else that permits the business enterprise to continue to function. (6) Price controls lead to a dissolution of moral standards. Willful and nonwillful violations are forced in order that economic enterprise can function. Through confusion and difficulty of interpretation, otherwise law-abiding producers and consumers unwittingly "break the law." Even those who "want to cooperate" must sometimes evade regulations in order to "live normally." (7) Price controls are ineffectual in fighting inflation by failing to address the causes.

Finally, price controls are at best a political rather than an economic device. They purport to promise "more goods for less money." They may do so, but only in an exceedingly short time frame. If government policy makers attempt to control prices over a longer period, holding prices below competitive equilibrium levels, they distort production, deteriorate product quality, and eventually result in consumers paying "more money for less consumable goods." The lessons of history on this point surely are not sufficiently encouraging for us to want to experience price controls again.

A better alternative to price controls is appropriate monetary and fiscal policies, as discussed in an earlier chapter, along with free trade policy and strong and effective antitrust prosecution wherever needed (business and labor), among other considerations. Also, in the long run, encouraged expansion of domestic output is required while reducing the forces causing excess demand. Price stability should ensue.

Topics for Discussion

1. Discuss the concept of "equilibrium price."
2. What are the expected effects on prices if demand shifts but supply does not? If both demand and supply shift?
3. What are the expected effects on prices if supply shifts but demand does not? If both supply and demand shift?
4. Discuss the nature of the cobweb theorem. Contrast a convergent cobweb model with a divergent model.
5. Discuss the four types of price cycles for farm products.
6. Contrast a simple index of prices with a weighted index.
7. How is the parity ratio calculated?

Problem Assignment

Select an agricultural commodity and prepare a price forecast for next year's crop, taking into consideration as many supply and demand factors as you can. Do not overlook the effects of carryover stocks, volume of imports, export conditions, and price inflation or deflation.

Recommended Readings

Colander, David C. *Economics,* 2nd ed. Chicago: Richard D. Irwin, Inc., 1995. Ch. 2 and 8.

Gwartney, James D., and Richard L. Stroup. *Economics: Private and Public Choices,* 7th ed. Orlando, FL: The Dryden Press, 1995. Ch. 3 and 23.

Lesser, William. *Marketing Livestock and Meat.* Binghamton, NY: The Haworth Press, 1993. Ch. 2.

Seitz, Wesley D., Gerald C. Nelson, and Harold G. Halcrow. *Economics of Resources, Agriculture, and Food.* New York: McGraw-Hill, Inc., 1994. Ch. 4.

18

Market Structures and Competition

Our discussion of price determination to this point has been based on a free market economy, or an economy in which there is pure competition between all market participants. Actually, many types of markets may exist. The purpose of this chapter is to describe the major types of market structure.

When selling goods or services, firms are usually classified into the following market types: (1) pure competition, (2) monopolistic competition, (3) oligopoly, and (4) monopoly. On the other hand, when buying goods or services, firms are usually classified into these types of markets: (1) pure competition, (2) monopolistic competition, (3) oligopsony, and (4) monopsony. A simplified portrayal of possible buyer and seller markets is shown in Table 18-1.

Table 18-1

Nine Market Structure Situations, Based on the Number of Buyers and Sellers

Buying Side	Selling Side		
	Many Sellers	Few Sellers	One Seller
Many Buyers	Pure competition	Oligopoly	Monoply
Few Buyers	Oligopsony	Bilateral oligopoly	Monopolistic oligopsony
One Buyer	Monopsony	Monopsonistic oligopoly	Bilateral monopoly

Pure Competition

Under **pure competition,** firms are so numerous that no one firm has any effect in the marketplace relative to controlling supply, influencing price, advertising, or differentiating its product. Each firm is a *price-taker* rather than a *price-maker.* For pure competition to exist, it is necessary that new firms have freedom to enter the business and that existing firms be free to discontinue business.

Depending on the commodity, production agriculture is probably as good an example of pure competition as exists. For instance, grain farmers produce a homogenous product; what one farmer produces cannot be distinguished from what another produces. Grain farmers compete with one another in selling their product. Because there are so many producers of the same commodity, no single producer is large enough to influence price in any way. This contrasts with the situation in many manufacturing and service industries, where price competition is less vigorous.

Of course, even agricultural markets are not purely competitive. It is difficult, because of the capital required in agriculture, for new firms to enter the industry. On the other hand, because of the specialized nature of many resources in agriculture, existing firms may remain in production even though it may be unprofitable to do so. There are simply few alternative uses for cotton pickers, for example. Farmers have modified the structure of markets somewhat by setting up cooperative associations. Also, big processors and distributors may contract for supplies of farm produce. Even more important, the federal government actively influences agricultural markets through such mechanisms as acreage controls, land retirement, crop loans, price supports, and marketing orders. Competition is still the dominant force in agricultural markets, however.

The Model of Pure Competition

Economists use pure competition as a model against which to judge all other models of market structure. Why? Under pure competition, the influence of each firm is so negligible that its individual effect on price is undiscernible. The output price, which equals average revenue and marginal revenue, is expressed as a horizontal line (Figure 18–1). Each firm will produce up to the point where marginal cost equals marginal revenue, the point where the additional revenue generated by producing the last unit of output just

Optimal output level in a purely competitive market

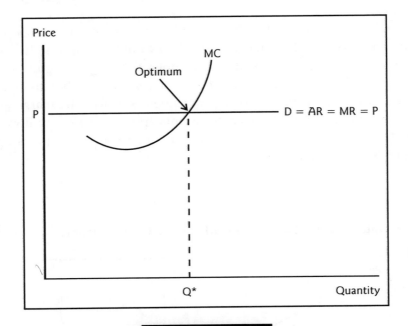

Figure 18-1

equals the cost of producing that unit of output. There is little the firm can do to bend or tilt this horizontal price line to its individual advantage. The firm cannot differentiate its product, which would have the effect of tilting the horizontal price line it faces, making the demand curve less than perfectly elastic.

Generally speaking, a more efficient allocation of resources is achieved under pure competition than under any other model, except for decreasing cost industries such as utilities. In the case of utilities, it is more advantageous for society to allow only one company to operate but to regulate its prices and services.

Imperfect Markets

In competitive structures that are imperfect, sellers' actions are not entirely independent and sellers have some degree of control over the price received and over how to adjust output to maximize profits. To a degree, sellers in this case are price-makers with a unique demand curve. They are unlike, say, grain

266 / ECONOMICS: APPLICATIONS TO AGRICULTURE AND AGRIBUSINESS

farmers who are, in general, price-takers with basically a perfectly elastic demand curve.

Generally, imperfect markets have an inherent tendency to restrict production and charge somewhat higher prices than would exist in purely competitive markets, although competitors still operate where marginal revenue equals marginal cost. The difference, however, comes about because marginal revenue is no longer equal to average revenue and price. The marginal revenue line lies to the left of the average revenue–price line, sloping downward from left to right (Figure 18–2). The allocation of resources is not as optimal and fluid as it might otherwise be in free markets.

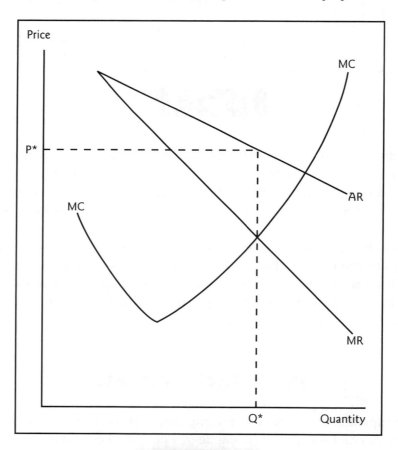

Figure 18–2

Factors Related to Imperfect Markets

Among the many factors associated with or causing markets to be imperfect are the following:

1. *Elimination of competitors by merger or takeover,* which results in fewer competitors.
2. *Licenses or business permits,* which restrict the entry of new competitors.
3. *Patents,* which give monopoly rights to holders for a period of years.
4. *Franchises,* which restrict the number of competitors in a given geographical area.
5. *Branding and copyrights,* which give monopoly privileges to their holders.
6. *Price maintenance agreements,* which assure certain markups for processors, distributors, etc.

We now proceed to a discussion of some of the types of imperfect markets.

Monopolistic Competition

As the term signifies, **monopolistic competition** contains elements of both monopoly and pure competition. The number of firms is relatively large, with each representing a fairly small share of the total market, able to exert only a limited amount of control over market price. However, the goods and services provided by these firms are not homogeneous, but differentiated in a variety of ways, such as by brand names, advertising, and packaging. This creates differences, real or perceived, in the minds of the users.

Consumers may be inclined to pay the differentiating firm more for its products or services than they would be willing to pay competing firms their products or services. In conjunction with this, the differentiating firm will tend to sustain its sales efforts through nonprice competition, such as by providing store displays, extending credit, and delivering packages, among a host of other techniques.

Since each firm in monopolistic competition considers itself relatively insignificant, some price cutting can be practiced with little fear of retaliation by its competitors. The demand curve facing an individual firm in monopolistic competition is not perfectly elastic, as it is under pure competition. Firms engaged in retailing of clothing, shoes, and furniture often compete in terms of monopolistic competition.

Oligopoly-Oligopsony

This type of market structure is one where a few firms dominate a certain line of business and where rival firms may observe each other closely relative to prices and pricing policies. When this market structure situation occurs on the selling side, we refer to it as **oligopoly,** and on the buying side, **oligopsony.**

There are also two subtypes of oligopoly-oligopsony: perfect and differentiated. *Perfect oligopoly-oligopsony* refers to situations where the products or services are identical and not differentiated or set apart. Perfect oligopoly is found, for example, in steel making. *Differentiated oligopoly-oligopsony* refers to situations where the products or services are similar but can be differentiated or set apart. Differentiated oligopoly is characteristic of auto and farm machinery manufacturing, for example.

The main characteristics of oligopoly-oligopsony are fewness and interdependence of firms, which, along with high capital requirements, increase the difficulty of new firms entering the business.

Oligopolists are also more interested in nonprice than price competition. They recognize that if one of their number reduces price, the others will more than likely have to do the same. On the other hand, by keeping prices steady, rivals can resort to competition in more subtle ways. Cigarette companies, for example, rely heavily on nonprice competition through branding and advertising programs.

While we usually think of oligopoly-oligopsony as being the creature of large firms, it may also characterize local markets, where the number of firms may be rather large, but each firm recognizes its rivals in setting price policies. Retail gasoline stations are a case in point, and periodic gasoline price wars are indicative of the effect a price cut by one participant can have on all sellers in the market.

Monopoly-Monopsony

In **monopoly-monopsony,** one firm controls all the supply or demand, respectively, and sets prices to suit itself, limited mainly by the availability of substitutes for its product or by public regulation, as in the case of utility companies. Also, when the federal government intervenes in a market situation and controls supply or acreage and sets prices, it is acting as a monopolist. The difference between a private monopolist and a public monopolist, such as the federal government, is that the latter is granted this privilege by elected representatives of the people. Private monopolies are illegal unless franchised or regulated by government. Utility companies represent one type of monopoly that is sanctioned because it is more efficient (less costly) to have only one firm supplying utility needs rather than several. In decreasing cost industries, such as utilities, it is usually more desirable to have a monopoly or one firm serve a given area because it eliminates the duplication of poles, lines, cables, etc., that would result in greater costs if pure competition were to prevail.

Theoretical Pricing and Output Under Monopoly

Suppose there was only one corn farmer in the United States who had absolute monopoly control over corn output and sales. How much should the farmer produce? What should the market price for corn be? Some answers are provided in the hypothetical situation in Table 18-2.

Table 18–2

Profit-Maximizing Input and Output Levels for a Monopoly with Corn Prices Falling from $4 to $1 per Bushel as Yields Increase with Additional Nitrogen

(1)	(2)	(3)	(4)	(5)	(6)	(7)	(8)	(9)	(10)	(11)	(12)
Nitrogen	Corn Yield	Price per Bushel	Total Revenue	Total Cost	Average Total Cost	Net Profit	Net Profit	Marginal Revenue	Marginal Cost	Marginal Revenue	Marginal Cost
Lbs./ac.	Bu./ac.	Average revenue	$/ac.	$/ac.	$/bu.	$/ac.	$/bu.	$/ac.	$/ac.	$/bu.	$/bu.
0	43	3.60	154.80	68.00	1.58	86.80	2.02				
								49.20	30.75	2.89	1.81
20	60	3.40	204.00	98.75	1.65	105.25	1.75				
								29.60	12.05	2.28	0.93
40	73	3.20	233.60	110.80	1.52	122.80	1.68				
								18.40	11.20	1.08	0.66
60	90	2.80	252.00	122.00	1.36	130.00	1.44				
								13.20	13.15	1.10	1.10
80	102	2.60	265.20	135.15	1.33	130.05	1.28				
								4.30	9.50	0.54	1.19
100	110	2.45	269.50	144.65	1.32	124.85	1.14				
								−16.50	11.00	−3.30	2.20
120	115	2.20	253.00	155.65	1.35	97.35	0.85				
								−20.95	7.90	−5.24	1.98
140	119	1.95	232.05	163.55	1.37	68.50	0.58				
								−24.65	10.60	−8.22	3.53
160	122	1.70	207.40	174.15	1.43	33.25	0.27				
								−29.05	12.55	−29.05	12.55
180	123	1.45	178.35	186.70	1.52	−8.35	−0.07				

A farmer under monopoly would produce at the point where marginal cost is equal to marginal revenue, as would the farmer under pure competition. As shown in Table 18–2, the optimum output level for the monopolist would be 102 bushels per acre, using 80 pounds of nitrogen per acre. The optimum price the monopolist would charge is $2.60. At this price, marginal revenue ($/bu.) equals marginal cost ($/bu.). At these assumed output and price levels, the monopolist would maximize profit at $130.05 per acre. The monopolist farmer faces a negatively sloping demand curve or an industry demand curve for corn, unlike farmers in pure competition, who face a perfectly elastic demand curve (a horizontal demand line) because they individually represent such a small part of the total market.

Had a farmer under pure competition faced a horizontal marginal revenue line of $2.00 per bushel, the optimum input level would have been at 140 pounds of nitrogen, producing 119 bushels of corn per acre, since, at that point, marginal cost equals marginal revenue of $2.00. This comparison is not entirely accurate, however, because the two farmers are dissimilar. One farmer would be producing all the corn (monopoly), while the other would be producing only a small part of total corn output (pure competition).

Suppose the federal government decided to regulate our monopolist corn farmer. Where would the government set the price? Very likely where marginal cost intersects average revenue, or at an input level of 120 pounds of nitrogen per acre, producing 115 bushels of corn per acre. The government-pegged price ($2.20) is considerably lower than what the monopolist would prefer ($2.60). At a price of $2.20, profit per bushel to the monopolist would be zero—that is, total returns would just cover total costs.

Another factor of concern to any monopolist corn farmer would be the production of other feed grains, especially those easily substituted for corn, like grain sorghum. With a high corn price set by our monopolist, sorghum producers would attempt to expand production and capture the feed grain market. To stay competitive, the corn monopolist would have to lower the corn price.

Other Aspects of Competition

Market competition is a broad subject. In this section, only a few selected aspects are discussed, including various laws and regulations that define, enforce, and govern competition in the marketplace.

Sherman Act

The Sherman Antitrust Act of 1890 is essentially an antimonopoly law. It was enacted in the era when trusts and combinations exercised power considered dangerous to the public welfare. The fundamental purpose of the Act is to prevent restraints to free competition that tend to limit production, raise prices, or otherwise control the market. It also seeks to secure equality of opportunity for businesses and to protect purchasers of goods and services.

Clayton Act

The Clayton Act of 1914 is designed to reach devices or practices that are discriminating and that might, under certain circumstances, lead to the formation of trusts. It is intended to supplement the purpose and effect of the Sherman Act.

The Clayton Act prohibits price discrimination, the intentional inducement or receipt of a discrimination in price, exclusive dealing arrangements and tying contracts, mergers, interlocking directorates, and intercorporate stockholding—any of the preceding having the requisite adverse effect on competition. Additionally, it forbids discrimination in the payments for, and the furnishing of, services and facilities and prohibits illegal brokerage payments.

Robinson-Patman Act

The Robinson-Patman Act, which became law on June 19, 1936, has as its purpose the prevention of discrimination in price and the prevention of other discriminatory practices injuriously affecting free competitive enterprise. The purpose of this act may be reduced to four basic provisions relating to interstate commerce:

1. Discriminatory prices that would injure competition may not be given to any buyer.

2. Commissions normally paid to brokers may not be received by any buyer.

3. Unfair discrimination between customers may not be made in regard to allowances or payments for advertising and promotion work.

4. Discrimination may not be made between customers in regard to special services, such as the extension of credit.

However, the Act does permit a seller to discriminate in price between buyers if the purpose is to meet the legitimately lower price of a competitor.

Capper-Volstead Act

The Capper-Volstead Act, enacted in 1922, clarified the status of farmer cooperatives under the Sherman and Clayton Acts. It makes legal an association of farmers for marketing purposes in interstate trade, provided such an association:

1. Is operated for the mutual benefits of its members.
2. Conforms to one or both of the following:
   a. No member is allowed more than one vote.
   b. Dividends on capital stock or membership capital do not exceed 8 percent per year.
3. Does not deal in products of nonmembers to an amount greater in value than that of products it handles for members.

Webb-Pomerene Act

The Webb-Pomerene Act of 1918 was passed for the purpose of allowing domestic firms to form export associations in order to compete more effectively in international trade.

The heart of the Webb-Pomerene Act is a provision that protects legally formed export associations from antitrust restrictions of the Sherman Act. Such export associations, which are only to engage in export trade, cannot restrain the export trade of any of their domestic competitors.

Federal Trade Commission

The Federal Trade Commission is charged with enforcement of the provisions of the Clayton Act, the Robinson-Patman Act, the Webb-Pomerene Act, the Federal Trade Commission Act, and other similar regulatory acts.

To safeguard further the operation of competition, Congress in 1914 established the Federal Trade Commission and charged it with the duty of stopping the use in interstate commerce, by persons, partnerships, or corporations, of unfair methods to suppress or injure competition.

The powers of the Commission were extended under the Wheeler-Lea Act of 1938, which also declared unlawful "unfair or deceptive acts or practices in commerce," whether or not in competition. In addition, the Wheeler-Lea Act expressly provides that the dissemination of any false advertising by United States mails, or in commerce by any means, that is designed to induce, or that is likely to induce, the purchase of food, drugs, devices, or cosmetics constitutes an unfair or deceptive practice.

Local Versus National Markets

It has been customary in economics to focus attention on competition in national markets, such as for auto and steel manufacturing. Unfortunately, the status of market competition in local markets is often overlooked or ignored. Contrary to popular belief, local markets are often less competitive than national markets. Before automobiles arrived on the American scene, the one locally owned general merchandise store was quite monopolistic. With no real competitors, it could charge almost any price desired. In addition, by extending credit, it was able to keep its customers in debt for long periods of time, especially when the store supplied both farm and home consumption goods. It was only with improved transportation and communication that the monopoly power of the local general storekeeper was broken. In early America, there were other instances of local monopoly-monopsony power, such as the single railroad, the cotton gin, the livestock buyer, the produce buyer, the grain buyer, etc. Presently, local monopoly-monopsony power is rare because of the availability of substitute stores, plants, and products and the increasing business skill of the American people.

On a national basis, also, monopoly-monopsony power may have been weakened, not only by technology, but also by antitrust legislation, such as the Sherman and Clayton Acts, and by various other governmental regulating authorities.

Price Leadership

In oligopolistic markets, where firms recognize their interdependence, price leadership is often dominant. In certain industries supplying farmers with inputs (fertilizer, feed, chemicals), price leadership is common. That is, the largest firm in the group feels confident that its smaller competitors will follow its lead in setting prices, either higher or lower. If they do not follow it, the largest firm may inflict certain punishments on its smaller rivals, such as dropping prices so low that the smaller firms are unable to compete. This threat and others that may be used tend to force all oligopoly participants to act in unison. This unity of action is not always thorough or longstanding, however, and some types of oligopoly behavior may be declared illegal under antitrust law.

Figure 18–3 shows a certain type of price behavior characteristic of oligopolies. The a–b portion of the curve illustrates a situation where one firm decides to lower its price. Its rivals follow the price cut.

Pricing behavior among oligopolists and a kinked demand curve (d–a–b)

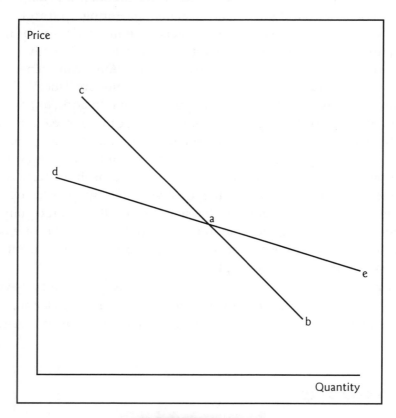

Figure 18–3

In the a–c portion of the curve, one firm decides to raise its price. Its rivals also raise their prices.

In the a–d portion of the curve, not all oligopolists go along with a price increase by one oligopolist. Sales are lost by the one price-raising firm, forcing it to return to point "a" if its rivals do not follow suit.

In the a–e portion of the curve, one oligopolist lowers its price, but the competitors do not follow. The price-lowering firm gains sales at the expense of its rivals.

What is more likely to happen, however, in an oligopoly market situation is a curve composed of d–a–b, or what economists call a *kinked demand curve*. This refers to a situation where rivals do not follow an oligopoly firm if it raises its price but do follow it if it lowers its price.

Considering all the ramifications presented by Figure 18–3, it is likely that price tends to remain at point "a" or moves up along portion a–c of the curve rather than a–d. Price cuts, or going down on portion a–b of the curve, are a rather rare and temporary instance among oligopolists.

Promotion Expenses

Promotion expenses are selling costs incurred to stimulate sales in two general ways: first, by informing potential buyers of the availability, characteristics, and prices of the products and, second, by persuading potential customers to buy. Thus, promotion expenses are both informational and persuasive. Costs devoted to informational purposes are functionally justified and are essential to the effective working of a free market system. It is necessary to inform potential buyers of the availability of goods and of their specifications, qualities, and prices. Promotion activities with a persuasive orientation, however, are not usually justified from the standpoint of public welfare. They reflect a diversion to sales promotion of productive resources that could otherwise be devoted to producing and distributing a larger volume of useful goods and services at lower prices.

While promotion expenses devoted to providing information are to be encouraged, those devoted to persuasion should not be, especially when persuasion is fictional.

Market Entry

The entry of new firms into a market is fundamental to a free enterprise economy. Barriers to entry are numerous. Some are erected purposely by firms already in the market, while other barriers result from technology, government, etc. In any case, market entry should be made easier, with restrictions held to a minimum.

Among the factors that could lead to easier market entry are these: (1) preventing issuance of patents to firms that have developed new technology with government funds, (2) expanding sources of capital for new businesses, (3) preventing monopoly over raw-material supplies, (4) discouraging excessive advertising and product differentiation, and (5) reducing political activity associated with the awarding of government franchises, licenses, and permits.

Evaluating the Markets

Markets can be evaluated in terms of their *structure*, the *conduct* of the firms, and the end results or market *performance*.

Obviously, market performance is the most important criterion, because no matter how markets are organized or conducted, the end results are what count. The mere classification of a market does not provide sufficient appraisal of the market performance of the firms in that market. Answering these questions will help in determining market performance:

1. How do prices charged relate to average production costs, and how large are the profits?
2. How do the size and efficiency of firms compare to the size and efficiency of the most efficient firm, and what is the extent of excess capacity?
3. What is the extent of sales promotion costs relative to production costs?
4. What about the character of the product, including its design, quality, and variety?
5. What is the rate of technological progress in the industry?

Market Performance of Farmers

Because farmers operate in a relatively free market on the selling side (with the exception of government intervention), prices of farm products approximate average total cost, with no excess or above-normal profits. In fact, in many cases, farm prices stay below average total cost for long periods of time.

Although the size of farms varies from very small to very large, farms are still sufficiently numerous at all sizes to preclude collusion and action in concert in most cases. Government farm programs are an exception to the otherwise lack of concerted action. Excess capacity in agriculture is largely a result of government farm programs, not the cause of them.

Farming has little or no promotion costs, product differentiation, or advertising at the farm level. Most farm products are simply graded, sized, and sold as homogeneous, standardized products.

The design, quality, and variety of farm products are, in general, in response to consumer desires transmitted through the market system. The huge public investment in agricultural research and technology precludes any one firm or farm from monopolizing technology. As a result, the rate of technology diffusion and adoption in agriculture is rapid as the new technology becomes available equally to all. The adoption and use of technology on a broad scale preclude excess profits, as supply pressures bear down continually on prices, driving them down to normal rates.

Creating More Market Competition

If a free enterprise economy is to survive, more, rather than less, competition has to exist between competing firms. This is essential if new ideas, products, and techniques are to be developed and applied. Also, vigorous price competition is essential. Prices should not be any higher than necessary for the judicious conduct of business, adequacy of financial reward to investors and risk-takers, and the sustenance of the firm over the long run. Price relationships are also important. That is, if one industry is more competitive than others price-wise, then it suffers as a result. A distortion is created in the command and utilization of resources. Labor, capital, and management will be attracted to the industry with higher-than-normal prices and profits. Therefore, it is in the general interest of all to maintain a price-competitive economy.

The following are a few suggestions for achieving a more price-competitive economy:

1. Encourage market entry, and eliminate or reduce barriers to entry.
2. Discourage excessive merger activity, especially where the end results are increased market power rather than marketing efficiency or cost savings.

3. Supply ample credit to encourage new businesses and to enhance small businesses already established.

4. Improve the management skills of small business owners and managers.

5. Curtail excessive market power wielded by either labor or management, or both.

These and other steps in themselves would help farmers by lessening the burdens they bear in dealing with firms that do not operate in purely competitive market structures.

Competition Farmers Have to Face

Concentration in industry and competitive conditions in agriculture result in the following inequalities:

1. *Industry can control production to prevent excessive price fluctuations.* Production control is possible in industry because:

    a. Authority is centralized.

    b. Costs can be cut on short notice (e.g., labor can be laid off).

    c. The productive process is such that it can be adjusted quickly to changing demand.

    In contrast, in farming:

    a. Command is not unified. Thousands of scattered farmers cannot get together at a moment's notice to decide on and pursue a common course of action.

    b. Even if command could be unified, many farm costs (taxes, interest on land, family labor) are fixed and would continue in spite of production cuts.

    c. The very nature of the productive process prohibits fast response to changes in market demand. Farmers plant their crops many months before they know what the demand and price situation will be. They cannot stop their grain growing and then resume it at a later date. All they could do would be to plow it under and suffer the loss of any investment made to that point.

2. *Industry can control marketing.* Industry is able to produce and market a steady stream of goods day after day. With its control over production, it can interrupt this flow to keep prices steady.

   In contrast, many farm products are marketable only at harvest times. Storage of nonperishable commodities helps some in regulating market flow, but this again involves the difficult job of agreement between thousands of farmers to maintain price stability. Perishables cannot be handled in this way.

3. *Industry can buy equipment and supplies at wholesale prices.* Because of their large-scale operations, industrial enterprises can make big savings in buying raw materials by carloads.

   Small-business farmers, on the other hand, must buy their machinery and supplies at retail and in small lots.

4. *Industry can finance its operation with little risk.* By issuing common stock, it can obtain money for an indefinite period and pay dividends at its own discretion. Instead of being paid out in dividends, profits may be reinvested in the business.

   Most farm operations, however, must borrow money and pay interest to be able to purchase operating inputs and capital assets. Even in difficult times, they are obligated to meet these debt obligations. Farm loans are usually secured with machinery and farm land as collateral, carrying the risk of foreclosure in bad times.

Regulating Licensed Monopolies

Licensed monopolies are found in abundance, including those companies permitted to furnish water, electricity, natural gas, transportation, communication, and local telephone services.

The problem of regulating these licensed monopolies in the public interest is a formidable one. Left to their own devices, licensed monopolies would prefer to supply Q* and charge price P*, where marginal cost intersects marginal revenue and their profits are maximized (Figure 18–4). However, a regulatory body may insist that a licensed monopoly operate where marginal cost intersects average revenue, supplying Q_1 and charging P_1. More output at a

Regulating a licensed monopoly

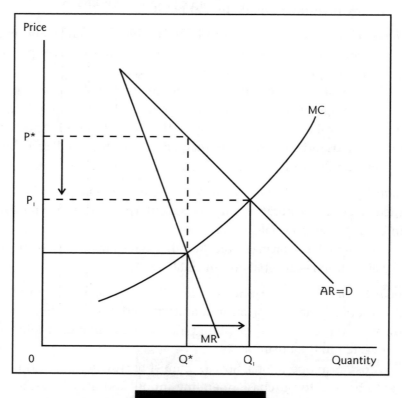

Figure 18-4

lower price would result, which should be the objective of an administrative body regulating a monopoly.

Topics for Discussion

1. Discuss the nature and characteristics of each of the four types of buyer and seller market structures.

2. Why is perfect competition used as the model to evaluate other types of market structures?

3. How does pricing under monopoly differ from pricing under perfect competition?

4. Discuss the laws pertaining to the maintenance of competition.

5. Discuss the forms of price leadership under oligopoly.

Problem Assignment

Select one agribusiness-type industry and prepare an essay on its competitive status, taking into account such factors as number and size of firms, utilization of present capacity, price competition, price leadership, technological advancements, promotion expenses, market entry, and other factors of your choosing.

Recommended Readings

Bailey, Dee Von, B. Wade Brorsen, and Michael R. Thomsen. "Identifying Buyer Market Areas and the Impact of Buyer Concentration in Feeder Cattle Markets Using Mapping and Spatial Statistics," *American Journal of Agricultural Economics,* 77(2):309–318 (1995).

Barkema, Alan, Mark Drabenstott, and Kelly Welch. "The Quiet Revolution in the U.S. Food Market," *Economic Review,* 76(3):25–41 (1991).

Colander, David C. *Economics,* 2nd ed. Chicago: Richard D. Irwin, Inc., 1995. Ch. 25–29.

Gwartney, James D., and Richard L. Stroup. *Economics: Private and Public Choices,* 7th ed. Orlando, FL: The Dryden Press, 1995. Ch. 19–22.

Hyde, Charles E., and Jeffrey M. Perloff. "Can Monopsony Power Be Estimated?," *American Journal of Agricultural Economics,* 76(5):1151–1155 (1994).

Purcell, Wayne (ed.). *Structural Change in Livestock: Causes, Implications, Alternatives.* Blacksburg, VA: Research Institute on Livestock Pricing, 1990.

Rogers, Richard T., and Richard J. Sexton. "Assessing the Importance of Oligopsony Power in Agricultural Markets," *American Journal of Agricultural Economics,* 76(5):1143–1150 (1994).

Seitz, Wesley D., Gerald C. Nelson, and Harold G. Halcrow. *Economics of Resources, Agriculture, and Food.* New York: McGraw-Hill, Inc., 1994. Ch. 7.

Shugart, William F., II, William F. Chappell, and Rex L. Cottle. *Modern Managerial Economics: Economic Theory for Business Decisions.* Cincinnati, OH: South-Western Publishing Co., 1994. Ch. 2.

PART FOUR

19 Economics Aspects of the Farm Supply Business

20 Food and Fiber Marketing Functions

21 Marketing Channels for Farm Products

22 Costs of Marketing Food and Fiber Products

23 Consumption of Agricultural Products

24 Economic Setting for U.S. Agricultural Policy

25 Achieving the Goals of Agricultural Policy

19

Economic Aspects of the Farm Supply Business

Modern agriculture consumes vast quantities of farm supplies, machinery, and equipment. In 1993, U.S. farmers spent the following amounts on selected production inputs:

Item	Amount (Million dollars)
Purchased feed	$21,400
Livestock and poultry	16,200
Farm services (excluding rent)	21,300
Rent	14,500
Fertilizer and lime	8,370
Agricultural chemicals	6,700
Petroleum products	5,540

As agriculture has become more commercial and specialized, the quantities of purchased inputs have increased, a trend expected to continue in the future.

In the past, farmers relied primarily on home-grown inputs for their production supplies. For example, chicks were hatched from eggs set at the farm rather than purchased from commercial hatcheries. Farmers used cottonseed meal and barnyard manure as fertilizer materials rather than relying on commercial fertilizers. Seed corn was obtained directly from the corn crib. Horses and mules were used instead of tractors. Many other examples could be cited.

In general, commercial farmers today are far different. They rely almost entirely on purchased inputs and very little on home-grown ones.

Scope of the Farm Supply Business

The scope of the farm supply business is extensive. Farmers spend annually about $155.5 billion for farm supply inputs and for fixed costs. Feed purchased for livestock and poultry is the most important single item, followed by capital replacement. The share attributed to feed purchases has increased over the years as commercialized confinement feeding has expanded. The intermediate product expenses can be broken out as follows:

Intermediate Product Expense Items	*Percent*
Farm origin	
Feed purchases	21.9
Livestock and poultry purchases	15.3
Seed purchases	5.3
Manufactured inputs	
Fertilizer and lime	8.6
Pesticides	6.9
Fuel and oil	5.5
Electricity	2.7
Other	
Repair and maintenance	9.3
Miscellaneous	24.6
Total	100.0

When we consider land, labor, and capital as the prime factors in farm inputs, relative production expenses associated with land (i.e., net rent, real estate interest, and property taxes) have remained relatively constant. Labor has dropped significantly, accounting for only about 10 percent of total production expenses at present, compared to about 75 percent in 1910. In contrast, the share of total production expenses made up by capital items (buildings, livestock, machinery, equipment, credit funds, operating inputs, etc.) has increased dramatically, from only 17 percent in 1910 to almost 80 percent at present. This indicates that farmers are substituting capital for labor, while land remains about constant.

Price Trends for Farm Inputs

Price trends for farm inputs have varied. On a per unit basis, prices for most inputs of farm origin—for example, seed, feed, and poultry and livestock—have been relatively stable. On the other hand, those input items that are of nonfarm origin cost more per unit now than before. Such items include fertilizer, motor vehicles and supplies, machinery and equipment, petroleum products, interest cost, hired labor, chemicals, insurance premiums, lumber and building materials, and taxes of all kinds.

Agricultural production is becoming more capital intensive. Expenditures for manufactured inputs, like fertilizer, pesticides, fuel, etc., increased 858 percent from 1950 to 1993. Expenditures for replacement of motor vehicles and of equipment and machinery increased 606 percent during the same period. In contrast, hired and contract labor expenses increased only 434 percent. These changes reflect the historical trends of (1) improved technology (machinery and equipment) being substituted for hired and family labor, (2) land becoming less important as a factor in total production, and (3) other forms of advanced technology being adopted, including biotechnology.

Farm input use and prices

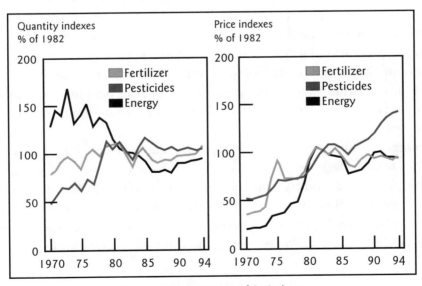

Source: U.S. Department of Agriculture

Figure 19-1

Distributing Farm Supplies

Each farm supply item has a distribution system unique to it, accounting for the wide diversity and complexity of the farm supply business. However, this diversity may offer opportunities for farmers and others to effect economies and improvements in the farm supply distribution system itself. It also offers many employment opportunities.

Some of the most important farm supply items are (1) feeds, (2) fertilizers, (3) farm machinery and equipment, and (4) pesticides. Each is briefly discussed, as is the role of farm supply stores in distributing these inputs.[1]

Feeds

Feed has historically been the major component (currently about 14 percent) of total farm expenses. U.S. farmers spend, on average, over $21 billion per year on purchased feed. The importance of the feed industry in its dual role as a major supplier of production inputs and as a major producer of farm products exceeds that of all other farm industries.

The feed industry has changed significantly over the past several decades. There are fewer feed manufacturers now, but existing firms are larger than in the past.

The most significant change affecting the manufactured feed industry is the emergence of on-farm feed mixing. About 60 percent of the feed used, based on its value, can be attributed to commercial sources. Most of the rest is mixed on the farm or within the particular livestock operation. Livestock operations have become larger, enabling them to justify investing in feed mixing equipment.

Changes in livestock and poultry production over the last 20 years have also had major effects on the feed industry. The average size of swine, cattle feeding, dairy, and poultry operations has increased considerably, while the number of operations has decreased. The large producers are often more specialized and sophisticated managers. The greatest changes have occurred in the poultry industry. Poultry production is almost all vertically integrated

---

[1] Parts are adapted from Kevin Kimle and Marvin Hayenga, *Agricultural Input and Processing Industries,* Report RD-05 (Ames: Department of Economics, Iowa State University, 1992).

through ownership or contractual arrangement. The integrated firms mix their own feed.

Advancements in livestock biotechnology will probably have profound effects on the feed industry of the future. Improved genetic lines of livestock resulting from the use of new biotechnological methods and the development of growth promotants will affect feed efficiency and nutrient requirements of feed-lot animals with the potential to significantly change the structure of feed product markets.

Manufactured feed products can be broadly categorized as complete feeds, supplements, or premixes. Complete feeds require no additional preparation. Supplements are formula feeds requiring the addition of grain to make a complete ration. They contain nutrients needed to supplement the often insufficient levels of protein, vitamins, and minerals found in most feed grains. A premix contains only vitamins and minerals and is used at the rate of less

Channels of U.S. commercial feeds distribution

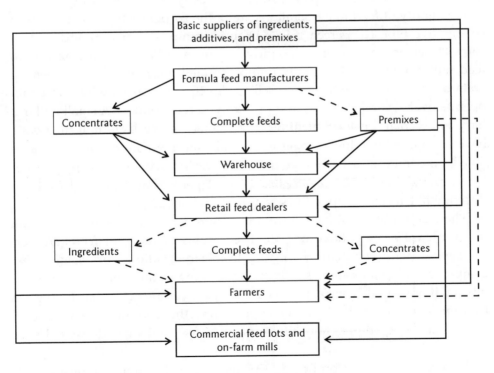

Source: U.S. Department of Agriculture

Figure 19-2

than 100 pounds per ton of finished feed. It is typically mixed with feed grains and a protein source, such as soybean meal, to provide a complete ration.

Complete feeds are the most important kind of feed manufactured, accounting for just under 80 percent of the feed produced by feed plants in the United States. Supplement production accounts for almost 20 percent of feed produced, while premixes account for about 1.5 percent. The percentage share of complete feeds has decreased since the 1970s. In contrast, use of supplements and premixes has increased dramatically. These changes can be attributed to the increase in the number of large beef, dairy, and swine operations that have expanded their capacities to mix their own feed by using premixes and supplements.

There are about 6,700 feed mill and mixer establishments in the United States. These mills range from small custom mills, where local grain brought by farmers is mixed with purchased concentrates, to large-volume mills, which produce a wide range of feed products. Individual feed companies offer product mixes that reflect demand in their specific geographical markets. Most of the large feed companies offer all three types of feed mixes.

A majority of feed mill operations are of the corporate type (59 percent) or cooperative type (28 percent). Sole proprietorships (9 percent) and partnerships (4 percent) are much less typical as a form of business organization for feed mills. Smaller mills, usually unincorporated, normally confine their operations to one or a few counties, selling directly to farmers, while large mills operate over several states, selling primarily to retail outlets. A small mill may have an average investment of $100,000, while a very large mill may easily have several million dollars invested in grain and ingredient storage facilities. Small mills may also be operated in conjunction with farm supply stores. Incorporated feed mills account for about 71 percent of volume of feed produced, while farmer cooperatives account for about 22 percent.

The feed industry is characterized by a number of diverse firms operating in an intensely competitive market. The various markets for feeds are very segmented, with competitors ranging from large national or regional feed companies to small local mills. With the increasing trend toward on-farm feed mixing, there has also been an increase in excess production capacity in many feed mills, which has intensified competition in the feed industry. As a result of this increased competition, the number of feed mills has decreased since the 1970s.

Factors affecting net profits of feed millers are (1) volume of business, (2) labor efficiency in the mill, (3) transport costs, (4) type of feed customers, (5) balancing of feed sales with mill capacity, and (6) ability to operate year-round instead of seasonally.

Problems of concern to feed mill operators include (1) growth of contract farming and vertical integration, (2) the increasing trend in on-farm feed mixing, (3) declining numbers of small livestock operations, (4) government regulations on labeling, manufacturing practices, and work-place and environmental safety, (5) technological advances in feed milling and distribution, specifically computerized feed formulation, (6) seasonality of operations, and (7) low profit margins.

Fertilizers

U.S. farmers spend about $8 billion per year on fertilizers. Fertilizer usage increased dramatically through the 1960s and 1970s, then peaked in 1981. It has been somewhat stable since the mid-1980s, ranging from 19.1 to 21.8 million tons through 1993. Fertilizer is an important farm input because of the large marginal or additional returns from fertilization relative to its cost.

Fertilizer prices, while stable or declining during the 1960s, have varied widely since 1973. After increasing significantly during the mid-1970s and again in 1980, farm prices of most fertilizer products fell during the period 1981 to 1986. Prices increased in 1987 and 1988, then remained somewhat stable through the early to mid-1990s. Currently, however, fertilizer prices have increased dramatically, due in part to a strong world demand for fertilizer and to increasing industrial demand for nitrogen.

The fertilizer industry is composed of basic manufacturers and fertilizer mixing firms that also distribute fertilizer, as well as a series of retailers. The basic manufacturers produce nitrogen fertilizer in the form of ammonia or an ammonia derivative, phosphorus fertilizer products from phosphate rock, and potassium fertilizer from potash. These materials may then be processed into a variety of forms for use as fertilizer. They are blended by the intermediate mixing firms and then distributed. The United States has more than 13,000 bulk blenders, fluid mixers, ammoniation-granulation plants, and retail stores.

Nitrogen is normally used as a form of ammonia, which has the chemical configuration NH_3, 1 molecule of nitrogen to 3 molecules of hydrogen. It is produced by a natural gas-driven compressor that extracts nitrogen from the air and combines it with hydrogen from natural gas. The United States is a net nitrogen importer. Oil companies have been in the forefront in ammonia production.

The phosphate industry is generally located close to phosphate rock deposits. In the United States, about 80 percent of phosphate rock supplies come from Florida, with other deposits located in North Carolina and in some western states.

U.S. potash production is small compared to U.S. potassium usage. About 80 percent of the potash used in the United States is imported from Canada. Most of the production in the United States comes from potash deposits near Carlsbad, New Mexico. These deposits mainly meet fertilizer needs in the southwestern United States.

There are about 11,000 large blending/mixing plants in the United States. The average size fertilizer plant in the United States has a storage capacity of 6,041 tons and an annual distribution of 15,490 tons. Almost half the fertilizer plants in the United States are privately owned, while almost 42 percent are cooperatives and about 9 percent are corporations.

The U.S. fertilizer market is expected to be somewhat flat throughout the remainder of the 1990s. Farm output will likely remain stable, and the intensity of fertilizer use on a per acre basis will also remain stable or even diminish slightly. Farmers are adopting management practices, specifically fertilizer programs, that more efficiently use fertilizer materials. Improved and more precise application techniques, along with more accurate assessment of the availability of nutrients already in the soil, will likely result in less intensive fertilizer use.

The fertilizer industry is also likely to be affected by environmental policy. Environmental regulations on certain aspects of fertilizer production and use will become more important through the 1990s. For example, local regulations concerning groundwater contamination by nitrogen fertilizers will affect the nitrogen fertilizer industry. If contamination of underground water supplies worsens, regulations or educational programs to decrease use of nitrogen fertilizers can be expected.

Farm Machinery and Equipment

The farm machinery and equipment industry has played a critical role in increasing the productivity of the U.S. agricultural sector. Currently, U.S. farmers spend about $14.6 billion each year on tractors and other farm machinery. Average production expenditures for farm machinery in 1993 were just over $7,000 per farm. Tractors and self-propelled vehicles account for about 30 percent of total machinery and equipment sales.

During the 1970s, there were seven full-line farm machinery manufacturers: Allis-Chalmers, J. I. Case, Deere & Co., Ford, International Harvester,

Massey-Ferguson, and White. Full-line companies produce and sell a complete line of major farm machinery and equipment items.

However, widespread reorganization and restructuring in the farm equipment industry took place in the 1980s. This was a result of the extended recession in the agricultural sector in the early part of the decade. Financial instability and depressed incomes, coupled with declining land values, forced many farmers out of the machinery market, severely reducing the profitability of farm machinery manufacturers and dealers. The adjustment took the form of plant closings, sharp employment reductions, and a number of mergers among firms in the industry. Small firms that did not survive the depressed sales environment in the early 1980s either went out of business altogether or saw their viable product lines acquired by others.

Today, the farm equipment market is dominated by a few manufacturers selling a broad, diversified line of equipment that meets the needs of a wide variety of agricultural producers. Short-line companies typically produce for full-line companies under some type of production agreement. Full-line companies acquire parts and assembled machines from short-line companies, foreign-based subsidiaries, and foreign manufacturers.

The changes in the farm machinery industry of the 1980s helped the industry make the transition from being primarily domestic to being competitive internationally. For example, the United States and Canada were basically exclusive markets for large farm machinery, while the rest of the world was a market primarily for small machinery. Domestic manufacturers decided to specialize in the production of large equipment to meet domestic demand. A growing market for small and midsize tractors in the United States and Western Europe, however, motivated domestic producers to enter into agreements in which foreign firms produced these smaller tractors under the domestic producers' nameplates.

Much of the success of farm equipment manufacturers can be attributed to their network of dealerships. Farmers make their purchasing decisions based on perceptions of product quality, price, and service. The service offered by a dealership is perhaps the most important factor influencing purchasing decisions.

The retail sector of the farm equipment industry is made up primarily of independent dealerships. There has been a major consolidation of farm machinery dealerships over the last several decades. The number of dealerships fell from 16,700 in 1967 to 9,500 in 1972, then to 7,900 in 1989. Average annual equipment sales per dealer increased from about $2 million in 1979 to over $3 million in 1988.

A number of machinery dealers are now carrying more than one line of farm equipment, a trend that has emerged as a result of the consolidation the farm equipment industry experienced throughout the 1980s. A dealer who, in the past, may have held only one franchise may now have two or three to generate sufficient volume to maintain a full service and parts operation and to provide adequate income for the dealer.

Pesticides

Agricultural chemicals are another important agricultural input, representing over 4 percent of total farm production expenditures. Farmers spend about $6.7 billion on agricultural chemicals, including herbicides, insecticides, and fungicides, to assure that crop production is not lost to pests. Sharp increases in pesticide use and subsequent concerns over environmental effects have triggered changes in the pesticide industry that are important to the agricultural sector.

The pesticide production process involves two stages: basic production and pesticide formulation. The main economic activity is carried out by firms in basic pesticide production. These firms conduct the research and development that results in pesticides being synthesized with various raw materials.

The second stage of pesticide production involves formulation of the pesticide products. The formulators mix inert materials with the technical pesticides to make them suitable for transportation and use. In the past, for the most part, the formulators were independent of the basic manufacturers. However, the basic producers have integrated into the formulation phase either by formulating their own pesticide products or by contracting with independent firms to formulate the products for them.

Pesticides are distributed by a diverse number of firms that are usually not considered a primary part of the pesticide industry. The retail market is widely dispersed, as are the firms selling pesticides. Basic producers are involved in distribution and retailing only to a limited extent. Instead, private firms, including cooperatives, carry out these functions. Competition between these distributors and retailers increased through the 1980s, with services provided becoming more important. Services, such as application and insect scouting, are an important tool for attracting farmers' patronage.

Growth in the market for pesticides in the United States will likely be less in the future than it has been in the past, because the market is largely saturated. Competition between pesticide manufacturers for market share is likely to remain intense. Environmental concerns are driving development of new types of pesticides, such as biopesticides, which will be safer to use.

Growth in pesticide use will be much greater in developing countries, as they work to increase food supplies. U.S. exports are expected to increase because U.S. pesticide production is expected to serve this expanding world market.

Farm Supply Stores

Farm supply store sales consist primarily of feed, seed, fertilizer, chemicals, veterinary supplies, and hardware. The majority of small stores are organized as either sole proprietorships or partnerships, while larger stores are either corporate or cooperative entities. Operators of farm supply stores tend to locate within successful farming areas, and in many cases there are more than one such establishment per locality. Stores located in larger urban areas have tended to shift to home and garden sales in recent years as the number of farmers has declined. Dollar sales per supply store vary considerably from as low as $100,000 to over several million dollars per year. The smaller stores are usually family managed, with some help during the busy spring season. The larger stores usually depend more on hired management and labor. Net profits among farm supply stores also vary considerably. Markups on merchandise have been decreasing due to more intense competition, while costs of operating stores have been increasing. Therefore, net profits per dollar of sales have trended downward. Smaller stores have suffered relatively more than larger stores.

One increasingly important concept in marketing supplies to farmers is the "one-stop shopping center," where a complete line of farm supplies and services is available. For example, bagged and bulk fertilizers are offered, together with bulk-blending services. The center may provide a feed mill operated to prepare complete feeds and/or a concentrate-grain mix, or it may offer a custom mixing service. Technical information is dispensed, credit is furnished, and other services are provided. Often, soil tests, seeds, fertilizers, and chemicals are combined and offered in one complete farm supply program. Gasoline, oil, tires, batteries, and accessories are usually sold also. This method of farm supply distribution is expected to continue to increase in importance, especially in areas with intense competition.

While most farm supply operators own only one store, there is a tendency to open branch supply stores. In other cases a firm may be operating a chain of retail stores over one or several states. Local co-op supply stores are increasingly federated into large wholesale supply cooperatives.

Another factor in the retail farm supply business is the increasing importance of marketing raw farm products. Having supplied the inputs for farm production, the store operator often finds that customers need assistance in marketing farm produce.

Still another factor in farm supply retailing is a tendency for some store operators to engage in livestock and poultry feeding themselves or by use of contracts.

Operators of farm supply stores are increasingly concerned with the trend to fewer but larger farms; direct selling to farmers by manufacturers and wholesalers; the growing volume of credit sales; the need to obtain well-trained personnel; more complete and tailored product and service programs; contract farming and vertical integration; and the implications of various federal farm programs on farm supply sales.

In order to hold or obtain the farmers' supply business, more and more store operators are faced with providing discounts to farmers who buy in large quantities. Farmers are now generally as concerned with services provided and product quality as they are with prices of farm supplies.

Role of Farmer Cooperatives in the Farm Supply Business

Cooperatives are business organizations owned and controlled by farmers. They have at least three basic features that distinguish them from regular profit-type corporations: (1) Voting is usually done on a one-person, one-vote basis rather than on a one-share, one-vote basis, as in profit-type corporations. (2) The amount of interest or dividends paid on equity capital is usually limited by law not to exceed 8 percent. (3) At the end of a fiscal period, any accrued net margins or savings are allocated back to patrons according to the proportion of their individual business with the cooperative relative to the total business done by the cooperative.

Farmer cooperatives engaged in farm supply business provide about 28 percent of the supplies, including machinery and equipment, that farmers purchase. Farmer cooperatives are especially significant in the areas of fertilizer and lime, agricultural chemicals, petroleum, seed, and feed manufacturing and distribution. For example, farmers purchase about 45 percent of the fertilizer and lime, 13 percent of the seed, 23 percent of the feed, and 32 percent of the agricultural chemicals they use from farmer cooperatives.

Pricing Farm Supplies to Farmers

Firms engaged in the manufacturing and selling of farm supplies usually compete in an oligopoly market structure. Firms retailing farm supplies to farmers might be characterized generally under monopolistic competition. In the latter case, such firms may price their products and services with some regard to their competitors, but since their offerings of products number into the thousands, they can price independently on many items and not lose sales to their competitors. Furthermore, product branding and locational differentiation insulate them from competition, to a certain degree.

Differential or Discount Pricing

One aspect of price competition between farm supply retailers concerns *differential pricing*. There are at least five types of legitimate pricing differentials: (1) quantity or volume discounts given according to size of purchases, either in terms of physical quantities or dollar volume; (2) spatial or transportation discounts or charges made according to whether the product is picked up at the plant or delivered; (3) temporal or time-of-payment discounts or charges made according to whether the customer pays cash or on time and/or whether early order bookings are made; (4) form-of-product discounts or charges made according to the form in which the product exists at time of sale (such as bulk versus bagged feeds and fertilizers); and (5) some combination of the above.

The ideal situation is to provide price differentials to farmers based on legitimate cost differences in handling and marketing supplies. Illegitimate pricing differentials, on the other hand, result when firms charge different prices to various buyers although the conditions and costs surrounding the sales are substantially similar. Thus, a legitimate pricing differential becomes illegitimate when it does not reflect true differences in the cost of making or selling the product. The Robinson-Patman Act, for one, prohibits illegitimate price differentials.

Agribusiness managers may maximize profits or minimize costs by following differential pricing. Thus, if certain customers purchase multiple numbers of one item (large-scale buying), it usually costs less per unit to process and deliver such orders. A lower price per unit is then charged for such sales. The increased sales revenue obtained by selling more units decreases fixed costs per unit and contributes to a firm's greater overall profit. Of course, differential pricing may reduce net profits if price reductions on large-volume sales exceed the reduced cost from handling and marketing them.

Market Performance of Farm Supply Agribusinesses

The number of agribusinesses manufacturing farm supply inputs, such as feed, fertilizer, pesticides, and farm machinery, is relatively small, although the number of firms retailing these supplies to farmers is large. Therefore, farmer-purchasers are more likely to be impressed with competitive conditions at the local retail level than with those at the manufacturing level. In this discussion, however, we refer primarily to the manufacturing level.

Positive market performance is evident in the manufacturing of farm supplies relative to product development and adoption of new technology. Also, industry profit levels and return on equity capital do not appear excessive.

Poor market performance is evidenced by some unused plant capacity, or overcapacity; by many plants that are too small, resulting in high costs; and by excessive advertising that is often somewhat misleading.

However, it appears generally that market performance in farm supply industries is improving when compared with performance in the past several years. The fertilizer industry, in particular, has recently undergone changes geared to the interest of farmers (high analysis and bulk fertilizer, for example). In feed milling, local mills geared to farmers' feeding needs have been started, with far less emphasis on brand names. In farm machinery, mergers and consolidations have produced fewer firms, but on the whole, the surviving firms are more vigorous in their competition than before.

Projected Trends in Use of Farm Inputs

Labor requirements are expected to continue decreasing while machinery and equipment inputs continue increasing. Sophisticated technology geared toward meeting site-specific nutrient and chemical needs to increase production efficiency will be among new technologies adopted by U.S. farmers. Land input will remain about constant, but increased yields per acre are anticipated.

Expected yield increases are linked to advancing biotechnology and, to a lesser extent, to increased use of fertilizers, lime, and pesticides. Feed expenditures will increase, as will the conversion of feed into meat, eggs, and their products. Property taxes, insurance, and interest costs will increase.

Further discussion of management of agribusiness is provided in Chapter 28.

Topics for Discussion

1. Discuss the relative importance of various farm supply items.
2. Discuss the changing role of land, labor, and capital in agricultural production inputs.
3. Discuss the various price trends associated with farm supply items.
4. Discuss the distribution of feeds, fertilizers, and farm machinery and equipment.
5. Discuss the role of cooperatives in the farm supply business.
6. Evaluate the market performance of farm supply agribusinesses.

Problem Assignment

Interview a farm supply store operator regarding items handled, their relative profitability, and trends in their use on farms. Prepare a report of this interview.

Recommended Readings

Ash, Mark, William Lin, and Mae Dean Johnson. *The U.S. Feed Manufacturing Industry, 1984.* Statistical Bulletin No. 768. Washington, DC: Economic Research Service, USDA, 1988.

Barse, Joseph R. (ed.). *Seven Farm Input Industries.* Agricultural Economics Report No. 635. Washington, DC: Economic Research Service, USDA, 1990.

Beierlein, James G., Kenneth C. Schneeberger, and Donald D. Osburn. *Principles of Agribusiness Management,* 2nd ed. Prospect Heights, IL: Waveland Press, Inc., 1995. Ch. 1 and 8.

Beierlein, James G. and Michael W. Woolverton. *Agribusiness Marketing: The Management Perspective.* Englewood Cliffs, NJ: Prentice-Hall, Inc., 1991. Ch. 1, 8, and 13.

Cook, Mike. "Cooperatives and Group Action." In *Food and Agricultural Marketing Issues for the 21st Century,* Daniel I. Padberg (ed.). FAMC 93-1. Food and Agricultural Marketing Consortium. College Station: Texas A&M University, 1993.

Coffey, Joseph D. "Implications for Farm Supply Cooperatives of the Industrialization of Agriculture," *American Journal of Agricultural Economics,* 75(5):1132–1136 (1993).

Kimle, Kevin, and Marvin Hayenga. *Agricultural Input and Processing Industries.* Report RD-05. Ames: Department of Economics, Iowa State University, 1992.

Osteen, Craig D., and Philip I. Szmedra. *Agricultural Pesticide Use Trends and Policy Issues.* Agricultural Economic Report No. 622. Washington, DC: Economic Research Service, USDA, 1989.

20

Food and Fiber Marketing Functions

Marketing can be defined as "those processes, functions, and services performed in connection with handling food and fiber from securing production inputs through production and until delivery into the hands of the consumer." Many activities are involved in marketing: assembling, storing, warehousing, sorting, conditioning, grading, packing, processing, packaging, pricing, financing, insuring, risk taking, wholesaling, transporting, promoting, advertising, communicating, merchandising, selling, distributing, retailing, inspecting, and regulating.

Agricultural marketing is a vital link between farmers and consumers, actually beginning with the decision to produce a product in response to consumer demand signals and continuing with production practices that will result in a consumer product with salable, quality characteristics. Marketing, therefore, is the creation of *form, time, and place utilities.*

Form utility refers to changing the form of a product, i.e., through processing, into another product that provides consumers with more utility (more satisfaction). **Time utility** is created primarily through storage—for example, providing the product to consumers over an extended period rather than flooding the market at harvest. **Place utility** refers the transportation of goods to overcome geographic differences between consumers and the sources of the goods they desire.

Approaches to the Study of Marketing

The purpose of this section is to introduce some of the key functions involved in the marketing process. The *functional* approach to the study of marketing considers the various activities performed to increase the utility of food and fiber products from producer to consumer. This approach is in contrast with the *commodity* and *channel* approaches, which focus on commodity marketing and the agencies involved in the movement of goods from producer to consumer. The commodity approach to marketing is considered more fully in the next chapter.

Assembly

Assemblers perform the useful function of *aggregating* or concentrating many small lots of a product into larger lots with more desirable features. For example, an assembler may sort among small lots of a product to provide large lots that are consistent. However, an assembler does not normally process or materially change the form of the product. Assemblers perform a very useful function in helping to link producing with consuming areas. Examples of assembling agencies include fruit and vegetable packing and sorting sheds, grain elevators, and livestock auctions.

Standardizing and Grading

Quality helps differentiate a product for price advantage. Current grades and standards for U.S. products provide a guide to quality assurance. Consumers know that their food is of a consistent and uniform quality. **Standardization,** as the term implies, includes the establishment of recognized standards for such characteristics or features as quality, quantity, size, weight, and color. Standardization makes grading possible, for once criteria are determined, products can be properly classified. In food markets, for example, certain eggs can be graded as Grade A and certain meat can be graded as Prime, Choice, and Select.

The function of standardization and grading increases marketing efficiency. A buyer can order from a supplier without having to physically visit the seller's storage facilities to inspect the products intended for purchase. This is so because the mention of a specific grade implies that the good has distinctive characteristics that are understood and accepted by both buyer and seller. Standardization and grading, therefore, make for easier contractual and exchange agreements between marketers.

U.S. standards are used extensively in marketing agricultural products in this country. For example, during fiscal year 1993, the USDA graded 37 percent of the shell eggs, 95 percent of the butter, and 55 percent of the frozen fruits and vegetables produced in the United States, plus 74.6 billion pounds of fresh fruits and vegetables and 8.8 billion pounds of processed fruits and vegetables. In addition, of all federally inspected meat and poultry production in the United States, the USDA inspected 81.2 percent of the beef, 80 percent of the turkeys, and 56 percent of the broilers and other poultry. The USDA also classed more than 97 percent of the cotton and inspected 97 percent of the tobacco produced in the United States.

The use of quality standards is, in most cases, voluntary. The standards are used in many instances by manufacturers or by packers in quality control work and as the basis of the federal and federal-state grading services.

Some states and communities have passed laws making the use of certain U.S. standards mandatory in their areas. Federal marketing agreements and orders often require the use of U.S. standards. USDA standardization specialists have developed grade standards for a wide variety of farm products. They develop new standards as new products are introduced or as older products find increased use and need standards to facilitate trade.

In developing or revising standards, specialists must consider the range of quality it is possible to attain, especially when producing large quantities. It would hardly be practical to set standards for the top grade of any product so high as to represent an ideal rather than an actuality. Standards must be easily understood and uniformly applied, because grading is often a subjective process. Objective methods of grading are being developed and put into use for a number of products.

Storing

Storage provides *time utility* by making goods available at the time they are needed. Storage has a beneficial effect on prices, since it removes a portion of the supply from the market at harvest time and allows it to enter the market later when prices are more favorable to farmers. Thus, storage leads to greater price stability.

Agricultural products that are stored may consist of two types: (1) unprocessed or raw farm products and (2) processed products. Types of storage facilities, among others, include elevators, warehouses, bins, sheds, open storage, kilns, tanks, ships, barges, coolers, and freezer lockers.

Storage of farm products may be accomplished (1) on-farm (e.g., the storage of grain in bins on the farm) or (2) off-farm in various types of commercial facilities. Also, off-farm storage may be classified as private or public warehouses; bonded, licensed or nonbonded, nonlicensed warehouses; and refrigerated or unrefrigerated storage. Some agricultural products may be also stored in bulk, sacked, or loose form.

Storage facilities have undergone many improvements in recent years, including (1) higher-quality construction materials; (2) better structural designs, leading to increased storage efficiency; (3) improved machinery and equipment, such as lifters, power trucks, pallets, conveyers, and unitized loads; (4) higher in-plant efficiency, including computerized inventory management and quality control; (5) more precise control of the internal environment (i.e., humidity and temperature); and (6) better pest and insect control to reduce damage and storage losses.

Processing

Processing is defined as any activity that alters the condition or nature of a product substantially or provides *form utility*. More farm products are being processed to some extent before reaching the consumer than in the past. Every indication suggests increased processing in the future, as consumers demand more convenience in food and fiber products.

Generally, the number of food processing plants has been declining, while the size of remaining plants has been increasing. Plant size has been getting larger because firms are able to lower average costs by expanding. Another trend is more **horizontal integration,** where several plants operating at the same stage of production are owned by one firm, and **vertical integration,** where processors attempt to control the production of their inputs.

Packaging

Packaging, i.e., designing and producing the container or wrapper, is an increasingly important function in marketing, serving not only to size and preserve contents, but also to merchandise products. With the dominance of self-service retailing, packaging characteristics, such as size, color, shape, and labeling, can influence product sales.

The multi-million-dollar packaging industry offers many innovations, such as "see-through" containers, vacuum sealing, and "boil-in-a-bag" and microwavable containers. These developments, funded largely through research and development monies, are necessary to provide products that consumers will purchase.

Whenever feasible, considerably more food packaging is being done at processing plants located near producing areas. Thus, fewer unsalable products are transported, ensuring better product quality. Innovations in automated packaging equipment have been dramatic.

Transportation

Transportation primarily provides *place utility*. Products must move from where they are produced to where they are consumed. Transportation also provides time utility because of the time required to move products from their origin to their destination.

The general pattern of transportation of farm products is in three steps: first, from the farm to the local market, which serves as an *assembly* point for the area's produce; second, from this local market to one in the consuming area; and third, from this market in the consuming area to the retailer.

The first and last phases of this movement are almost always performed by trucks. The longest movement, the haul from the local assembly point to the terminal market, may be by truck, rail, water, or air. Thus, we find the most common carriers, motor and rail, supplementing each other in the transportation process in some situations and competing with each other for the same traffic in others. Transportation is provided through networks of highways, railroads, waterways, and airlines. Each mode of transportation has unique characteristics giving it an inherent advantage over the others in certain situations. A carrier may be a trucking firm or a trucker with a single, owner-operated vehicle; a railroad; a water transport service; or an airline. Sometimes the owner of the cargo also owns the shipping vehicle or vessel and is then referred to as a "private carrier."

The government, at both state and federal levels, has played a regulatory role in the freight transportation industry, deciding such matters as how a firm may enter the business, what routes it may serve, and what rates it may charge. Such control applies to railroad and air transport and to a great deal of transportation on the highways and waterways. The federal agencies that exercise such control include the Federal Aviation Administration (FAA) and the Department of Transportation (DOT) for airlines, the Federal Maritime Commission (FMC) for certain water transport services, and the Interstate Commerce Commission (ICC) for other carriers. The Federal Energy Regulatory Commission is responsible for setting rates or charges for the transportation and sale of natural gas and for establishing rates or charges for transportation.

In 1887, the U.S. Congress passed the Interstate Commerce Act creating the ICC to regulate transportation in interstate commerce. The law declared that unreasonable rail rate discrimination must cease, rates must be published, and fares must be reasonable. The ICC has jurisdiction over railroads, trucking companies, bus lines, freight forwarders, water carriers, coal slurry pipelines, and transportation brokers.

Federal promotion and regulation of civil aviation have been carried out by the FAA and the Civil Aeronautics Board (CAB). The CAB promoted and regulated the civil air transportation industry within the United States and between the United States and foreign countries. The CAB granted licenses to provide air transportation service, approved or disapproved proposed rates and fares, and approved or disapproved proposed agreements and corporate relationships involving air carriers. After the CAB ceased to exist as an agency in 1984, some of its functions were transferred to the DOT.

Deregulation of the transportation industry began in 1977 and continued through 1980. Before deregulation, government regulatory agencies essentially controlled carrier rates and the service offerings available to the public. There was little incentive for carriers to innovate. However, with the removal of much of the regulatory control, shippers and carriers are now free to creatively negotiate rate and service packages that best meet the needs of both parties. Also, transportation rates have decreased substantially as a result of deregulation, especially in the air and trucking industries.[1]

Nearly every state government has a public utilities commission that exercises a degree of control over intrastate transportation. However, a large share of the farm products hauled by truck and water for farmers is exempt from control by federal and state governments. The federal government, in its control, is concerned mainly with transportation from one state to another or between the United States and foreign countries. On the other hand, a state, in its control, is concerned primarily with hauling from one place to another within its own borders.

Of course, many products are grown a relatively short distance from the consuming market. In those cases, trucks are commonly used to haul the products from farm to market.

---

[1] Donald F. Wood and James C. Johnson, *Contemporary Transportation*, 4th ed. (New York: Macmillan Publishing Company, 1993).

Truck Transport

Truck transport has increased relative to other types of hauling for both raw and processed foods and fibers. Improved trucks, greater hauling capacity, and improved roads have contributed to this increase. Also, in hauling raw or unprocessed farm products, trucks are exempt from Interstate Commerce Commission regulations, except for safety and driving regulations. Trucks owned by farmer cooperatives are also exempt, allowing much more flexibility and lower hauling costs than otherwise.

However, truck owners are still controlled in other ways, such as through licenses and taxes, load limits, regulations governing length and height of trucks, and state and interstate commerce commissions wherever applicable.

The basic advantages of truck transport over other forms are in rapid movement of perishable foods; flexible schedules; lower costs, especially for small loads; dependability; speed; easier back-hauling; less loss in transit; and faster, more efficient loading and unloading. Grain, livestock, milk, vegetables, and fruits account for about 85 percent of the truck haulings of farm produce in the United States. Improvements in highways (such as the interstate highway system) and the use of refrigerated trucks and piggyback arrangements with railroads have increased the advantages of trucks over other carriers. Truck transport, however, continues to be hampered by varying and intricate regulations and restrictions imposed by different states.

Rail Transport

Rail has long been a primary form of mass transportation. The advantages of rail transport are in movement of large volumes of bulk commodities, safety, and economy of hauling over long distances.

Initially, railroads were given land grants to speed their development. Later, their rates and scheduling became highly regulated under both the Interstate Commerce Commission and state agencies. This regulation limited their ability to compete, and railroads lost much of their share of the transportation business. However, since deregulation of rail transport in the early 1980s, the volume of goods moved by rail has stabilized. About 20 percent of all carloads transported contain agricultural products, including farm products, food and kindred products, lumber and wood products, and pulp and paper products.

Railroads, over the years, have made progress in improving transportation by using larger freight cars; specially designed cars; piggyback systems; improved rail communications; unit trains; and computerized central traffic stations, yards, and makeup of trains. Other improvements have included railroad mergers to reduce overhead costs, lower labor costs, and improved services.

Water Transport

Water transport, by both inland waters and ocean, has increased its share of the total business. This method is especially adapted to movement of grains and other large-volume, bulk commodities with which speed is not critical. It is a relatively low-cost method. Interstate water traffic is exempt from regulation when bulk commodities, such as grains, are shipped. Also, public development of waterways has helped water traffic become more competitive.

Air Transport

Air transport is especially adapted to conditions requiring speed and dependability and to commodities having a high value per unit, such as flowers, nursery items, fruits, seafood, and specialty crops. Otherwise, it is too costly for bulky commodities.

Air cargo transport has had most of its development since World War II. Although air transport did not appear too promising in the beginning, technological developments provided it with significant potential. Planes built especially for cargo reduced transportation costs, permitting lower rates and wider markets. Though the proportion of intercity ton-miles[2] of freight traffic on domestic airways remains small, the rate of increase in freight traffic volume on commercial airways has been dramatic. A wider variety of commodities are being transported by air.

Financing

In the movement and exchange of goods from producer to consumer, there are lags between the time of production and the time of sale and between the time of sale and the time of payment. Thus, financing must be provided to ensure that firms have adequate liquidity to continue normal business operations. If financing were not provided, business firms could not operate the scale of plants or perform the variety of functions necessary for efficient marketing.

---

[2] A ton-mile is the movement of 1 ton (2,000 pounds) of freight the distance of 1 mile.

Market Communications

Principal types of communication media are newspapers, magazines, direct mailings, telephone (including facsimiles), word of mouth, radio, television, and computers.

In farm product marketing activities, agribusinesses employ all types of media. Newspapers are used for commodity reports, market news, advertising, and farm news. In general, magazines are used to reach consumers of processed farm products and farmers as buyers of inputs. Direct mailings have a number of uses, primarily advertising. Telephones serve a number of purposes. Satellite communications are involved in conveying market news reports and up-to-the minute price information. Fax machines are often used in executing and confirming sales transactions. Radio and television, moreover, are used for advertising and for market and farm news programming. Newsgroups on the Internet, which are increasing in popularity, provide a unique opportunity for producers and others to discuss a variety of pertinent issues.

The U.S. Department of Agriculture maintains more than 170 year-round market news offices to furnish daily and weekly reports on market prices, supplies, and other market conditions related to all major farm products. Many of the offices are operated cooperatively with the states. Market information is collected from all the major trading centers and disseminated through press associations, newspapers, and radio and television stations.

Trade associations deserve special note in the area of agribusiness communications. Not only do trade associations represent and educate their members, but they also inform the general public concerning trade activities, problems, and responsibilities. With dues and assessments collected from members, executives of trade associations, usually with small staffs, wield considerable influence in getting trade group policy integrated into local, state, and federal legislation. Practically every phase of the food and fiber sector is represented by one or more trade associations. It is expected that their roles will increase in importance over time.

Advertising and Promoting

Food marketing firms spend about $11 million annually in direct consumer advertising conducted through electronic and print media and through coupons. Consumers are the primary target of advertisers. The share of advertising in newspapers, about 24 percent of all advertising expenditures, has fallen slightly over the last decade. Television, radio, and magazine advertising has been relatively stable at 22 percent, 7 percent, and 5 percent, respectively, of all advertising expenditures. Advertising for food and food products makes up about 16 percent of all television and 7 percent of all magazine advertising expenditures.

Advertising programs may be conducted to increase consumption of branded products, or they may be "generic" in the sense of promoting a category of unbranded products, such as the "Drink milk—It shows," "Beef—It's what's for dinner," or "Pork—The Other White Meat" slogans. Almost every U.S. commodity group is involved in some type of generic advertising or promotion. The various producer promotion groups (National Dairy Promotion and Research Board, Beef Promotion and Research Board, National Pork Board, and National Potato Promotion Board) seek, as their ultimate aim, to influence consumers to use their respective generic products. Over 312 federal- and state-legislated producer promotion programs cover over 80 farm commodities.

Advertising expenditure is less intensive and advertising content more informative for food products that are less differentiable, such as milk and meat. In contrast, more persuasive advertising content is used for products that are differentiable, requiring large advertising expenditures.

The effectiveness of advertising in influencing consumers depends on the media used, the educational and income levels of the targeted consumer group, and the appeal of the advertising technique itself. Some advertisers develop their own ads, while others hire advertising agencies.

Advertising, or specifically "truth-in-advertising," is regulated by trade associations, the advertising media, the Federal Trade Commission and other government agencies, Better Business Bureaus, and consumers themselves through failure to purchase the advertised products.

Pricing

The *pricing* function serves to equate the exchange of goods and services between buyers and sellers. Pricing allocates goods from sellers to buyers and acts as a signal to producers, intermediaries, and consumers in guiding their market choices.

Although prices are determined by supply and demand, pricing decisions are made within different institutional environments. This analysis provides another viewpoint of market price behavior. Six of the possible market environments are discussed below.

1. *Open markets.* In an open market (such as a livestock or produce auction, a stock exchange, or a commodity futures market), the interaction of supply and demand—the meeting of minds of numerous individuals who represent the supplies and the demands—gives rise to price determination. Major farm commodities commonly traded in open markets include livestock and fresh vegetables, among others.

2. *Legislated markets.* In legislated markets, state and/or federal legislation may decree either or both the minimum and maximum prices that can exist. Examples are price supports for certain farm products, minimum wage laws, and fair trade laws. It does not necessarily follow that those prices that are legislated are good and fair. Since they may be set too high or too low, such legislated prices may create distortions in the marketplace, causing their short-run and long-run effects to vary. Major farm commodities traded within legislated markets to some degree include cotton, rice, wheat, feed grains, peanuts, and tobacco.

3. *Formula pricing markets.* In certain markets, prices are arrived at by a formula, usually under some control by a public or quasi-public body. Examples include state and federal milk marketing orders. Major farm commodities traded within formula pricing markets include milk and sugar.

4. *Negotiated markets.* In negotiated markets, prices are arrived at by formal negotiations between buyer and seller, sometimes under rules specified by state and/or federal legislation. Collective bargaining by labor unions under the Wagner Act is an example. Or, in a less formal and more individual fashion, broiler and egg contracts between a farmer and an integrator are negotiated. Major farm commodities included in negotiated markets are poultry and eggs, hogs, and vegetables and fruits for processing, all produced under contract, and some forestry products.

5. *Group-disciplined markets.* In certain markets, prices for services are set by a professional or occupational group and are binding on all members of that group (members are penalized for deviating from the established prices). Examples include medical societies, legal societies, barbers' unions, etc.

6. *Regulated markets.* The Interstate Commerce Commission and state public service commissions set and regulate the prices of freight, natural gas, water, and electricity, as well as bus fares and railroad fares. Thus, regulated markets are quite important in the national economy, the agricultural sector included.

Risk Sharing

One economic tool for sharing risks is hedging in the futures market, which helps reduce price risk. A brief explanation of hedging will illustrate the mechanics of this risk management tool.

At planting, producers are, of course, uncertain about what commodity prices will be at harvest. Hedging is a tool farmers can use to shift some of the risk resulting from this price uncertainty to speculators trading in the futures market. Hedging involves taking an equal but opposite position in the futures market to what one has in the cash market. Should prices fall, a producer who placed a hedge will be protected.

For example, soybean producers plant soybeans in the spring, with plans to sell the resulting crop at harvest in the fall. To hedge against adverse price movements, a soybean producer sells an October futures contract in May for October delivery. By selling the futures contract, the producer is said to have taken a **short** position in the market. In contrast, a **long** position refers to buying a futures contract. When the delivery time arrives in October, the producer may either deliver the beans or buy back the contract at some price, canceling the October delivery contract. Whether the producer will deliver the beans or buy an offsetting contract will depend largely upon the price of beans in the cash market in October and the price of October futures contracts.

Assume the following situation:

(A) *May*—Farmer X sells 10,000 bushels of beans for October delivery at $5.50 per bushel. $55,000

(B) *October*—Farmer X buys back his October futures contract at $5.30 per bushel to offset previous contract. $53,000

Gain on futures transactions $2,000

(C) *October*—Farmer X sells soybeans for cash on local market at $5.30 per bushel. $53,000

Farmer X could have delivered the beans for $5.50 and not bothered to buy an offsetting contract.

Farmer X protected his soybean price at $5.50 per bushel ($5.30 cash plus $.20 gain on futures transactions). Had he speculated, his price would have been $5.30 per bushel, the October cash market price. However, the $.20 per bushel gain on this particular futures transaction is a gross amount, because he must pay the costs of futures trading, including possible commissions and margin calls. Commissions are charges by the brokerage firm for completing transactions in the futures market. Margin calls are money that must be sent to the brokerage firm to maintain a futures market position when the market is moving against the trader's position. Margin calls may be required any time a trader, whether hedging or not, holds a futures contract.

Farmer X protected his May sale of beans for October delivery at $5.50 per bushel, which he had considered satisfactory at planting time. If he had not sold futures, his profit might have been greater, but his loss could have been also very large. For example, if the bean price had fallen to $5.00 per bushel at harvest, Farmer X would have incurred a $5,000 loss by speculating on the cash price at harvest. By hedging, Farmer X removed himself from speculation in the cash grain market. Possible large gains from speculation were sacrificed, but possible losses from speculation were reduced also. Hedging, then, is an economic tool for producers to use to reduce price risks.

Let's consider another example, one where the cash and futures prices of beans rise between the time the hedge is set in the spring and the time the crop is harvested in the fall.

(A) *May*—Farmer Y sells 10,000 bushels of beans for October delivery at $5.50 per bushel $55,000

(B) *October*—Farmer Y buys back October futures contract at $5.70 per bushel to offset previous contract $57,000

   Loss on futures transactions $2,000

(C) *October*—Farmer Y sells beans for cash on local market at $5.70 per bushel $57,000

Farmer Y obtained $57,000 for cash beans minus $2,000 loss on futures, for a net of $55,000, or $5.50 per bushel. Thus, Farmer Y was successful in locking in her bean price at $5.50 per bushel. While it is true she would have profited by $2,000 had she not hedged, it is also true that she could have lost by not hedging. Her margin of success, therefore, depended largely upon the spread between cash and futures prices.

Commodity **options** are another tool for managing price risk. Purchasing an option gives the holder the *right*, but without obligation, to a position in the underlying futures market. Options potentially eliminate two major barriers to the use of commodity futures in forward pricing agricultural commodities. First is the fear that the price set when the hedge is placed is too low. Second is the need to manage a margin account and answer margin calls as the market rallies against the short position in the futures.

Options can eliminate the need for a margin account, and they leave open the potential for higher prices to the producer. They leave open the potential for lower costs to the user (a food processor, for example) who wants protection from higher input prices.

A **put** option gives the right to a short position in the futures market. A **call** option gives the right to a long position in the futures market. In our soybean producer example, the producer would buy a put option to obtain the right to a short position in the futures market. The producer is essentially buying "price insurance," providing protection against price decreases. Should prices rise, the option is simply abandoned.

A food processor or cattle feeder will be interested in the "cost insurance" provided by a call option to obtain the right to a long position in the futures market. Food processors or cattle feeders who purchase food grains or feeder cattle as inputs for their production processes will want protection from price increases, which is what call options provide.

Should prices increase for the commodity producer or decrease for the input user, each will simply not exercise the right to a particular position in the futures market. By failing to exercise the option, each will lose the **premium** paid for the option, which is the market-determined value of an option for a particular futures contract and for a particular price level or **strike price**. Strike prices are price levels designated by the commodity exchange, such as the Chicago Board of Trade or the Chicago Mercantile Exchange, for which put and call options are traded. Premiums vary, depending on:

1. The strike price being considered relative to the underlying futures

2. The level of volatility in the underlying futures

3. The time left before the option expires

4. The prevailing interest rate[3]

Other commodities eligible for futures trading besides soybeans are corn, oats, cotton, wheat, broilers, live cattle and hogs, among others.

Futures markets are regulated by the Commodity Futures Trading Commission in the U.S. Department of Agriculture and by the commodity exchanges themselves.

---

[3] Wayne D. Purcell, *Agricultural Futures and Options: Principles and Strategies* (New York: Macmillan Publishing Company, 1991), pp. 225–227.

Distribution

Distribution is a marketing function. The term refers to the wholesaling and retailing of products to the final consumer. Included are all the efforts devoted to getting products from the processor or manufacturer to the point of final purchase for consumption. Therefore, distribution also involves such marketing functions as transportation and communications. Distribution has at least two facets: physical and economic. The former refers to the physical handling and movement of goods from one place to another, while the latter refers to the process of exchange of ownership and to costs, margins, selling prices, and net returns, among other factors. The analysis of distribution agencies and channels is an important element of marketing.

Product Development

An increasingly significant marketing function is the research and development involving new food and fiber products and the modification of existing products. These efforts are conducted for several reasons: (1) Firms view research and development (R&D) as a weapon to gain market share from competitors. (2) Affluent consumers tire of existing products, and their buying appetites must be rekindled periodically with new products or new forms of existing products. (3) Innovation and creation are fundamental traits of human endeavor that are reflected in an almost endless stream of technology. As consumers, we enjoy product variety.

Marketing Research

Attempting to determine what consumers want and prefer has become an important marketing function. If producers and marketers are to produce and market effectively, they must have more precise knowledge about their customers. Marketing research attempts to provide such information through consumer interviews, taste panels, store experiments, control displays, and other methods.

Marketing Regulations and Services

Various federal acts have an important bearing on regulating certain market activities. Some of the more important ones are discussed briefly.

Food, Drug, and Cosmetics Act

The adulteration of food products can reduce their wholesomeness to the extent of presenting health and safety hazards. Early government actions for consumer protection focused on food safety. The Pure Food and Drug Act, passed in 1906, then amended in 1930 and again in 1938 to become the Federal Food, Drug, and Cosmetics Act, provides the basis for food regulations in the United States.

Agricultural Marketing Agreements Act

Marketing agreements and orders are designed to improve returns to growers through orderly marketing. These are self-help programs through which growers can work together to solve marketing problems they cannot solve individually.

A marketing order may be issued by the Secretary of Agriculture only after a public hearing on the proposed order and after it has been approved by growers voting in a referendum.

Under a marketing agreement and order program, an industry can regulate the handling and marketing of its crops so as to prevent erratic flow to market, reduce the total supply in primary market channels, prevent low-quality produce from depressing prices, standardize containers and packs, or prevent unfair trading practices. Such a program also provides a means of financing marketing research and development and collecting statistics and shipping information necessary for effective program operation.

Federal Seed Act

This act requires truthful labeling of seed shipped in interstate commerce, prohibits false advertising, and prohibits importation of low-quality seed and screenings.

Packers and Stockyards Act

The Packers and Stockyards Act regulates the business practices of those engaged in interstate and foreign commerce in livestock and poultry marketing as well as meat and poultry packing. It sets out rules for fair business practices and for free, open, competitive markets.

Perishable Agricultural Commodities Act

This act prohibits unfair and fraudulent practices in the marketing of fresh or frozen fruits and vegetables and requires that dealers, commission merchants, brokers, shippers, and growers' agents handling these commodities in interstate or foreign commerce be licensed.

Standard Containers Acts

Containers subject to these laws must have specific capacities to prevent the use of deceptive containers in marketing produce.

U.S. Warehouse Act

The U.S. Warehouse Act authorizes licensing and bonding of public warehouses for storage of agricultural products. It provides for periodic inspection of warehouses and products to insure safekeeping.

Commodity Futures Trading Commission

This authority strives to assure correct registration of prices, protects the "hedging" services of the futures markets, and assures fair practices in futures trading. Futures prices must be protected against unfair or manipulative trading because they are used as guides in the buying and selling of "cash" wheat, corn, soybeans, and other crops at country points and at terminal markets. Hedging, which is the nonspeculative buying or selling of futures to offset or diminish price risk in handling actual commodities, is commonly engaged in by merchants, processors, farmer cooperatives, and individual farmers. It is an operation that obviously depends for its effectiveness on fair trading practices.

Agricultural Marketing Act of 1946

This act provides the basic authority for many functions of the Consumer and Marketing Service. In enacting this law, Congress declared its intent to provide for a scientific approach to the marketing of farm products and to establish an integrated administration of all federal laws aimed at improving the distribution of farm produce through research, marketing aids and services, and regulatory activities.

U.S. Cotton Standards Act

This act provides for (1) the establishment, preparation, distribution, and use of official standards for cotton and (2) a cotton classing service, on a fee basis, for shippers, spinners, and the general public.

U.S. Grain Standards Act

The U.S. Grain Standards Act (1) authorizes official standards for grain, (2) requires an inspection, based on these standards, of grain sold by grade in interstate or foreign commerce when shipped from or to a designated inspection point, and (3) prohibits deceptive handling and inspection practices.

Tobacco Inspection Act

This act provides for establishing and promoting the use of standards of classification for tobacco and maintaining official tobacco inspection and market news services.

Wool Standards Act

This act authorizes the use of certain funds for wool standardization and grading work.

Nutrition Labeling and Education Act

This act, passed in 1990, requires nutrition labeling for most foods offered for sale that are regulated by the Food and Drug Administration (FDA). The nutrition label, which follows an established standard format, is required to include information on total calories and calories from fat and on amounts of total fat, saturated fat, cholesterol, sodium, total carbohydrates, dietary fiber, sugars, protein, vitamins A and C, calcium, and iron, in that order.

Evaluation of Performance of Functions

It is recognized that certain functions must be performed to get products from the farm to the consuming household. Some of these, however, can be combined with others for greater efficiency, such as assembly, standardization, grading, and processing. Also, costs of performing certain functions might be reduced if firms doing them were of larger size, if labor were more productive, or if management were better. Although we rely on competition between firms for the efficient performance of these functions, the net result is not always optimum because of market imperfections, such as lack of knowledge, poor communication, monopolistic tendencies, government intervention, and other factors.

Sometimes functions may be overperformed—for example, packaging. Is it necessary to wrap certain packages several times with expensive materials? Is this done to preserve the contents or to differentiate the product even more?

Another question concerns the labor used in performing marketing functions. Could we save on labor for handling and rehandling? For example, as cans of processed vegetables leave the processor, they are loaded into trucks and then perhaps unloaded at a central warehouse; loaded into a smaller truck for delivery to the retail food store, unloaded, and stacked in the back of the store; loaded into carts and unloaded onto food store shelves; loaded by the consumer into a shopping cart and unloaded at the checkout counter; loaded into a shopping bag, put into a car, and unloaded at home; loaded onto a shelf and, lastly, unloaded from that shelf for final preparation. Obviously, the labor employed in doing all this loading and unloading will cost more than the farm product that is in the can. Could we improve on this system?

Topics for Discussion

1. Define "marketing."
2. What is meant by the functional approach to marketing?
3. Select one or more of the marketing functions presented, and discuss the role of the function(s) in the marketing of a selected farm product.
4. Discuss one or more of the regulatory laws as they affect the marketing of farm products.
5. Can you think of other marketing functions besides those discussed?

Problem Assignment

Select a farm commodity that is eligible for futures trading, and illustrate how a producer or agribusiness manager might use hedging to reduce price risk.

Recommended Readings

Beierlein, James G., Kenneth C. Schneeberger, and Donald D. Osburn. *Principles of Agribusiness Management,* 2nd ed. Prospect Heights, IL: Waveland Press, Inc., 1995. Ch. 3.

Beierlein, James G., and Michael W. Woolverton. *Agribusiness Marketing: The Management Perspective.* Englewood Cliffs, NJ: Prentice-Hall, Inc., 1991. Ch. 10, 11, 12.

Chafin, Donald G., and Paul H. Hoepner. *Commodity Marketing from a Producer's Perspective.* Danville, IL: Interstate Publishers, Inc., 1989.

Cramer, Gail L., and Clarence W. Jensen. *Agricultural Economics and Agribusiness,* 6th ed. New York: John Wiley & Sons, Inc., 1994. Ch. 12.

Economic Research Service. *Food Marketing Review.* Washington, DC: Economic Research Service, USDA. Annual issues.

Futrell, Gene A., and Robert N. Wisner (eds.). *Marketing for Farmers,* 2nd ed. St. Louis: Doane Information Services, 1987.

Kinnucan, Henry W., and Cynda R. Clary. "Brand Versus Generic Advertising: A Conceptual Framework with an Application to Cheese," *Agribusiness: An International Journal,* 11(4):355–369 (1995).

Kinnucan, H. W., O. D. Forker, J. P. Nichols, and R. W. Ward. "Research and Marketing Issues Facing Commodity Promotion Programs." In *Food and Agricultural Marketing Issues for the 21st Century,* Daniel I. Padberg (ed.). FAMC 93-1. The Food and Agricultural Marketing Consortium. College Station: Texas A&M University, 1993.

Lee, Jasper S., James G. Leising, and David E. Lawver. *AgriMarketing Technology: Selling and Distribution in the Agricultural Industry.* Danville, IL: Interstate Publishers, Inc., 1994.

Lesser, William. *Marketing Livestock and Meat.* Binghamton, NY: The Haworth Press, 1993. Ch. 10, 11, 12.

Manchester, Alden C. *Rearranging the Landscape: The Food Marketing Revolution, 1950–91.* Agricultural Economics Report No. 660. Washington, DC: Economic Research Service, USDA, September 1992.

Meier, Kenneth J., and E. Thomas Garman. *Regulation and Consumer Protection,* 2nd ed. Houston, TX: Dame Publications, Inc., 1995. Ch. 10.

Newman, Michael E., and Walter J. Wills. *Agribusiness Management and Entrepreneurship,* 3rd ed. Danville, IL: Interstate Publishers, Inc., 1994.

Purcell, Wayne D. *Agricultural Futures and Options: Principles and Strategies.* New York: Macmillan Publishing Company, 1991.

Schap, Keith. *Commodity Marketing: A Lender's and Producer's Guide to Better Risk Management.* Chicago: American Bankers Association and the Chicago Board of Trade, 1993.

Smith, Deborah T. (ed.). *Marketing U.S. Agriculture—1988 Yearbook of Agriculture.* Washington, DC: U.S. Department of Agriculture, 1988.

21

Marketing Channels for Farm Products

In the previous chapter, we discussed marketing functions and we outlined several of the steps involved in the movement and exchange of products from farms to consumers. Associated with these marketing functions are firms of different types (sole proprietorships, partnerships, corporations, cooperatives) that execute or conduct these functions as business operations either for profit or not for profit, as the case may be. A variety of government entities are also involved. By grouping these various firms according to the jobs or tasks they perform in the marketing system, we may develop another ap-

Marketing channels for farm products

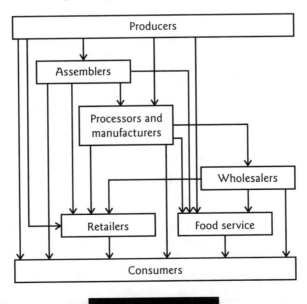

Figure 21-1

proach to marketing, namely, one dealing with the channels through which products flow as they move from farms to consumers. The focus of this chapter, then, is on *marketing channels* (Figure 21-1), which involve assemblers, processors and manufacturers, wholesalers, and retailers of food and fiber products. Included in retail marketing channels is food service.

Types of Marketing Channels

The movement of food and fiber from farms and ranches to processing plants and then on to consumers is a complex process involving many different marketing channels, kinds of facilities, and types of skilled personnel. Table 21–1 presents some typical marketing channels for various farm commodities as they move from the farm to the first handler, buyer, or receiver and, subsequently, to others in the marketing chain.

Table 21–1

Processing and Marketing Outlets (Except Retailing Outlets) for Various Types of Farm Commodities

Type of Commodity	Main Type of Market Outlets	
	First Handler	Subsequent Handlers

Cotton	Gins, cotton buyers, cotton merchandisers	Lint:
		  Compresses, warehouses, textile mills, apparel mfrs., exporters
		Seed:
		  Oil mills, exporters, linters

Sugar cane, sugar beets, maple, and honey	Raw sugar factories, processors	Refiners, molasses users, sugar by-product plants, processors, confectioners, beverage bottlers
Fruits, nuts, and vegetables:		
Fresh	Local farmers' markets, shippers, packing sheds, auction markets, terminal markets	Brokers, distributors, wholesalers
Processed	Driers, shellers, freezers, canneries, dehydrators, processors, distillers	Brokers, wholesalers, distributors

(Continued)

Table 21–1 (Continued)

Type of Commodity	Main Type of Market Outlets	
	First Handler	**Subsequent Handlers**
Beef cattle, hogs, and sheep	Auction markets, terminal markets, direct sales markets	Feedlots for further feeding, meat packers, meat slaughterers, renderers, wholesalers, brokers, curers, fabricators, meat locker plants, pharmaceutical mfrs.
Table eggs	Processors, packers	Wholesalers, direct sales markets, brokers
Broilers and turkeys	Processors, renderers	Wholesalers, brokers, distributors, exporters
Fish	Canners, freezers, processors, curers	Marine oil refiners, further processing, wholesalers, brokers, distributors, exporters, by-product plants
Feed grains (corn, oats, barley, soybeans, sorghum)	Country elevators, terminal elevators, driers, oil mills, other farmers	Feed mills, ingredient mills, refiners, manufacturers, exporters, seed houses, wholesalers, brokers
Seeds	Processors, cleaners, driers, handlers	Wholesalers, farm stores, home and garden stores
Hides, skins, pelts, mohair, and wool	Processors, curers, tanners, finishers	Mills, manufacturers, exporters
Milk and milk products	Haulers, processors, bottlers, manufacturers	Distributors, wholesalers, brokers, institutions
Tobacco	Curers, warehouses, dealers, stemmers	Manufacturers, exporters, distributors, wholesalers
Greenhous, nursery, and floral products	Shippers, distributors	Wholesalers
Wood and wood products	Loggers, veneer and plywood mills, lumber dealers/brokers/assemblers, pulpwood mills, paper mills, sawmills, by-product processors, planing mills, paperboard mills, naval stores dealers	Preservers, manufactureres, wholesalers, distributors, chippers, processors, brokers, exporters
Food grain (rice, rye, wheat, barley, oats, and corn)	Elevators, driers, brewers, malters	Bakeries, brokers, distributors, wholesalers, exporters, millers, processors, seed houses

In general, marketing channels are being consolidated; that is, more products are being moved more directly from farm to processor, wholesaler, retailer, and consumer. The functions still must be completed, but fewer and more consolidated firms are performing the tasks.

Assemblers

Because most processors have sought to capture economies of size, they need large quantities of the commodities they process. Assemblers pool small lots of farm commodities into the large lots that processors need. Assemblers increase the efficiency of the market system by providing commodities of specified quality, quantity, and volume at the right time, thus ensuring efficient processing, at least from the input procurement perspective.

Assemblers who move farm products on to wholesale markets include resident, traveling, and order buyers; auctions; and an increasing number of grower-shippers.

Resident buyers own or operate local marketing facilities, such as grain elevators, warehouses, and fruit and vegetable packing plants. They may operate independently or as representatives of other firms.

Traveling buyers, who operate independently or as representatives of terminal buyers or firms operating at local shipping points, go from one producing district to another as crops mature. For example, fruit buyers may purchase citrus fruits in Florida during the winter, then buy peaches from Georgia to South Carolina, moving from south to north as the season progresses. Purchases are generally made in truckloads or rail-car loads from the producers.

Order buyers function as shipping-point brokers or agents for wholesale buyers and may be salaried or paid on a commission basis. They are important in the livestock, fruit, and vegetable businesses.

Auctions have had an important influence on the operations of local markets, especially those that handle livestock, wool, fresh fruits, and vegetables. Some auctions are owned by wholesalers or processors. Ownership of others is specialized, and farmers using them pay fees for the services.

With fruits and vegetables, the *grower-shippers* have increased in prominence. Many are former cash buyers whose interests as producers have gradually increased. Besides their own production, they may also pack and market products produced by their neighbors, generally for a fixed packing charge and/or a marketing commission.

In general, the trend is toward fewer but larger assemblers somewhat more vertically integrated or coordinated with subsequent handlers and processors.

Processors and Manufacturers[1]

Most farm products are processed to some extent before they reach consumers. By processing, commodities are transformed, providing form utility. Some products, like eggs for table use, need relatively little processing, but others, like microwavable meals, need substantially more. Thus, processors of one kind or another are key participants in the marketing channel. Because of changing consumer needs and desires, processors and manufacturers have had to increase plant size, adopt new technology, and improve processing techniques. Also, they have developed a wide variety of "convenience" foods and new products besides expanding advertising and other product differentiation programs.

The food processing sector in the United States was generally quite profitable through the early 1990s. Processed food shipments, which total about $395 billion annually, account for about 13 percent of all U.S. manufacturing activity and represent the largest sector in the economy.

The structure of the food processing industries has changed considerably in recent decades. Many smaller processors have discontinued operations because they were unable to capitalize and manage needed changes. There were over 1,250 divestitures (sales of business holdings) between 1982 and 1992. Others have merged with stronger companies, capital-wise and management-wise, to survive. Almost 2,900 mergers took place between 1982 and 1992. The 50 largest food processors control 47 percent of sales, compared with an average of 27 percent for all other industries.

Despite the increased level of concentration by the larger firms, competition among the 16,000 firms in the 49 food processing industries is intense. Even in periods of economic stagnation, food processors use price and nonprice competition to gain both consumer acceptance and retail shelf space in the $160 billion name-brand retail food market and in the $240 billion food processing market.

Most consumer-directed nonprice competition consists of product differentiation in the form of advertising and new product introductions in the name-brand market. About 40 percent of U.S. food sales are branded products; the other 60 percent are either unbranded, undifferentiated products or products sold to food service or other food manufacturers. Sales to food serv-

---

[1] Much of the material in this and following sections was adapted from selected issues of *Food Marketing Review* (Washington, DC: Economic Research Service, USDA).

ice or other food manufacturers are only minimally affected by mass media advertising and other forms of product differentiation. Therefore, increasing or maintaining market share for food processors in these undifferentiated sectors is determined by price or contractual arrangements. However, the trend is more differentiation through new product introductions and advertising, even for traditionally undifferentiated products such as red meats, poultry, fish, and some dairy products.

American consumers have over 230,000 packaged food products from which to choose. Including new size introductions, over 16,000 new grocery products have been introduced in some years (Table 21–2). Industry estimates put the failure rate of new food products at 90 to 99 percent. The top 20 food processing firms account for only about 17 percent of new product introductions. Smaller or midsize firms account for the remainder, as they try to make inroads in these concentrated industries.

Table 21–2

Selected New Product Introductions

Item	1991	1992
	---- Number ----	
Baby foods	95	53
Bakery foods	1,631	1,508
Baking ingredients	335	346
Beverages	1,367	1,538
Breakfast cereal	108	122
Candy, gum snacks	1,885	2,068
Condiments	2,787	2,555
Dairy	1,111	1,320
Desserts	124	93
Entrees	808	698
Fruits and vegetables	356	276
Pet food	202	179
Processed meat	798	785
Side dishes	530	560
Soups	265	211

Source: Economic Research Service, USDA.

Cooperative Marketing

Farmers have at least three choices available for marketing farm products: (1) they can sell their produce to nonfarm businesses, letting them perform the marketing steps; (2) they can sell some produce (rice, cotton, wheat, peanuts, corn, etc.) to the federal government, such as under the various price support programs; or (3) they can organize cooperatives for marketing their own products.

There are just over 2,200 farmer marketing cooperatives in the United States, down from 3,650 in the early 1980s. The business volume of selected products marketed through farmer cooperatives is given in Table 21–3. Cooperatives are especially important in marketing milk and dairy products, grains, sugar products, rice, nuts, fruits, and vegetables.

Table 21-3

Business Volume of Product Marketing by Farmer Cooperatives, 1992

Product Marketed	Net Business
	1,000 dollars
Beans and peas (dry edible)	265,400
Cotton and cotton products	2,076,563
Dairy products	20,238,617
Fruits and vegetables	7,591,016
Grain, soybeans, and soybean meal and oil	15,223,325
Livestock and livestock products	4,938,452
Nuts	916,316
Poultry products	1,216,237
Rice	771,583
Sugar products	2,225,676
Tobacco	415,484
Wool and mohair	18,977
Miscellaneous	2,298,265
Total	58,195,911

Source: U.S. Department of Agriculture.

Wholesaling Farm Products

The movement of food and fiber from processing and manufacturing plants to wholesalers and then to retailers and others is an important link in the chain of agricultural distribution. Wholesalers are operators of firms engaged in the purchase, assembly, transportation, storage, and distribution of groceries and food products for sale to retailers, institutions, and business, industrial, and commercial users.

Wholesalers typically purchase goods in large quantities, rail-car loads or truckloads. They then break these large quantities into smaller lots to meet the needs of their clients, the retail food and food service establishments.

There are over 476,000 wholesale business establishments of all kinds in the United States. Annual sales total over $2,525 trillion. Employment exceeds 6.3 million persons, with an annual payroll of over $181 billion. Wholesaling grocery and related products and farm-related raw materials involves over 880,000 employees in some 54,700 firms with $500 billion of sales annually.

There are three basic types of wholesalers (Figure 21–2): (1) merchant wholesalers, operators of firms primarily engaged in buying and selling groceries and grocery products on their own account; (2) manufacturers' sales branches and offices, wholesale operations maintained by grocery manufac-

Wholesale food sales, 1992
Merchant wholesalers dominated the sector's $453 billion in sales.

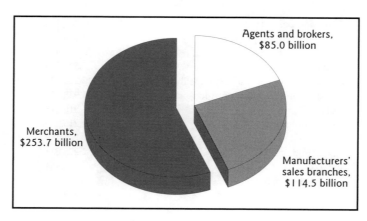

Source: Economic Research Service, USDA

Figure 21–2

turers (apart from their manufacturing plants) for sales and distribution purposes; and (3) agents and brokers, wholesaler operators who buy or sell as representatives of others for a commission and who usually do not store or physically handle products.

Merchant wholesalers can be divided into several types. General-line wholesaler merchants handle a broad line of dry groceries, health and beauty aids, and household products. Limited-line wholesaler merchants handle a narrow range of dry groceries dominated by canned foods, coffee, spices, bread, and soft drinks. Specialty wholesale merchants handle perishables, such as frozen foods, dairy products, poultry, meat, fish, fruit, and vegetables. Wholesale clubs are hybrid wholesale-retail establishments selling food, appliances, hardware, office supplies, and similar products to its individual and small-business members at prices slightly above wholesale.

Structural Changes

The wholesale food industry experienced two major structural changes in the 1980s and early 1990s. First, there was a surge of mergers in the early to mid-1980s. These mergers, which slowed by the end of the 1980s, produced national grocery and food distributors, replacing local and regional operations that had previously existed. Fewer but larger firms meant higher aggregate sales concentration.

Second, there was a rapid increase in the number of wholesale clubs in the late 1980s and early 1990s. This growth has intensified competition between wholesale clubs and between wholesale clubs and traditional supermarkets. Wholesale clubs have already saturated large-population areas. Sales in many of those stores have been level or declining because the number of customers per store declines as companies and their competitors increase the number of stores in a territory. Existing supermarkets have responded to the increase in wholesale clubs by offering their own club packs (for example, selling a 12-pack of paper towels at a competitive price), offering unadvertised specials, featuring attractive bakery and restaurant (deli) items, and putting special emphasis on the various services that supermarkets provide that clubs do not.

Wholesale club firms have begun moving into areas with lower population densities. There is concern whether the consumer traffic will be adequate to generate the volume of business necessary to operate profitably on the narrow profit margins common to wholesale clubs.

The increased involvement of U.S. wholesale firms in international markets is another important trend. Prompted initially by free trade agreements between the United States and other countries, many wholesalers have moved quickly to enter foreign markets through joint ventures with retail food and grocery firms in those countries.

Retailing Agricultural Products to Consumers

Agribusiness retailers include those establishments selling groceries, prepared foods, soft drinks, tobacco and tobacco products, alcoholic beverages, floral products, clothing, shoes, furniture, and home furnishings (from agriculturally derived products). There are over 588,000 retail food stores and eating and drinking places in the United States.

Retail Grocery Stores

The retail grocery is near the end of the marketing channel, where agricultural and other products, in a multitude of forms, are placed before customers. The basic task of food retailers is to anticipate and service customer needs and wants and to acquire and competitively price 3,400 to 100,000 items, depending on the type of store. A store must be convenient to its customers. Retailers must select, train, and supervise personnel. Sometimes they must oversee food processing—meat fabrication, deli foods, and salad bars—and repackaging—meat, fresh fruit, vegetables, and cheese. Often they must advertise. A major task of a retailer is to receive the merchandise into the store, prepare it for display, price it, place it on the shelves, and check it out. Finally, there is the important job of providing the capital, keeping records of transactions, and paying operating expenses. A retailer's efficiency in those tasks affects the prices customers must pay for food.

Table 21-4

Consumer Grocery Expenditures, 1992

Product	Share of Supermarket Sales
	----------- Percent -----------
Perishables	47.66
Dairy	6.74
Fresh and cured meat and poultry	15.32
Produce	10.18
Service deli	4.58
Other	10.84
Dry grocery (food)	28.51
Beer, wine, liquor	5.46
Breakfast foods	2.55
Condiments, dressings, and spreads	2.06
Soft drinks	3.59
Other	14.85
Total foods	76.17
Dry groceries (nonfoods)	12.44
Health and beauty aids	4.22
Prescriptions	1.15
General merchandise	4.22
Unclassified	1.80
Total	100.00

Source: Economic Research Service, USDA.

Types of Grocery Stores

There are about 165,000 grocery stores, with annual sales of about $360.8 billion and employment of about 3.2 million. Retail grocery stores are usually classified into three types: (1) supermarkets, (2) convenience stores, and (3) superettes (Figure 21–3).

Supermarkets represent just under 10 percent of all retail food stores but account for almost 72 percent of total grocery sales. Some key trends affecting supermarkets are discussed in the next section.

Convenience stores, which make up about 20 percent of retail food stores, account for 12.6 percent of grocery sales. Traditional convenience stores have experienced increased competition from petroleum retailers that have added convenience stores to their gasoline stations. These retail outlets are not counted as food stores because food product sales make up less than half of total sales.

Superettes and "mom and pop" stores represent about 37 percent of all retail food stores but account for only 10 percent of sales. These smaller grocery stores, which offer a variety of food and nonfood grocery products, are usually located in areas not adequately served by supermarkets, such as in densely populated urban areas or sparsely populated rural areas. As supermarkets explore new growth opportunities, these smaller grocery stores may face increased competition, which could limit their long-term survival.

Food sales in 1992
Sales reached $384 billion, with supermarkets accounting for more than 71 percent of total sales.

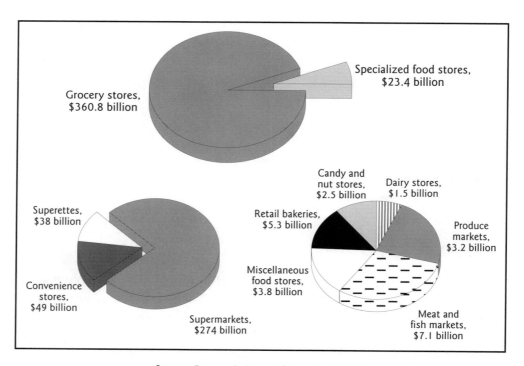

Source: Economic Research Service, USDA

Figure 21–3

The remaining 6 percent of retail grocery sales are made by specialized food stores. Specialized food stores are primarily engaged in the retail sale of a single food category. Examples are retail bakeries; meat and fish markets; candy and nut stores; natural and health food stores; coffee, tea, and spice stores; and ice cream stores.

Store Format

The share of conventional supermarket formats has been declining as a percentage of both sales and number of all supermarkets. Conventional supermarkets have been increasingly displaced by newer store formats, such as the superstore and the warehouse/limited-assortment supermarket. The superstore offers a wider variety. The warehouse/limited-assortment type offers larger sizes of fewer items at lower prices than the typical supermarket. Nonconventional formats now account for more than half of all supermarkets and more than two-thirds of supermarket sales. Superstores, which account for one-quarter of all supermarkets and nearly one-third of supermarket sales, have been increasing in importance since their inception in the 1980s.

The evolution of supermarket formats has had important implications for consumers and the food distribution and retailing industries. By the late 1970s, it became clear that the needs of consumers in a changing economic and demographic environment could no longer be met by a single supermarket format encompassing a common mix of products and services. Alternative store formats allowed retailers to better serve individual consumer segments. At the same time, these new formats provided retailers the opportunity to distinguish themselves from their competitors by offering a wider variety of products and services or by offering larger quantities of fewer items, but at a lower price per unit.

Other outlets have been developed that are also challenging traditional grocery store retailers. These nontraditional outlets offer grocery products (both food and nonfood) as part of a broader line of retail merchandise. As nontraditional outlets, such as warehouse wholesale club stores, deep discount drugstores, and mass merchandisers, offer more grocery products, they must compete with traditional grocery retailers, primarily supermarkets.

Grocery chains, those food retailers owning 11 or more stores or outlets, now account for about 65 percent of food store sales, up from 37 percent in 1948. The chain store's share of total grocery store sales was stable through the mid-1980s, due in part to improvements in purchasing and in operating efficiencies not available to non–chain store retailers (independents) because of their lack of affiliation with a full-service food wholesaler.

Technology and Innovation

A key trend among grocery retailers is acceptance of methods of payment beyond the traditional cash or personal check. Special incentives by credit card issuers have spurred grocery retailers to accept credit cards for purchases, opening up the last, and largest, nondurable retail credit card market. With this development, more progress is being made to facilitate the use of electronic payment methods, including debit cards and food stamp benefit cards. Electronic payment methods take less time than the personal check, the most common form of payment. They, along with optical scanning technology, have improved checkout, or *front end*, productivity.

Many grocery operators are also seeking ways, through new technology, to increase the productivity of inventory, or *back end*, operations. These include the use of electronic data interchange (EDI) for warehouse ordering and of inventory data exchange between the retailer and its suppliers. Efforts are also being made to extend the use of EDI for transmission of promotional information and discounts to retailers and for automated billing and payment transfers for customers.

Food Service

"Food service," the fastest growing part of the agribusiness sector, refers to the commercial and noncommercial establishments that prepare and dispense meals and snacks intended for on-premise or immediate consumption. This includes fast-food drive-through sales and sales of foods, such as pizza, delivered to homes. Candies, popcorn, pretzels, nuts, and drinks qualify as food service if other foods are not available. The share of the average consumer's personal dollars spent on food service has been increasing over the last several decades.

The rapid growth of food service can be attributed primarily to the changing life style of American consumers, who demand quickness and convenience. People are waiting longer to marry. The number of households in which both parents are employed outside the home or in which there are single parents has been steadily increasing over recent decades. Often in such households, no one has the time or the energy to prepare every meal and clean up afterwards, so there is the incentive to eat out more often, even if it is more costly.

Types of Establishments

Commercial or public establishments, freestanding or part of a host establishment like a hotel, prepare meals and snacks and then serve and sell them for profit to the general public. Commercial places include drinking places, eating places, lodging places, recreation and entertainment places, and retail hosts. Commercial food service sales of meals and snacks account for nearly $191 billion, about 76 percent of food service sales.

Food service sales in 1992
Sales reached $251 billion, with commercial sales accounting for 76 percent of total sales.

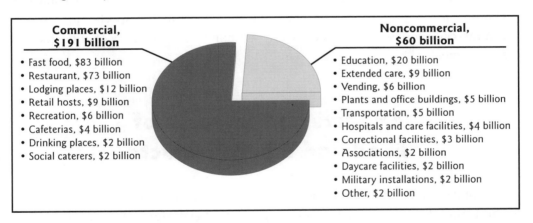

Commercial, $191 billion
- Fast food, $83 billion
- Restaurant, $73 billion
- Lodging places, $12 billion
- Retail hosts, $9 billion
- Recreation, $6 billion
- Cafeterias, $4 billion
- Drinking places, $2 billion
- Social caterers, $2 billion

Noncommercial, $60 billion
- Education, $20 billion
- Extended care, $9 billion
- Vending, $6 billion
- Plants and office buildings, $5 billion
- Transportation, $5 billion
- Hospitals and care facilities, $4 billion
- Correctional facilities, $3 billion
- Associations, $2 billion
- Daycare facilities, $2 billion
- Military installations, $2 billion
- Other, $2 billion

Source: Economic Research Service, USDA

Figure 21–4

Drinking places are those establishments that primarily serve alcoholic beverages for on-premise or immediate consumption but that also provide food service. These include bars, beer gardens, taverns, nightclubs, and saloons. At 57.5 percent, beer sales are the primary component of alcoholic beverage sales, followed by distilled spirits at 30.9 percent, and wine at 11.6 percent.

Eating places are those establishments that primarily serve prepared meals and snacks for on-premise or immediate consumption. These include restaurants, lunchrooms, fast-food outlets, and cafeterias.

Lodging places are those establishments that provide both lodging and food service to the general public. These include hotels, motels, and tourist courts.

Recreation and entertainment places include food service operations in theaters, in bowling alleys or billiard halls, in commercial sports establishments, in country clubs, at public golf courses, and in miscellaneous amusement and recreational establishments.

Retail hosts are food service operations that operate in conjunction with, or as part of, retail establishments, such as department stores, drugstores, and other miscellaneous retailers.

Noncommercial establishments are those in which meals and snacks are served as a supportive service rather than as the primary service. These include schools, colleges and universities, hospitals and extended-care facilities, vending areas, correctional facilities, the military, and transportation. Noncommercial food service accounts for $60 billion (24 percent) of total food service meal sales.

Exports and Imports of Food and Fiber Products

Domestic markets are not the only outlet for U.S. food and fiber products. The United States is the world's top agricultural exporter, with sales of $42.5 billion in 1993. Export markets, specifically developing countries, are the key source of growth potential for the agricultural sector.

Consumers in countries around the world have a large appetite for U.S. agricultural products, particularly high-value U.S. consumer foods and beverages. Rising per capita incomes, increased interest in Western-style foods, greater demand for convenience, and growing awareness of healthful foods among consumers in major overseas markets have helped make consumer foods the fastest growing segment of U.S. agricultural products sold abroad. Exports of processed foods total almost $23 billion, with meat products the largest component. In fact, the U.S. share of the global market for these products increased from 10 percent in 1985 to 16 percent in 1992.

The United States is the largest bulk commodity exporter. The estimated U.S. share of world trade for 1993 was 27 percent for wheat, 46 percent for feed grains, 18 percent for rice, 60 percent for soybeans, and 23 percent for cotton. Since the mid-1970s, the U.S. share of the global market for bulk commodities has remained fairly stable at about 30 percent, although it did fall to the low 20s in the mid-1980s.

U.S. agricultural exports generate employment and income in both the farm and nonfarm sectors. Agricultural exports create nearly 1 million jobs in the United States. Of those, 340,000 are created on the farm, which represents 10 percent of the farm labor force.

With imports of $24.4 billion annually, the United States is also one of the world's largest importers of agricultural products. Almost $22 billion worth of those imports are processed foods. However, agricultural imports make up only a small portion (5 percent) of total U.S. imports.

Imports provide consumers with agricultural products that are either not produced or not available in sufficient quantities in the United States. Major imports generally not produced in the United States include certain spices, coffee, teas, cocoa, rubber, and bananas. Domestic production of other goods, such as certain cheeses and tobaccos, is insufficient to satisfy domestic demand. There are also imports of seasonal fruits and vegetables during the off-season in the United States. Also, some agricultural products, such as sugar, are imported in their raw form for processing and packaging in the United States because foreign producers have a cost advantage over U.S. producers.

Agricultural imports provide jobs within the United States in transportation, storage, handling, processing, and distribution. Additionally, imports provide foreign countries with needed revenue in the form of U.S. dollars, which, in turn, can be used to purchase other U.S. products.

International trade, when no restrictions or barriers are in place, will take place between nations according to the law of comparative advantage, discussed previously. Nations will produce those commodities for which they have the greatest advantage or least disadvantage, comparatively. In total, all nations will be better off if they trade than if each attempts to be self-sufficient.

Another premise is that for any nation to export, it must likewise be willing to import. If a nation's exports exceed its imports, other nations are in debt to that nation. Conversely, if a nation's imports exceed its exports, that nation is in debt to others. A nation must then attempt to rectify this import-export deficit by transferring funds to nations with surplus positions.

Governments use various methods to restrict imports, including (1) tariffs, or taxes; (2) variable levies, or tariffs that are adjusted up or down depending on how high an import price is desired; (3) quotas, or quantity limitations; (4) licenses, or permits enabling an exporter to market a specified quantity of the goods; (5) foreign exchange controls, or regulations affecting how importers may convert their funds to other currencies; and (6) embargoes, or complete restriction over imports.

U.S. agriculture, as a whole, stands to benefit more by a freer trade policy than by one of restriction, because U.S. agricultural exports, including feed grain, food grain, oilseed, poultry, and cotton exports, are relatively more important than the smaller amount of agricultural products imported. Also, many agricultural products imported are noncompetitive products, i.e., products like bananas, coffee, and tea, which are not in direct competition with U.S. products.

Further discussion of international trade is presented in Chapter 29.

Industrial Uses of Farm Products

Increasing productivity in the agricultural sector has spurred interest in finding alternative uses for agricultural commodities. There have also been other factors that have heightened the interest in finding agricultural substitutes for industrial materials. One example was the sharp increases in petroleum prices by the Organization of Petroleum Exporting Countries (OPEC) in 1973 and 1979. Americans were made aware of their dependence on foreign oil, which kindled interest in organic substitutes for petroleum.

Farm products with the largest industrial usage are oils from various seeds, by-products of livestock slaughter and the dairy industry, cotton, and starch from corn. The proportion of the main agricultural commodities used for industrial purposes varies greatly but is increasing generally. The following paragraphs list a few examples.

There appears to be some potential for replacing part of the petrochemicals used to manufacture plastics with biodegradable cornstarch derivatives. This will increase demand for corn. Soy ink, made from soybeans, is beginning to supplant petroleum-based ink products for some uses.

Oils extracted from the seeds of soybean, corn, cotton, sunflower, flax, and rapeseed plants have a variety of industrial uses. They become components of chemicals such as plasticizers, which add pliability to plastics and other substances; stabilizers, which help other substances resist chemical change; emulsifiers, which enable the mixing of normally unmixable liquids; surfactants, which reduce the surface tension of liquids and are commonly used in detergents; and nylons and resins, which are basic ingredients of many industrial products.

Lactose (milk sugar) is the major carbohydrate found in milk. Lactose is used as an excipient and diluent for drugs. It dilutes the active ingredient in a medication to help ensure that a uniform dose is ingested in a form that is convenient and agreeable for the user. Fatty acid derivatives of lactose can be used as surface-active agents and emulsifiers in toothpaste and other toiletries because they are derived from natural ingredients and tend to be nontoxic.

Casein, the protein found in greatest quantity in milk, has had nonfood uses for some time, including specialty adhesives, premium paper coatings, and biodegradable plastics. Recently, "high clarity" casein has been used in the manufacture of television screens and as a component of the light-sensitive emulsion on some photographic film.

The industrial uses of agricultural products are obviously varied and extensive, the result of much public and private research. These new and alternative uses provide important and viable outlets for agricultural commodities.

Military Demand

U.S. military demand for food and fiber products consists of meeting the needs of personnel stationed domestically and abroad. In addition, various food aid programs to foreign nations are administered through military channels, especially in areas of military conflict. U.S. food and fiber products are an important military tool in many situations, to say nothing of the rehabilitation value of food and fiber products after military conflict ceases.

Compressing the Marketing Channels— A Model Channel

It is likely that we have too many channels between producers and consumers. If so, what would a "model" channel look like? Perhaps the broiler industry can give us a glimpse of a model marketing channel. First, through vertical integration, broiler production decisions, from hatching to slaughter and from processing to distribution, are made by the integrator, representing only one "profit center," not several segments as in other industries. Second, processed broilers, including whole birds, parts, and further processed products, move directly from processing plants to retail supermarkets, packaged in a variety of forms, from ice pack or frozen to individual portions, ready to eat. Third, consumers purchase the broilers, in forms ranging from whole birds to prepared entrees, from the supermarkets and complete necessary preparations before consumption. Obviously, this type of marketing channel is about as compact as possible. The ultimate would be for vertically integrated broiler production and processing firms to integrate further and own the retail outlets. From the other direction, supermarket chains could vertically integrate backward and gain control of broiler processing and production. Still another possibility, although far-fetched, would be for consumers to own both the production and processing firms and the supermarket chains, perhaps through cooperatives.

Aside from the channel ownership question, the efficiency of broiler production and marketing is evident. Pork production and marketing is following suit. Broiler production and marketing takes advantage of economies of size, eliminates several handling and rehandling operations, provides rapid movement from plant to store, and centralizes decision making and accountability. It is no wonder, then, that broilers are among the most competitively priced meats in supermarkets.

Topics for Discussion

1. Discuss the processing channel as one outlet for farm products.
2. Discuss the wholesale channel for farm products.
3. Discuss grocery stores as a retail channel for farm products.
4. Name and discuss other types of food retailers besides supermarkets.
5. Will industrial uses and military demand for farm products increase? Explain.

Problem Assignments

1. Select a farm commodity, and discuss the main types of market outlets for it up to the consumer level.
2. Visit a marketing facility in your area, interview the manager, and write a summary of his or her marketing channel problems.

Recommended Readings

Baker, Gregory A., Max S. Wortman, Jr., and Granger Macy. "Critical Success Factors for Managing Quality in Food Processing Firms," *Agribusiness: An International Journal*, 10(6):481–490 (1994).

Barkema, Alan. "Reaching Consumers in the Twenty-First Century: The Short Way Around the Barn," *American Journal of Agricultural Economics*, 75(5):1126–1131 (1993).

Barkema, Alan, Mark Drabenstott, and Kelly Welch. "The Quiet Revolution in the U.S. Food Market," *Economic Review*, 76(3):25–41 (1991).

Beierlein, James G., and Michael W. Woolverton. *Agribusiness Marketing: The Management Perspective*. Englewood Cliffs, NJ: Prentice-Hall, Inc., 1991. Ch. 10, 11, 12, 13, 23.

Cramer, Gail L., and Clarence W. Jensen. *Agricultural Economics and Agribusiness*, 6th ed. New York: John Wiley & Sons, Inc., 1994. Ch. 12.

Economic Research Service. *Food Marketing Review*. Washington, DC: Economic Research Service, USDA. Annual issues.

Goldberg, Ray A. "New International Linkages Shaping the U.S. Food System," *Choices*, 8(4):15–17 (1993).

Kuchler, Fred, Sarah Lynch, Katherine Ralston, and Laurian Unnevehr, "Changing Pesticide Policies," *Choices*, 9(2):15–19 (1994).

Manchester, Alden C. *Rearranging the Landscape: The Food Marketing Revolution, 1950–91*. Agricultural Economics Report No. 660. Washington, DC: Economic Research Service, USDA, September 1992.

McKinney, Luther C. "Workable Food Safety Regulation," *Choices*, 9(2):10–14 (1994).

McLaughlin, Edward W., and Peter Fredericks. "New Product Procurement Behavior of U.S. Supermarket Chains: Implications for Food and Agribusiness Suppliers," *Agribusiness: An International Journal*, 10(6):481–490 (1994).

Padberg, Daniel I. (ed.). *Food and Agricultural Marketing Issues for the 21st Century*. FAMC 93-1. Food and Agricultural Marketing Consortium. College Station: Texas A&M University, 1993.

Panyko, Frank. "New Directions in Grocery Retailing," *Journal of Food Distribution Research*, 26(1):98–103 (1995).

Russo, David M., and Edward W. McLaughlin. "The Year 2000: A Food Industry Forecast," *Agribusiness: An International Journal*, 8(6):493–506 (1992).

Smith, Deborah T. (ed.). *Marketing U.S. Agriculture—1988 Yearbook of Agriculture*. Washington, DC: U.S. Department of Agriculture, 1988.

Wilson, Paul N., James C. Wade, and Julie P. Leones. "The Economics of Commercializing New Crops," *Agribusiness: An International Journal*, 11(1):45–56 (1995).

22

Costs of Marketing Food and Fiber Products

All the various marketing functions between farms and retail outlets involve costs. Regardless of the channels, agencies, or persons performing the marketing tasks, there are fixed costs, such as for plants, warehouses, utilities, depreciation, and property taxes. There are also many variable or out-of-pocket costs. Possible additional costs include losses from spoilage and trimming, to name only a couple.

The purpose of this chapter is to discuss the costs of marketing, including payments or charges for the performance of all marketing steps, markups for profits, and also any services that add value to food and fiber. We discuss two approaches. The first involves using the value of an established market basket of foods as a basis of comparison and looks at the effects of changes in retail prices. The second considers how consumer food spending is distributed among major marketing functions, such as processing and retailing.

When consumers want additional services that can be provided efficiently, they will be willing to pay necessary marketing costs. Naturally, there are costs associated with value-adding services, and they are reflected in the retail price; but those services make foods more attractive and convenient to consumers, often reducing preparation time and costs. Likewise, higher wages may add to costs but also attract and retain more efficient workers. Greater efficiency may offset part or all of the increased wage costs. Most food and fiber products in their raw state are not yet ready for market in the sense of being ready to cook or wear. A long series of steps and processes are involved before raw products can be termed "marketable."

Importance of Individual Farm Commodities

The first step in the marketing process is the sale of farm commodities at the farm level. The share of the value of agricultural products sold held by major farm commodities is about as follows:

Commodity	Percentage of Total
Meat animals	28.8
Dairy products	11.3
Feed crops	11.3
Poultry and eggs	9.4
Oil crops	7.6
Vegetables	7.1
Fruit and tree nuts	5.8
Food grains	4.8
Cotton	2.9
Tobacco	2.9
All other crops	7.6
Miscellaneous livestock	1.5
Total	100.0

Livestock and livestock products, including poultry and eggs and dairy products, account for about 50 percent of the value of agricultural products sold, and field crops for about 50 percent.

Market Basket Prices[1]

To better determine why food prices change, the retail price of food can be divided into two categories: the farm value of retail food sold to consumers and the farm-to-retail price spread. By evaluating these categories separately we can determine what happened to the prices that farmers received for food commodities and what happened to the charges for marketing services. The

---

[1] Much of the material discussed in this chapter is adapted from Denis Dunham, *Food Cost Review, 1993*, Agricultural Economic Report No. 696 (Washington, DC: Economic Research Service, USDA, August 1994).

Food price components

Farm value of food products rose for the first time in 3 years, but the 1993 value was only 8 percent higher than the value a decade earlier.
1982-84=100

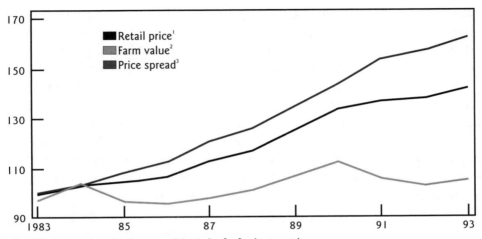

[1] Retail price based on the Consumer Price Index for food eaten at home.
[2] Farm value based on prices received by farms.
[3] Price spread represents processing and distribution charges.

Source: Economic Research Service, USDA

Figure 22-1

USDA uses a market basket concept to separate these two components of food prices. The market basket contains the average quantities of food that mainly originate on U.S. farms and are purchased for consumption at home in a base period.

Farm Value—The Market Basket Approach

Farm value, using the market basket approach, is a measure of the return, or payment, farmers received for the farm product equivalent of retail food sold to consumers. Farm value, expressed in dollars, is calculated by multiplying farm price times the quantity of farm product equivalent of food sold at retail. An allowance is made in farm value if by-products are obtained in processing. The farm value usually represents a larger quantity than the retail unit, because the foodstuffs that farmers produce lose weight through storage, processing, and distribution.

The farm product equivalent varies between foods. Only a slight amount of raw milk is lost, for example, as it is handled and processed for sale in cartons to consumers. Therefore, the farm value of a gallon of milk at the retail level is only slightly more than the price that milk producers receive for a gallon of milk. In contrast, because of shrink and dressing percentage, among other factors, nearly 2.4 pounds of live animal yield 1 pound of Choice beef at retail. The payment the cattle producer receives for that larger quantity of live animal is the gross farm value of the retail price of 1 pound of beef.

The farm value averages about 26 percent of the retail price of all foods in the market basket. Farm value has been trending downward because abundant food supplies depress farm prices while rising food processing and distributing costs increase retail prices. The farmer's share is higher for products requiring minimal processing. In contrast, the farmer's share is less for those goods that require considerable processing and handling. For example, the following products generally indicate this relationship:

Product Group	Farm Value Share of Retail Price (Percent)
Poultry	44
Meat products	40
Dairy products	34
Fresh fruits	22
Fresh vegetables	26
Processed fruits and vegetables	19
Fats and oils	22
Bakery and cereal products	7
Market basket of food products	26

The farm value share of the retail price varies greatly between commodities. Some of the factors influencing farm value share include the costs of transporting from farm to consumer, product perishability, seasonality of production, and charges for retailing. These factors help explain why the farm value share for fresh fruits and vegetables is relatively low.

Farm-to-Retail Price Spread

The farm-to-retail spread for the market basket of foods attempts to measure charges for performing services connected with a fixed quantity of foods of a constant type and quality. However, the types of services incorporated into foods sold in grocery stores have changed over time, a result of new product introductions, such as boneless meat and poultry products, and greater food preparation, such as fruits and vegetables sold at salad bars. Prices for these new and usually higher-value foods are incorporated into the market basket retail price calculations over time, thus changing the type and quality of foods in the market basket. These changes in foods marketed with added services may increase price spreads.

Distribution of Food Spending

Spending for domestically produced food represents the retail market value of food purchased by or for civilian consumers. Both the quantities of food bought and the prices paid affect spending levels. Consumers spend over $490 billion annually for food originating on U.S. farms. About 60 percent of consumers' food expenditures are made at retail grocery stores on food for use at home. The remaining 40 percent of these expenditures represent the retail value of food served at public eating places, hospitals, schools, and other institutions.

Sales data suggest that consumers are purchasing more food at public eating places. For example, expenditures for food eaten away from home increased 76 percent from 1983 to 1993, while expenditures at food stores increased only 45 percent during that period. Higher employment levels have increased household income and reduced the amount of time available for meal preparation and cleanup at home. As a result, consumers are purchasing a higher percentage of their meals at restaurants. A stronger economy encourages consumers to substitute meals in restaurants for meals at home.

Like the market basket approach, consumer food expenditures can be divided into two categories: farm share and the marketing bill. The purpose of this approach is to provide more information about the source of changes in consumer food expenditures (Figure 22–2).

Farm Value—Distribution of the Food Dollar

The farm value of the consumer food dollar is a measure of the payments farmers received for the raw commodities equivalent to food purchased by consumers at food stores and eating places. The farm value of food commodities represents about 21 percent of consumer food expenditures. The farm

Distribution of food expenditures

The marketing bill was 78 percent of 1993 food expenditures.

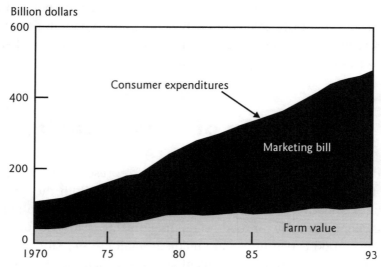

Data for foods of U.S. farm origin purchased by or for consumers for consumption both at home and away from home.

Source: Economic Research Service, USDA

Figure 22-2

value is a much smaller part of expenditures for food eaten away from home than for food bought at stores, because of the cost of preparing and serving food is a major part of the cost of food eaten away from home. The farm value accounts for only about 16 percent of expenditures for food consumed away from home, compared to 26 percent of expenditures for farm food in food stores.

Marketing Bill

The **marketing bill,** the difference in dollars between the farm value and what consumers spend for food produced on U.S. farms, is about $382 billion. As illustrated in Figure 22-2, marketing costs have been the major source of rising food expenditures over the last several decades. Detail of changes in the marketing bill can be evaluated in two ways: (1) by dividing the total marketing bill into the costs of key marketing functions, such as processing and retailing, and (2) by dividing the bill into costs of principal inputs, such as labor and packaging.

Costs of the functions performed are different for food bought in food stores than for meals and snacks purchased for consumption away from home. The farm value is about 26 cents of each dollar spent in food stores, and the marketing bill is about 74 cents. The marketing bill can be allocated as follows: processing, 33 cents; wholesaling, 10 cents; transportation, 6 cents; and retailing charges, the last 25 cents (Figure 22–3).

Marketing functions of the food dollar in 1993

Processing remained the most expensive marketing function for food eaten at home.

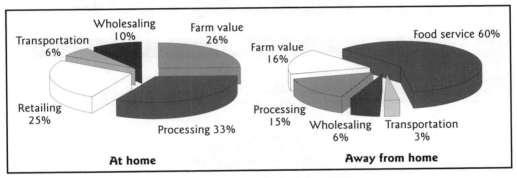

Source: Economic Research Service, USDA

Figure 22–3

For meals and snacks purchased away from home, the farm value is about 16 cents of each dollar spent, and the marketing bill is about 84 cents. The marketing bill can be itemized by function as follows: processing, 15 cents; wholesaling, 6 cents; transportation, 3 cents; and cost of food service, or preparation and serving of food eaten away from home, the remaining 60 cents (Figure 22–3).

The marketing bill can also be divided into its key specific-cost items. These include labor, packaging, transportation, depreciation, advertising, fuels and electricity, profits, rent, interest, and repairs (Figure 22–4).

Labor costs are the largest component of the marketing bill. Rising labor costs have accounted for almost half the total increase in the marketing bill over the last decade. Labor costs consist of wages and salaries; employee

What a dollar spent on food paid for in 1994

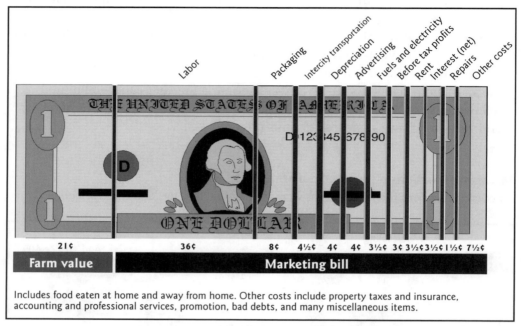

Includes food eaten at home and away from home. Other costs include property taxes and insurance, accounting and professional services, promotion, bad debts, and many miscellaneous items.

Source: Economic Research Service, USDA

Figure 22–4

benefits, such as group health insurance; estimated earnings of proprietors and family workers; and tips for food service.

Wage supplements have increased because of rising health insurance premiums and pensions. Health insurance benefits, the costs of which have been increasing rapidly because of the rising cost of medical care, have been a key issue in collective bargaining between workers and food companies.

At 8 percent of the food dollar, packaging is the second-largest component of the marketing bill. Paperboard boxes and containers account for the largest packaging cost. About 40 percent of the food industry's total packaging expenses are for paper and paperboard products, with cardboard boxes the primary containers used for shipping processed foods. Metal containers are the second-largest packaging expense, making up about 20 percent of total food packaging costs. Cans have become less important for food packaging because of the increased popularity of glass and plastic bottles, the year-round availability of fresh fruits and vegetables, and the increased use of microwavable dishes for frozen foods. Costs of plastic containers and wrapping materials account for nearly 20 percent of food packaging costs.

Transportation costs, about 4.5 percent of food expenditures, have risen due in large part to more expensive equipment and higher driver wages. The increase in trucking costs has pushed trucking rates up in most places.

The energy bill, about $17.3 billion, makes up about 3.5 percent of retail food expenditures. The energy bill includes only the costs of electricity, natural gas, and other fuels used in food processing, wholesaling, retailing, and food service. Public eating places and other food service facilities incur the greatest share, nearly 40 percent, of fuel and electricity expenses related to food marketing. These energy expenses have risen because of large growth in the away-from-home food market. Also, away-from-home food service has the highest energy cost per dollar of sales, about 3.1 percent. About 85 percent of this cost comes from the use of electricity. Energy costs of retailers are the second largest, at about 26 percent of the energy bill, also mainly for electricity. The food processing sector is responsible for another 20 percent of the total energy bill. Electric power accounts for 56 percent of food manufacturing energy costs, with natural gas making up the other 44 percent.

About 3 percent of food spending is designated as corporate profits of food industry firms. Corporate profits are influenced by internal factors, such as labor productivity, operating costs, and raw material costs, and by external factors, such as interest rates and the general condition of the economy as a whole.

Reasons for a Higher Marketing Bill

Consumer expenditures for food increased almost 200 percent from the mid-1970s to the mid-1990s. Most of that growth is reflected in the 243 percent increase in the marketing bill. The farm value share of food expenditures increased only 96 percent. A logical question to ask is, Why did the marketing bill increase so much?

The rise can be attributed to increases in the costs of the components of the marketing bill, primarily labor, the largest component. Labor costs increased 268 percent from the mid-1970s to the mid-1990s; packaging materials increased 205 percent; transportation, 151 percent; fuels and electricity, 276 percent; and corporate profits, 115 percent.

A common error when discussing what a dollar spent for food pays for is to assume that a farmer's small share of the consumer food dollar means low farm income and that a farmer's large share means high farm income. This assumption fails to consider the effect of quantity sold and the costs of a particular marketing method. Most egg producers, for example, have found that 100 percent of the consumer's dollar (i.e., deliveries to homes) generates less net income than 50 to 60 percent on a much larger volume of eggs sold to dealers.

Also, in providing value-adding services for a product, the added cost of convenience must be measured against the consumer's time in performing similar services. For example, suppose Joe and Jane Consumer want pizza for supper. They can purchase all the necessary ingredients, mix the dough to make the crust, chop the onions and peppers, brown and drain the sausage, etc., and make the pizza themselves. Another option is for them to purchase an already prepared frozen pizza and "pop" it into the oven. Depending on the value of a consumer's time, it is possible for some services (e.g., basic pizza preparation) to be performed more cheaply in a food processing plant than in the consumer's kitchen. As disposable incomes and demand for more leisure time increase, there is a greater opportunity to sell consumers such increased product services.

In contrast, the farm value share of the consumer food dollar has been declining relative to total food expenditures. Key factors, some unique to agricultural production processes, resulting in the low farm value share of the consumer food dollar include the following:

1. Most farm goods are bulky; that is, they have a relatively low value per unit of volume.

2. Most farm goods mature during a short period, thus intensifying the marketing tasks, such storing and processing.

3. Farm products are perishable, although perishability varies between commodities and by degrees.

4. Farm production is variable—there are years of abundance and of scarcity.

5. Farm products have various marketing dispositions, such as fresh, canned, and dried.

6. Farm products are often produced in relatively small quantities over a wide area and are heterogeneous. They must be assembled, sorted, and graded before final sale.

7. Consumers, being a diverse population, demand a wide variety of goods, packages, and services, including further-processed and ready-to-consume foods.

Appraising the Farmer's Share

By comparison, 36 percent of the consumer food dollar pays labor costs in the various sectors of the food industry, while the farmer's share is only 21 percent. The farmer has other production resources to compensate in addition to labor. The farmer's share of the consumer food dollar is only an estimate of the return for the farm product. This value does not provide a direct measure of the income position of farmers, nor does it provide any basis per se for judging the "fairness" of shares going to farmers or to other groups. Also, food marketing margins are not measures for appraising economic efficiency of the tasks being done. A product with high marketing margins may be handled with great efficiency, while one with low margins may be inefficiently handled or processed. Therefore, it is necessary to carefully appraise each task performed, as well as how it is performed, before drawing conclusions regarding its efficiency.

The number of marketing intermediaries needed to facilitate the movement of commodities from producers to consumers is decreasing. Thus, marketing requirements for these intermediaries and, subsequently, for producers will continue to become more specific, precise, and contractual. Farmers who are more numerous and smaller than the intermediaries dealt with will need to recognize the benefits of cooperative efforts on their part as a method of counteracting market power. Some level of government intervention may be required to protect producers and consumers from economic exploitation.

Opportunities in Marketing

Since marketing is so complex, many opportunities exist to simplify and improve the process. Some of these are (1) reducing the cost of performing necessary steps or services; (2) eliminating unnecessary steps or services; (3) developing more equitable bargaining power between parties to any given transaction; (4) improving communication of consumer desires through the marketing system to producers; (5) providing a wide variety of safe and wholesome food products to the consuming public; (6) consolidating small establishments to obtain greater economic efficiency; (7) improving marketing personnel, management, and resource use; (8) improving accounting and cost data; (9) increasing involvement in international markets; and (10) improving the gathering and dissemination of market news and information.

Retail Price Movements and Farm Level Price Movements

Why is it that retail food prices do not fall in proportion to the decline in prices farmers receive? There are many explanations for this question, among which are:

1. The competitive structure is different. Farmers compete in a more "perfect market" than retailers, the latter having some control over their prices while farmers do not.

2. The costs and margins of firms completing intermediate steps in the food production process are rather inflexible and slow to change, usually going upward.

3. Productivity per worker in processing, wholesaling, and retailing does not increase very rapidly, and higher wages simply raise the unit cost of labor.

4. Retailers, for example, prefer to keep a steady price on their merchandise rather than to mark up or mark down prices daily in line with fluctuations in farmers' daily prices.

5. Retailers and others in the marketing channel price their merchandise on the basis of an overall "target" return to stockholders' equity or invested capital. Therefore, their prices are less responsive to changes in farmers' product prices. Maximization of stockholders' returns is a primary goal, not maximization of the prices farmers receive.

6. Policies of a few companies with control over wide segments of the industry may have a greater influence on prices paid to farmers and prices charged to consumers than policies established by many smaller buyers and sellers. A company that sells in more than one product market need not sell every item at a profit. Rather, it may classify its products into such categories as profit-making items and loss leaders and may follow different policies in selling products of the different classes.

7. Because of the less elastic (more inelastic) demand for food at the farm level, the farmer's share of the consumer food dollar drops relatively more when supplies are increased. The converse happens when supplies are decreased. Food demand at retail, especially for specific food

products, is more elastic; thus, supply changes affect retailers' share relatively less.

Topics for Discussion

1. Discuss the trends of the farm value of consumers' food spending.
2. What are the principal inputs that make up the marketing bill?
3. Why is the marketing bill taking a larger share of the consumer food dollar?
4. Discuss some opportunities in marketing.

Problem Assignment

Select a farm commodity, and analyze the farm value share and the farm-to-retail spread of that commodity. Identify the source of changes in the farm value share and in the farm-to-retail spread.

Recommended Readings

Barkema, Alan, Mark Drabenstott, and Kelly Welch. "The Quiet Revolution in the U.S. Food Market," *Economic Review*, 76(3):25–41 (1991).

Brooker, John, David Eastwood, Morgan Gray, and Brian Carver. "Using Supermarket Scanner Data to Assess the Impact of Value-Added Produce on Purchases of Uncut Produce," *Journal of Food Distribution Research*, 26(1):138 (1995).

Buzby, Jean C., John T. Jones, and John M. Love. "The Farm-Retail Price Spread: The Case of Postharvest Pesticides in Fresh Grapefruit Packinghouses," *Agribusiness: An International Journal*, 10(6):521–528 (1994).

Dunham, Denis. *Food Cost Review, 1993*. Agricultural Economic Report No. 696. Washington, DC: Economic Research Service, USDA, August 1994.

Godwin, M. R., and L. L. Jones. "The Emerging Food and Fiber System," *American Journal of Agricultural Economics*, 53(5):806–815 (1971).

Kinsey, Jean. "Seven Trends Driving U.S. Food Demands," *Choices*, 7(3):26–28 (1992).

Manchester, Alden C. *Rearranging the Landscape: The Food Marketing Revolution, 1950–91*. Agricultural Economics Report No. 660. Washington, DC: Economic Research Service, USDA, September 1992.

23

Consumption of Agricultural Products

Food consumption and prices are determined by the interaction of supply and demand. In the short run, supplies of agricultural products are relatively fixed and inflexible, and prices adjust to clear the market. What is produced is consumed. When supplies are high, price goes down and consumers buy more. On the other hand, when supplies are tight, prices are high and individual purchases are smaller. Short-run changes in consumption are essentially the result of changes in supply rather than changes in tastes and preferences.

In the long run, producers respond to market signals transmitted through the system by shifting their resources to produce more of the higher-priced goods and less of lower-priced goods. Changing demographic factors, such as household size and age distribution of the population, and changing tastes and preferences can affect consumption patterns over the longer run.

The consumer, in a free market economy, is the object of many efforts and resources devoted to production and distribution of products and services. Consumers, with money in hand, essentially "hold the votes" by which goods are selected or rejected, and the market listens carefully to their votes. In the final analysis, therefore, consumers reign supreme in a free market economy. In a government-controlled economy, such as under communism, consumers are often ignored in favor of what the bureaucrats may decide consumers need. Obviously, the bureaucrats are often wrong, because consumers have varied, ever-changing, and complex needs and desires.

Over the years, U.S. consumers have evolved from being highly self-sufficient in the earlier days of the Republic to being very dependent on others. Because of the complex interrelationships that have developed in the consumer sector of the economy, consumers are largely dependent on one another for their welfare. Modern consumers are also affected by diverse influences, not the least of which is telecommunications. Television, for example, continually introduces consumers visually to new foods, clothing, housing, and other goods and services. Human wants are stimulated, creating greater expectations (perceived needs) in the minds of consumers.

Profile of the U.S. Consuming Population

The United States is primarily urban. People who live in large cities and their suburbs and in towns of at least 2,500 account for three-fourths of the total population.

The level of education is increasing in the United States. Coupled with that, the percentage of persons employed in managerial and professional positions and in technical, sales, and administrative support positions has been increasing. In contrast, the percentage employed in precision production, craft, and repair positions and as operators, laborers, and farm workers has fallen during the last decade.

Improved education and rising incomes create not only an expanding demand for goods and services, but they also allow consumers to be more selective and have more discerning tastes and preferences. Rising incomes increase expenditures on more further-processed and therefore more expensive foods, as consumers demand more convenience and quality.

Each year the U.S. population grows by between 2 and 3 million people. The annual growth rate in the early 1990s was about 1.1 percent, slightly higher than the average annual growth rate of 1 percent observed through the 1970s and 1980s. The population, as a whole, is growing older. The median age increased from 28 years in 1970 to 33.1 years in 1991. In the last few years, the largest increase has occurred in the 25 to 39 age group. The elderly population has also been increasing, a reflection of improving medical technology and health care.

Consumer wants are changing largely because (1) incomes have risen; (2) more households have two wage earners; (3) work structures permit more leisure time; (4) eating-out expenditures increase as leisure and travel increase and as the number of dual-wage-earning households increases; (5) knowledge about nutrition has expanded; and (6) new products offer wider variety and more convenient preparation. These changes will likely continue in the years ahead, causing even more modifications in diets and family living and spending.

Consumer Expenditures for Food and Fiber Products

The breakdown of consumer personal consumption expenditures by general category is given in Table 23–1. If we use food, tobacco, clothing and shoes, and furniture as estimates of the consumption of food and fiber products, we can see that expenditures for those products amount to about 22 percent of disposable income. However, there are other uses of agricultural products not reflected in this estimate. These include chemical and industrial uses, lumber and other wood products used for construction, and many paper products.

Table 23–1

Personal Consumption Expenditures, by Type, as a Percentage of Disposable Income, 1980 and 1992

Expenditures	1980 % of Pers. Cons. Expend.	1980 % of Disp. Income	1992 % of Pers. Cons. Expend.	1992 % of Disp. Income
Food	19.5	17.5	15.3	14.1
For off-premise consumption	13.8	12.4	10.1	9.2
Purchased meals and beverages	5.3	4.8	4.9	4.5
Other	0.4	0.3	0.3	0.4
Tobacco products	1.2	1.1	1.2	1.1
Clothing and shoes	7.5	6.7	6.8	6.3
Housing	14.6	13.1	14.5	13.3
Household operation	12.2	10.9	10.5	9.7
Furniture	1.2	1.1	1.0	0.9
Medical care	11.9	10.6	17.0	15.7
Personal business	5.8	5.2	8.6	7.9
Transportation	13.5	12.1	11.2	10.3
Recreation	6.7	6.0	7.7	7.1
Education and research	1.9	1.7	2.4	2.2
Religious and welfare activities	2.2	2.0	2.8	2.6
Other	1.8	1.5	1.0	0.8
Totals	100.0	89.5	100.0	92.0

Source: U.S. Bureau of the Census, Statistical Abstract of the United States, 114th ed., Washington, DC, 1994.

Consumer expenditures for food can be divided among major food groups about as follows:

Item	Percentage of Total
Meat	28.3
Fruits and vegetables	23.7
Dairy products	13.3
Bakery products	10.8
Poultry	6.8
Grain mill products	3.7
Eggs	1.2
Other foods (includes fats and oils, sugar, tree nuts, peanuts, and miscellaneous foods)	12.2
Total	100.0

Trends in Food and Fiber Expenditures

Over the past several decades, the percentage of disposable income spent for food and clothing by U.S. consumers has fallen slowly. This is because consumers are spending relatively less of their higher incomes on food and clothing and more on other goods and services. For example, in 1970, consumers spent about 20 percent of their disposable income for food. That amount had dropped to just over 15 percent by 1991. Also, the retail price per unit for food and clothing has not risen as rapidly as for other goods and services. Although a smaller proportion of consumers' disposable income is now expended for food and fiber than formerly, aggregate dollar expenditures have risen and doubtless will continue to rise. For comparative purposes, it is noted that consumers now spend relatively more of their income for medical care, housing, recreation, personal care, and education than for food and clothing.

While no two wage earners spend their money in exactly the same way, there are predictable patterns of expenditures for families in various income categories. Studies have shown that families with low incomes spend most or all of their money for necessities—food, shelter, and clothing. Families with high incomes spend more total dollars for these necessities. They buy more expensive food, larger homes, and more clothes and services. They eat out more and consume less bulky carbohydrates and more high-protein foods, fruits, and vegetables. Those with higher incomes probably "waste" more food than those with lower incomes. Higher-income families spend a smaller proportion of their total income on food than do lower-income families.

A nineteenth century Prussian statistician, Ernest Engel, was the first to describe this behavior pattern in what is now known as "Engel's Law of Consumption." This law states that as family incomes rise, the proportion spent for food declines. Or, as incomes rise, food purchases increase but not as fast as income. If incomes fall, the proportion spent for food increases.

Not all food products are affected in the same way by changing income levels. Consumption of some foods increases as incomes rise. These items are called **superior** foods, which has nothing to do with whether they are nutritionally good. The term merely indicates how a change in income affects consumption. Other foods are termed **inferior,** which means that consumption decreases as incomes rise. Those foods whose consumption is not affected by changes in income are termed **neutral.**

Since family disposable income has been trending upward for a good many years, there is a growing proportion of the population in the middle-income group. This change means that those who produce and sell inferior foods find it increasingly difficult to expand the market for their products. Of course, there are other factors that affect the market for food and fiber. One of these is the location of consumers—whether they are in urban or rural areas or in warm or cold climates.

Family size and composition by age and sex are additional factors. The educational level of consumers is still another factor, although it is often correlated with income.

National origins and religion are also significant. For example, Jewish and Catholic days of obligation affect the demand for particular foods.

Farm Family Spending

Farm families increasingly spend their money in approximation of the ways urban families spend theirs. For example, farm families are producing less of their own foodstuffs. Instead, they are buying more groceries from stores. Rural consumers want and obtain household appliances, electronics, and furnishings that urban consumers demand. Farm children have educational and recreational pursuits similar to those of urban children. Increasingly, the farm family budget provides for life insurance, health insurance, medical care, recreation allowances, educational materials, and a variety of personal services. Farm families now spend increasingly more on housing than before. Expenditures for transportation by farm families are also on the rise.

Profile of Food Consumption

Assuming each person consumes three meals a day, a U.S. population of around 258.2 million requires a total of 774.6 million meals daily, or 282.7 billion meals annually. On a weight basis, the average American consumes about 1,500 pounds of food per year. This equates to a total of some 387.3 billion pounds for the entire U.S. population. These data indicate the tremendous size of the food sector and the enormous responsibility that rests upon food producers, processors, and marketers in the United States.

U.S. consumers choose their diets from a national food supply adequate to meet the recommended allowances of the National Research Council. However, the nutritive value of the food supply has changed in several respects over the years. The per capita food energy level of food consumed is about 3,700 calories daily, compared to 3,300 calories in 1970 and 3,530 in 1910. Sources of the energy from food have also changed:

Sources of Calories	1910	1970	1990
	--------- (Percent) ---------		
Carbohydrates	57	60	63
Fat	32	25	23
Protein	11	15	14

The total calories and the percentage of calories from carbohydrates in diets have increased, while the kinds of carbohydrates consumed have changed. In general, more consumers are eating healthier diets, with a higher percentage of complex carbohydrates and a lower percentage of fats.

The kinds of fats in the food supply have also shifted significantly, with the consumption of animal fats largely replaced by the consumption of vegetable fats. Fewer calories, relatively, are derived from fat consumption than in the past.

Per capita consumption of beef has declined from its peak of 88.8 pounds (boneless, trimmed equivalent) in 1976 to about 63 pounds now. Per capita egg consumption has also been declining, from 310 eggs per person in the early 1970s to about 233 now. Per capita pork consumption has remained fairly steady at just under 50 pounds annually. In contrast, per capita poultry (chicken and turkey) consumption, currently about 61 pounds, has been increasing steadily.

In recent years the consumption of whole and 2 percent milk has been declining, while the consumption of 1 percent and skim milk has been increasing. Ice cream consumption has fallen slightly, as consumers substitute frozen yogurt. Total cheese consumption has increased slowly. Mozzarella cheese has shown the greatest increase in per capita consumption because of its use as a key ingredient on pizza.

Per capita consumption of fresh fruits and vegetables has been increasing at a slightly faster rate than consumption of processed fruits and vegetables.

Per capita consumption of flour and cereal products has been increasing steadily since the early 1970s. Much of the increase can be attributed to rising per capita consumption of durum flour, used in making pasta and spaghetti, and of rice and oat products.

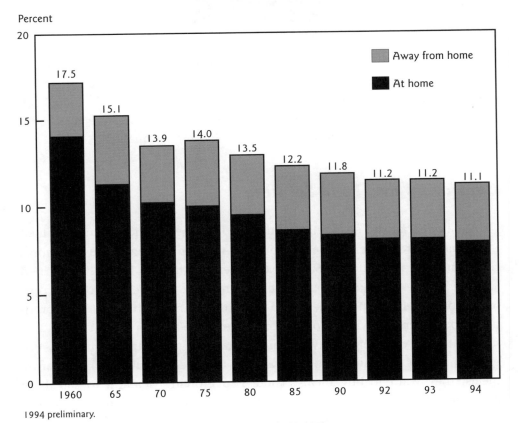

Figure 23-1

Food expenditures as a percentage of total private consumption expenditures, selected countries, 1990

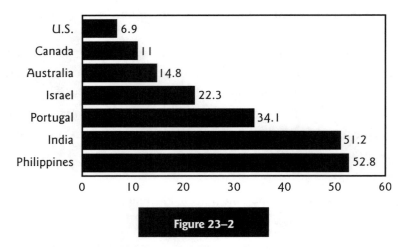

Figure 23-2

American consumers probably get the best buy in the world in food and fibers. Most people of the world spend from one-third to one-half their disposable income for food. Americans, who spend only about 7 percent, on average, of their private consumption expenditures for food, receive a better diet with greater variety, more convenience, and higher quality.

Consumers and Food Services

A major social and economic trend is the increased participation of women in the labor force. Overall, the participation rate for women in the labor force rose from 34.8 percent in 1960 to 56.6 percent in the late 1980s. For married women 35 to 44 years old, the rate went from 36.2 to 72.7 percent. Not surprisingly, convenience is now one of the most important attributes of food products. Consumers are willing to pay for convenience and service.

Tight time constraints have created a rapidly growing convenience-oriented industry. Consumers do not want to purchase ingredients for preparing meals; they want to buy meals. There is a growing demand for meals in containers that can go from the freezer to the microwave to the garbage, eliminating much cooking and cleanup.

But someone somewhere between the farm and the retail store has provided all the services necessary for convenience—the peeling, washing, cooking, and packaging. These services add somewhat to the cost of most convenience foods, compared with the cost of unprepared foods. However, some convenience foods are actually cheaper than the fresh forms because they are more compact for shipment and storage and have a longer shelf life. For example, frozen and canned peas are usually less expensive than fresh peas.

It is likely that if "value" or "price" were attached to a person's work in the kitchen (i.e., opportunity cost), most convenience foods would be cheaper to serve than those bought unprepared. This is an important consideration for consumers with better alternative uses for their labor.

Food Spending According to Family Cycle

Food and fiber spending levels are not necessarily constant. For example, when a young couple marry, their total food budget is likely quite modest. With the arrival of the first child, baby food expenditures are encountered. The food budget increases sharply as the child reaches adolescence. With other additions to the family, the cycle is repeated. After all the children leave home, the middle-aged parents cut the food budget considerably and reduce their own food consumption for dietary reasons. As the parents become more elderly, food use declines still more as physical activity lessens, but the inclusion of nutritious foods in the diet should not be reduced.

Therefore, food budgeting over a life cycle is dynamic, although plateaus in food expenditures are sustained over long periods of time, especially before arrival of children and after they leave.

In addition, food expenditures per capita decline as family size increases. It is generally cheaper per person to feed large families than small families. For example, for two-person households, weekly food expenditures are about $39.00 per person; for three-person households, $35.10 per person; for four-person households, $32.70 per person; and for five-person households, $28.20 per person. Reasons for this include:

1. There are economies of size in buying larger quantities of food and less waste in preparing meals for more people.

2. Most larger families include young children, who generally consume less food than adults.

3. Budget limitations—with no added income, a family can afford to buy only so much more food.

Table 23–2

Weekly Food Cost, Moderate Cost Plan, for Different Types of Families, 1990 and 1993

Type of Family	1990	1993
Couple, 20 to 50 years old	$ 74.70	$ 78.30
Couple with children:		
One child, 1 to 5 years old	93.00	97.50
One child, 6 to 14 years old	102.50	107.60
One child, 15 to 19 years old	108.00	113.20
Two children, 1 to 5 years old	108.20	113.40
Two children, 6 to 11 years old	130.10	136.40
Two children, 12 to 19 years old	136.70	143.30
Couple, 51 years old an over	71.80	75.30

Factors Surrounding Consumer Decision Making

Prices

In the food and fiber marketing chain, consumers are the last link. As such, they represent the final demand for food and fiber products. Other links in the chain represent the intermediate steps, or what economists call "derived demand." Since overall consumer demand for food and fiber products is rather stable, except for growth in population, the supplies coming to market greatly affect prices at the consumer level. Large supplies tend to move prices sharply down, while short supplies have the opposite effect.

However, consumer demand for food and fiber must not be viewed only as a lump sum. Individual products that make up consumer demand differ widely in their price response to supply and demand conditions. Demand for broilers, whose supplies have been expanding rapidly in recent years, responds greatly to price decreases, while demand for table eggs does not. This indicates differing price elasticities of demand for broilers and table eggs.

Retail prices for selected livestock products, 1970 to 1993

Prices expressed as a percent of 1970 prices.

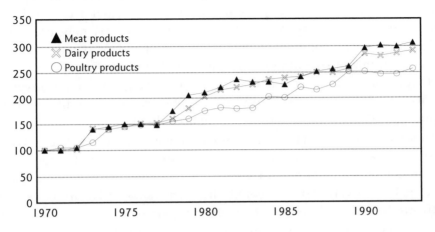

Source: U.S. Department of Agriculture

Figure 23–3

Retail prices for selected crop products, 1970 to 1993

Prices expressed as a percent of 1970 prices.

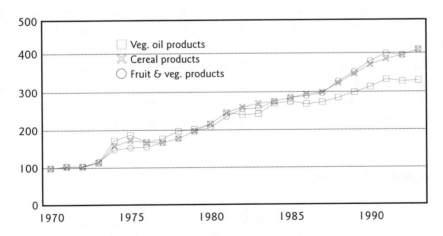

Source: U.S. Department of Agriculture

Figure 23–4

There is another aspect of demand that warrants mention, however. This is a situation where demand for a product increases or decreases independent of price changes. For example, a shift in consumer demand to the right, an increase in demand, may result from several factors: (1) new uses for old products, (2) new products, (3) increased use of old or new products, (4) increased income or spending power, particularly for low-income groups who would purchase more food if they had the money, and/or (5) advertising appeals, which might cause consumers to shift some expenditures to food products, away from other items in the household budget.

Another important factor to consumers is marketing costs. These costs are rather sticky and inflexible. Hence, if retail prices drop, the farmer gets less because marketing costs remain somewhat constant. If retail prices rise, the farmer gets relatively more. Theoretically, farmers could ensure better prices for themselves if they could act together to successfully limit the quantity supplied to consumers. Realistically, for most commodities this is an impossible task because there are so many producers and they are scattered widely throughout the country. Threats of imports and synthetic products also negate the wisdom of such an undertaking. Furthermore, actions to limit supplies to raise prices would need extension to all food and fiber products because of the substitution of one product for another.

One possibility for farmers to raise prices is to practice price discrimination by separating the demand for their product between elastic and inelastic responses. For milk, as an example, producers could restrain the sale of fluid whole milk, which has an inelastic demand, and profit by doing so, while lowering the price of other milk going into elastic uses, such as cheese and butter. By separating the markets and allocating milk accordingly, total net profits from milk are enhanced.

Reducing the marketing spread by improving market efficiency is another possible source of increased profits to farmers, provided they can obtain a share of the savings in marketing costs. Farmers may accomplish this by organizing cooperatives or by bargaining with processors, for example.

Consumer Price Index

A **price index** is a ratio showing the price of a basket of goods and services in various years in relation to the price of the basket in a base year. The indicator of price changes most important to the general public is the Consumer Price Index (CPI). This index, calculated by the Bureau of Labor Statistics of the U.S. Department of Labor, measures cost of living for a typical urban family. It is based on the prices of about 400 items selected to represent all goods and services purchased. Prices for food, fuels, and certain other goods and services are obtained every month in cities chosen to represent all urban places in the United States. Prices of most other commodities and services are collected every month in the five largest cities and every three months in the other cities.

In building the Index, a weighted average of price changes for all items in each urban location is calculated first. The weight for any item or group of items is the portion of total spending, as determined by periodic consumer surveys, used for the purchase that item or group. For example, suppose that the average urban family spends a total of $3,000 during a one-month period, of which $600 is spent on food. Then, food would have a weight of 0.20 (600/3,000 = 0.20) in the CPI. The data from all locations are combined, using the populations of the locations as weights. Finally, the CPI is computed so as to express current prices as a percentage of average prices prevailing during some base period.

Other Aspects of Consumer Decision Making

Consumers are motivated in their purchasing decisions not only by price but also by habit. Once the habits are established, they are rather difficult to change. Even as consumers contemplate the purchase of a new product, "old" habits form a frame of reference within which decisions are made. Of course, as income levels and residence change and community influences become structured, these "old" habits are modified somewhat. Economic wants and needs change too in response to family size and age, occupation(s), etc. Advances in communication technology, especially with respect to advertising media, have also served to modify or change tastes, preferences, and consumption habits. Easy access to consumer credit, through widespread credit card use, has also resulted in significant change in consumer spending habits.

Modern-day consumers are becoming more educated, sophisticated, and complex individuals. Since most are considerably above the poverty level, they have more discretion in buying and may view a product not only in terms of economic factors but also in terms of social, psychological, and cultural ones. It is no longer a question of buying soap just to clean away dirt, but one of buying it also to beautify and to gain social acceptance. Thus, consumers' goals or objectives may change over time from only need-satisfaction to ego-satisfaction as well.

It is likely that consumers base purchasing decisions not only on economic factors, such as price, but also on emotional or social factors, such as keeping up with the neighbors. The "facts" that consumers may consider when making decisions can just as often be "wrong" as "right." This is especially true with regard to nutrition, where consumers are, on the whole, poorly informed, often believing erroneously that certain foods have nutritive properties that they do not have. To become a "wise" consumer requires certainly more education, study, and training than in years past. Add to this the fact that consumers have many more purchasing decisions and a much wider array of alternative goods and services than ever before. And, the modern consumer is less concerned with food and relatively more concerned with other consumer goods and services.

While modern communication media do offer more alternatives and make consumers more aware of them, the information is generally so fragmented and often so vague that consumers have difficulty synthesizing these many fragments of information into an understandable whole. Thus, it is questionable whether decision making is greatly enhanced at all. Also, some of this information may be incorrect, misleading, or misinterpreted, consequently providing a faulty base on which to make decisions.

Economics of Food Buying

There are many misconceptions about food buying and the value of food. Simply because a food item is expensive does not ensure that it is also of high quality or highly nutritious. For example, cooked oatmeal, which is quite nutritious, costs far less per serving than widely advertised cold cereals. However, because consumers are often pressed for time in the morning, they choose a quickly eaten bowl of cold cereal as a typical breakfast, if they have breakfast at all. Also, price specials and so-called bargains are not always bargains. Whether a food represents a bargain is dependent on many factors:

1. *Is it offered at a conveniently located store?* Money saved on a food item at a store some distance away might be less than the transportation cost (i.e., the cost of gasoline) incurred getting it. Even if bargains represent a real saving in money, a busy consumer may find shopping around too costly in time.

2. *Will the family eat and enjoy it?* No food is a bargain if the family will not eat it. However, creativity and persistence on the part of the meal plan-

ner may turn a disliked food into a family favorite. A different method of preparation or special care in serving may change attitudes.

3. *Is it packaged in a quantity that meets family needs?* Large cans and packages may represent a saving over small containers. If, however, a large container means leftovers that are eventually discarded, it is no bargain.

4. *Can it be properly stored at home until used?* Very large quantities—a quarter of beef, a bushel of apples, a case of green beans—can often be purchased at low per unit cost. If such items can be stored properly to prevent spoilage and are not in such large quantities that the family will tire of them before they are consumed, they represent a real saving.

5. *Does the consumer have time and skill to prepare it?* Few are interested in preparing all foods from "scratch," even if money is saved. Making bread at home, for example, may be too time-consuming to be worth the pennies saved. In response to increasing demand for convenience, stores are offering more and more foods that are prepared or partially prepared. The cost of this preparation sometimes, but not always, increases the selling price of the food item. Consumers who know how much more they pay for frozen French fries than for potatoes to be prepared at home are better able to decide if the time saved is worth the extra cost.

6. *How does its cost compare with that of other foods of similar food value?* The money-wise shopper knows which groups of foods go together to make up a good diet and economizes by selecting best buys from each of these groups.

Some consumer guides to wise buying of food and fiber products are the following:

1. *Plan purchases.* Purchases that are planned—by developing a menu for a several-week period, for example—are usually wiser than are impulse purchases. Examination of food ads also helps in planning purchases.

2. *Compare prices.* Prices among competing brands, for example, should each be converted into price per unit, such as per ounce, pound, quart, etc., to ascertain the true cost per given quantity, everything else being equal. This technique provides consumers a more rational basis in purchasing.

3. *Examine labels*. Familiarity with labels, grades, etc., will help consumers to purchase more wisely than if they base their buying decision on appearance alone.

4. *Use house brands*. In many cases, house brands or those owned or controlled by the retailer may be better buys than nationally advertised brands. In many cases, house brands are manufactured and packaged by national food companies and sold to retailers at a lower price.

5. *Products in season*. For some commodities, prices are lower when the bulk of the supply comes to the market. Buying certain products out-of-season will cost more. Consumer information agencies have a regular advisory in most newspapers that indicates what foods are in plentiful supply and lower priced.

Food Purchasing Decisions

Consumers may approach the problem of purchasing food in one or a combination of several ways, depending on their objectives. For example, a *maximization* objective in terms of maximizing *calories* purchased for a given sum of money is one method. The consumer tries to obtain the maximum number of calories by purchasing potatoes, beans, rice, spaghetti, etc. Low-income consumers often follow this method.

The opposite of maximization is *minimization*. That is, the consumer tries to purchase a given number of *calories* at *least cost*. This is often referred to as the *least-cost* method of food buying. Institutions such as hospitals and nursing homes use this method. Menu planning in institutions of all kinds is typically done using appropriate computer programs. Computerized meal planning begins by ascertaining the food preferences of the consumers to be served. A food list is prepared indicating the number of times a food is preferred during the week, for example. Next, the caloric content and nutritive elements of each food are calculated and costs per unit of serving are estimated. Subsequently, the amount budgeted for food is entered into the computer, together with certain restrictions on menu plans, such as minimum calories and minimum vitamin, mineral, and protein levels. The computer then provides an acceptable menu at least cost. As prices of food products change, the computer is instructed to calculate new sets of least-cost but acceptable menus. Care is exercised to produce menus that provide variety and palatability, among other considerations.

Another *minimization* objective might be minimizing *preparation time*: consumers decide to purchase a given amount of food that requires the least amount of preparation time. This method is especially attractive to dual wage-earning families, employed individuals, or those who, because of high incomes, prefer to spend a minimum amount of time in the kitchen. Minimizing preparation time necessarily requires a larger expenditure for a given number of calories because much of the kitchen work is already built into the food product itself.

Domestic Food Assistance Programs

Federal and state governments, through their domestic food assistance programs, influence food consumption, especially among lower-income persons. A few of those food assistance programs are discussed briefly in this section.

The U.S. Department of Agriculture administers a number of federal food assistance programs through the Food and Nutrition Service (FNS). Most of these programs are directed at low-income Americans. Included are the Food Stamp Program; the National School Lunch Program; the School Breakfast Program; the Special Supplemental Food Program for Women, Infants, and Children (WIC); the Nutrition Program for the Elderly; and the Food Distribution Program on Indian Reservations. Many of the programs are designed to give children access to a more nutritious diet, with the goal of improving their eating habits through nutrition education, and to encourage consumption of domestically produced foods.

The Food Stamp Program, which was begun as a pilot program in 1961 and then made permanent in 1964, issues monthly allotments of coupons that are redeemable at participating retail food stores. Increasingly, the paper coupons, or food stamps, are being replaced with Electronic Benefit Transfer (EDT), usually an on-line system in which participants use magnetic-strip cards (similar to debit cards) to access their food stamp accounts at the point of sale. The motivation behind the new electronic system is to eliminate the printing of coupons. This, in turn, is expected to significantly reduce street trafficking and fraud. The Food Stamp Program serves over 27 million persons per month. Average monthly benefits are about $68 per person.

The National School Lunch Program is a federally assisted meal program operating in about 93,000 public and nonprofit schools and residential care institutions. It provides nutritionally balanced, low-cost or free lunches to about 25 million children each day. Children from families with incomes at or

USDA costs for food assistance

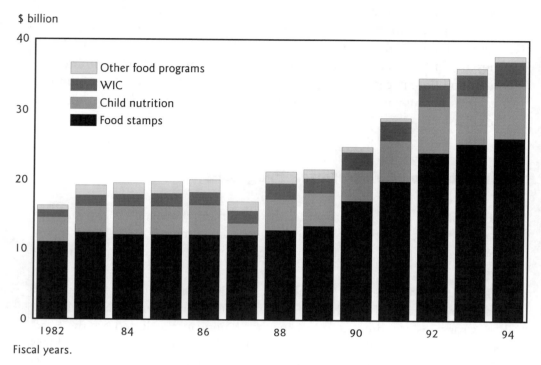

Source: U.S. Department of Agriculture

Figure 23-5

below 130 percent of poverty ($18,655 for a family of four) are eligible to receive free meals. Children from families with incomes between 130 and 185 percent of poverty ($26,548 for a family of four) are eligible to receive reduced-price meals. Children from families with incomes above 185 percent of poverty pay the full, locally established price.

The School Breakfast Program grants cash assistance to states to provide nonprofit breakfast programs in eligible schools and residential child-care institutions. The programs operate in almost 55,000 schools and institutions, serving a daily average of 5.7 million children. Eligibility requirements are the same as for the National School Lunch Program.

The WIC Program is a grant program whose goal is to improve the health of pregnant, breast-feeding, and non-breast-feeding postpartum women and of infants and children up to five years old by providing supplemental foods, nutrition education, and access to health care. Each month, more than 5 million WIC participants receive vouchers, redeemable at retail food stores, that allow them to purchase a monthly food package specially designed to supplement their diets. The foods provided are high in protein, calcium, iron, and vitamins A and C.

Consumer Demand for Fibers

Aside from the food needs of the population, clothing and shelter are also derived from agriculture. U.S. fiber consumption represents a substantial proportion of the world's total fiber consumption. For example, in the United States, 4.6 billion pounds of cotton, or 18 pounds per person, are used annually. Also, over 150 million pounds of apparel and carpet wool are used.

Per capita consumption of lumber is about 240 board feet. About 82 percent is from softwoods, and the remainder from hardwoods. New housing construction is the primary end-use of lumber. Building an 1,800-square-foot home uses about 10,000 board feet of lumber. Per capita consumption of paper and paper products is about 680 pounds each year.

As is true for food, the demand for fibers is closely related to population, incomes, style trends, and standard of living. For example, a large increase in spare time, travel, and recreation causes greater demand for casual and sport clothes but less demand for more formal attire. The increasing proportion of both young children and elderly people in the U.S. population causes more demand for children's clothing and for less-formal attire for the elderly.

Development of Substitute Products

The development of synthetic food and fiber products is progressing rapidly. Synthetic fibers have taken over some of the natural fiber markets. The growing use of synthetic fibers, such as nylon, orlon, dacron and others, has reduced the demand for natural fibers. Powdered milk, artificial citrus juices, and imitation meat products have also been developed. Synthetic sweeteners are commonly substituted for sugar, and synthetic leathers for hides and skins. Other examples could be cited.

Consumers may readily accept synthetic food and fiber products if price, quality, and other characteristics are competitive with those of natural agricultural products. With improved technology, synthetic products may continue to infringe upon traditional natural food and fiber markets. However, in some cases synthetic foods and fibers are combinations of natural and synthetic products. In other cases, one agricultural product becomes a replacement for another, such as soybean oil for cottonseed oil.

What are the reasons for the rapid rise of synthetics? (1) Supplies of natural products are not always dependable and consistent. (2) The prices of synthetics generally have declined with improved technology and lower production cost. (3) Synthetics are produced under industrial-type conditions, which can be better controlled. (4) The quality of synthetic products has improved. (5) Research and development expenditures on synthetic products are massive. (6) Synthetic products sometimes create new and expanding markets in response to consumer needs. (7) Large firms with greater marketing power than agricultural producers are often the developers and marketers of synthetic products.

Consumer Protection

Consumers in the United States are protected from adulterated and poor-quality foods and fibers by a series of city, state, and federal regulatory and inspection laws. Food safety regulation is much more complex today than it was in the past, requiring a broad range of controls. Herbicide and pesticide residues, irradiation, *E. coli* bacteria, toxicity of additives and supplements, and sanitary conditions of processing facilities are examples of areas addressed by current regulations.

Local and state governments often have laws that were passed with the goal of protecting consumers. These may include checking of scales to insure proper weights and measures; health checkups of food establishment workers; sanitation regulations over food processing and dispensing establishments; inspection of meats, milk, and certain other foods for wholesomeness; and water and air pollution regulations over businesses such as paper mills, feed mills, and fertilizer plants. Municipal inspection may include surveys and examinations of supply warehouses; retail stores; delivery trucks; refrigeration systems; and restaurants, hotel dining rooms, and other public eating places.

At the federal level, a host of consumer protection laws are in effect. Consumers benefit in many ways from the services and safeguards provided by the U.S. government. Some services are direct, such as inspection of meat and poultry for wholesomeness. The government inspects the cleanliness and sanitation of plants in which meat, poultry, and other products are processed and checks the honesty of labels as well. USDA stamps and labels on food inform the shopper.

The Pure Food and Drug Act and the Meat Inspection Act were passed in 1906, establishing that the federal government has a responsibility and the legal right to ensure a safe and wholesome food supply. These two laws gave the Department of Agriculture broad jurisdiction over food in interstate commerce. The Pure Food and Drug Act was designed to ensure better, purer food for consumers by addressing a wide range of problems, including filth, decomposition, and adulteration of food. It also protects consumers by forbidding misbranding, mislabeling, and the use of misleading containers and requires that the weight, measure, or count be specified and the ingredients listed. The Meat Inspection Act set up a federal meat inspection system.

Other laws that have had an effect on food safety include the 1960 Color Additives Amendment, which regulates coloring additives; the Fair Packaging and Labeling Act of 1966, which regulates size, contents, shapes, and labels of packages; the Egg Products Inspection Act of 1970, which regulates egg processing plants; and the 1971 Freedom of Information Act, which opens government actions to public scrutiny.[1] The Nutrition Labeling and Education Act, passed in 1990, requires mandatory nutritional labeling on nearly all processed foods regulated by the Food and Drug Administration.

In addition to food protection, consumers are safeguarded in textile purchases by the Wool, Fur, and Textile Fiber Products Acts. Also, protection of household furniture purchases and regulation of furniture advertising come under the trade practice rules of the Federal Trade Commission.

While these and other laws are designed to provide consumer protection, they can be costly to implement. Often they are not funded adequately or administered properly. Consumer protection laws that have insufficient funds for enforcement fall short of their intended objective.

Consumer Problems

Consumers are not without problems, even in an affluent society. Because generally they have discretionary income, consumers attract greater sales ef-

---

[1] Ben Senauer, Elaine Asp, and Jean Kinsey, *Food Trends and the Changing Consumer* (St. Paul, MN: Eagan Press, 1991), p. 243.

forts in the form of fancier packages, advertising, frills, and merchandising gimmicks. Accurate labeling and grading of consumer products are another problem.

Consumers, in general, are neither articulate nor organized, whereas producers and merchants combine with others to further their interests, making their demands effective through both the price system and legislation. As individuals, consumers may loudly protest against price or quality; but collectively, they do very little except occasionally achieve some limited success with localized boycotts against alleged high retail prices.

The organization of local, state, and national consumer information councils, such as the Consumers Union and the Consumer Federation of America, appears to provide consumers with a more active voice in economic affairs. What these councils particularly need, however, are more sources of accurate, unbiased information about consumer goods. The federal government could, if it chose to do so, release vast amounts of data it has collected on the quality and reliability of consumer products, such as appliances, clothing, medical products, etc. The government could also sponsor more research on consumer products and disseminate the results widely.

Many consumers also want more products sold at retail to be quoted on a price-per-unit basis, such as per ounce, pound, etc. This would facilitate comparison shopping and place greater emphasis on price competition. One disadvantage is the added labor cost of performing this task in the retail stores.

It is important also to realize that consumers possess no yardstick with which to measure their wants or the efficiency of their "plant." The home is too small a unit to operate with anything like the same efficiency as that of an industrial plant. In a factory, machines may operate 24 hours a day, while in the home, a piece of equipment—the vacuum cleaner or the washing machine—may be used but a few minutes each day. Under such limitations, the household cannot hope to have precise guides and measurements to assist in the study of consumption as does a factory.

Consumers are faced with so many choices over such a wide range of products and prices that misallocation of resources is likely in many households. Therefore, the need for accurate, timely, unbiased information on foods and fibers, among other consumer items, is critical. In one shopping test, 33 college-trained consumers tried to select the least expensive of 20 common items in a familiar supermarket. They were wrong in 43 percent of their decisions, and their miscalculations cost them 9 percent more than they needed to pay.

Several steps seem appropriate in achieving more and better consumer information:

1. Provide better consumer education for youth in high schools. Although most high schools offer some consumer education, it is often inadequate.
2. Devote more public resources to consumer education. By and large, most public resources are devoted to producer interests rather than to consumer interests.
3. Provide for adequate grading and labeling of foods and fibers at the retail level, where consumer decisions are actually made.
4. Provide recognition of consumer interests at various levels of government, from the city council to the federal government.
5. Increase consumer-interest communications, such as expanded food sections in newspapers and food programs on radio and television, among other techniques.

Consumer Trends

Several trends in food and fiber consumption, together with consumer behavior, are apparent:

1. The desire to spend less time in the kitchen is likely to continue. The demand for convenience will remain strong.
2. As they become more affluent and the proportion of income needed for food falls, American consumers will become more concerned about taste, convenience, variety, safety, and nutrition.
3. Americans are consuming an increasingly larger share of their meals outside the home. Among other factors, this results from families with both parents working who eat out for the convenience and from those with more leisure time—the retired, for example—who eat out to enjoy the slower pace.
4. Diversity will increase in importance. Ideas about what is good, healthy, or affordable vary among consumers. Varied life styles require creativity in food presentation and delivery.[2]

---

[2] Ibid., Ch. 3.

5. Consumers have less allegiance to conventional food and fiber products and are more apt to buy synthetically produced items.

Labeling

Ingredient and nutrition labeling, which was controversial in the 1970s, is again widely debated. Food companies argue that such regulations would be too costly, while consumer advocates argue that they would help consumers make better-informed and nutritionally sound food choices.

Beginning January 1, 1975, food manufacturers became responsible for proper nutritional labeling if their products were enriched or fortified or if the manufacturers made nutritional claims. Otherwise, nutrition labeling was voluntary.

The Food and Drug Administration introduced new regulations in 1990 that generally make labeling mandatory. Nutrition labeling is required on every packaged food product that is a "meaningful source of calories."

The new regulations require information be included on levels of cholesterol, saturated fat, and dietary fiber per serving and also on the percentage of calories from fat. With these four additions, the need to provide nutritional information on thiamin, riboflavin, niacin, and one of the two protein listings has been dropped. The purpose is to give more attention to nutrients and other food components that have been linked to chronic diseases and less to those associated with particular nutritional deficiencies that are increasingly rare in the United States.[3]

Consumer Group Action

Consumers, many of whom lack expertise in buying, purchase goods and services mostly on an individual basis. There are several alternatives available that may assist consumers in purchasing goods and services more economically: (1) consumer information periodicals that provide technical information and may grade and rate products by consumer tests and evaluation; (2) consumer buying clubs or food cooperatives that organize consumers into group

---

[3] Ibid., p. 170.

purchasing units through which purchases of fresh produce and other items are often made at wholesale prices; (3) consumer cooperatives, in which several hundred to several thousand consumers organize themselves formally to conduct a retail store operation by cooperative principles mentioned in an earlier chapter; (4) class action lawsuits, in which one or more consumers, on behalf of themselves and others having the same grounds, institute legal proceedings against manufacturers or sellers to redress an injury or grievance; (5) government information agencies, such as the Cooperative Extension Service, which provide various types of information on consumer products and services and on purchasing; (6) government regulatory agencies that initiate and/or receive complaints against businesses relative to faulty or misrepresented products and services.

Topics for Discussion

1. Present a profile of the U.S. consuming population.
2. Discuss the trends in food and fiber expenditures.
3. Present a profile of food consumption.
4. How are food prices determined at the consumer level?
5. Discuss the principal factors in consumer decision making and food purchasing decisions.
6. Discuss the nature of federal and state food assistance programs.
7. Discuss some of the laws designed to protect consumers.

Problem Assignment

Prepare a budget for a typical family of four on both a monthly and an annual basis, including necessary minimum expenditures for food, clothing, shelter, and other consumption needs.

Recommended Readings

Adrian, John, and R. Daniel. "Impact of Socioeconomic Factors on Consumption of Selected Food Nutrients in the U.S.," *American Journal of Agricultural Economics,* 58(1):32–38 (1976).

Barkema, Alan. "Reaching Consumers in the Twenty-First Century: The Short Way Around the Barn," *American Journal of Agricultural Economics,* 75(5):1126–1131 (1993).

Barkema, Alan, Mark Drabenstott, and Kelly Welch. "The Quiet Revolution in the U.S. Food Market," *Economic Review,* 76(3):25–41 (1991).

Kinsey, Jean. "Changing Societal Demands: Consumerism." In *Food and Agricultural Marketing Issues for the 21st Century,* Daniel I. Padberg (ed.). FAMC 93-1. Food and Agricultural Marketing Consortium. College Station: Texas A&M University, 1993.

Lutz, Steven M., David M. Smallwood, James R. Blaylock, and Mary Y. Hama. *Changes in Food Consumption and Expenditures in American Households During the 1980s.* Statistical Bulletin No. 849. Washington, DC: Economic Research Service, USDA, December 1992.

Manchester, Alden C. *Rearranging the Landscape: The Food Marketing Revolution, 1950–91.* Agricultural Economics Report No. 660. Washington, DC: Economic Research Service, USDA, September 1992.

Meier, Kenneth J., and E. Thomas Garman. *Regulation and Consumer Protection,* 2nd ed. Houston, TX: Dame Publications, Inc. 1995. Ch. 10, 11, 17.

Putnam, Judith J., and Jane E. Allshouse. *Food Consumption, Prices, and Expenditures, 1970–1993.* Statistical Bulletin No. 915. Washington, DC: Economic Research Service, USDA, December 1994.

Senauer, Ben, Elaine Asp, and Jean Kinsey. *Food Trends and the Changing Consumer.* St. Paul, MN: Eagan Press, 1991.

U.S. Department of Agriculture. *Nutrition: Eating for Good Health.* Agriculture Information Bulletin No. 685. 1992.

24

Economic Setting for U.S. Agricultural Policy

Much of the discussion about economic theory to this point has focused on the decision making of firms and individuals in a free market environment. However, there are no truly unhindered free markets. At a fundamental level, the government must establish and enforce property rights before markets can exist. Thus, there is at least some government involvement in the marketplace. The extent of the government's intervention is guided by established public policy. The agricultural sector is certainly no exception, although the extent of the involvement is expected to decrease in the near future.

Agricultural policy is a course of action designed to achieve certain objectives in the food and fiber economy. The purpose of this chapter is to introduce agricultural policy and to set the stage for making farm business decisions that take agricultural policy into account.

In the next chapter, we will examine specific farm policy goals and actions. Here we describe the general policy-making process, provide a brief history of early U.S. agricultural policy, and then discuss key characteristics and trends in the agricultural sector that influence the objectives of agricultural policy. The economic setting must be understood before agricultural policy can be effectively used to achieve established goals.

Public Policy

The purpose of government is to conduct the public's business. A wide variety of areas and issues—civil liberties, civil rights, economics, welfare, international affairs, security and defense, etc.—make up public business. Individuals and groups concerned with each of these areas generate ideas about what should be done to achieve desired results in a given area. These ideas are based on ideological preferences, self-interest, and assumptions about what is expected to happen if certain actions are taken. This collection of ideas and assumptions forms the basis of public policy.

Policy can be defined as a definite course of action, given current conditions, to guide and determine present and future decisions in the accomplishment of agreed-upon goals and objectives. **Public policy,** then, defines a course of action or guiding principle for governments to follow to achieve stated goals or solve problems of public concern. Public policies involve the laws, rules, or regulations designed to move the course of action from a conceptual stage to implementation.

The Policy-Making Process

The process of making public policy in a democracy involves answering these questions: (1) What is the problem? (2) What is the "best" solution to the problem? (3) How should a policy based on this solution be implemented? (4) How should the policy be evaluated?

Identifying a problem that must be addressed involves a number of steps. There must be a felt need, perhaps arising from dissatisfaction with how things are or from a vision of how to improve some aspect of society. This felt need is then defined and transformed into a goal by an individual, a legislator, or an interest group. The next step is cultivating public awareness—educating the public about the merits of this goal and its benefits in addressing the identified felt need. Following public awareness is public acceptance—general agreement that there is a need for action.

After identifying a problem comes deciding how it should be handled. This involves determining the importance of reaching the goal, based on values, and then analyzing alternative policy plans for addressing the problem. The analysis should involve determining expected costs and benefits of the proposed policies. A variety of individuals and groups are involved at this point. Based on the analysis, the "best" plan can be written into a bill, introduced into the legislative process, and enacted.

After the policy has been drafted and enacted, it must be implemented, or carried out. Often, there are many layers of bureaucracy to penetrate in the implementation process. Different agencies must agree that the project satisfies certain requirements before it can proceed through the process. These various checkpoints represent the successful efforts of various interests over time to be included in the policy-making process.

The policy-making process also includes determining ways to evaluate the policy's effectiveness in accomplishing its stated purpose. To evaluate a policy

The policy-making process

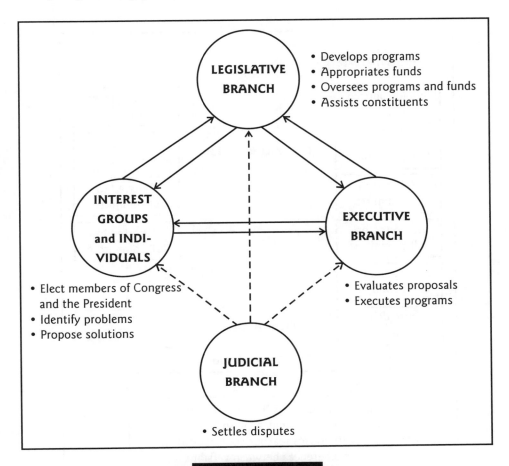

Figure 24-1

properly, there must be a well-defined set of criteria to use in measuring results. The more precise and specific these criteria are, the more useful is the evaluation. Therefore, a policy should always clearly state its goals and expected results. This does not always occur. Often, vague and general wording is used to accommodate a variety of interest groups. When this happens, useful evaluation is difficult.

Once a policy has been decided, enacted, implemented, and evaluated, there is typically mixed public reaction to it. Those whom it was designed to help may be dissatisfied with it and want it revised. Others may want it repealed, perhaps because it is imposing an excessive burden. Public reaction

The legislative process: How a bill becomes law

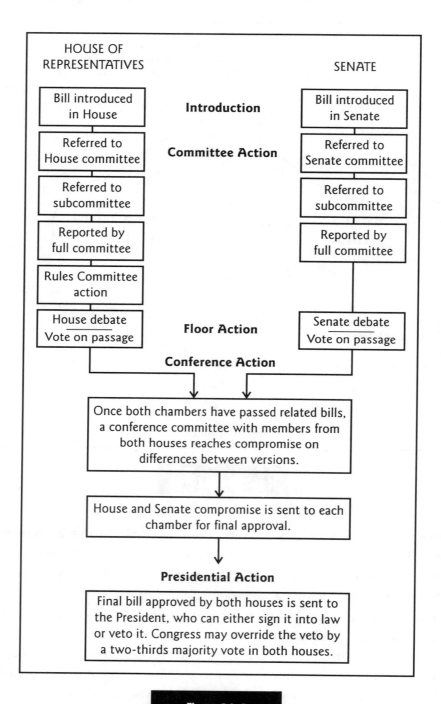

Figure 24-2

usually identifies new problems and new felt needs, and the process starts all over again.

Understanding Government Intervention in the Agricultural Sector

A variety of factors provide the political motivation for and public acceptance of government involvement in the otherwise free market agricultural sector. We discuss some of these in this section, starting with a brief historical synopsis of the government's role in the agricultural sector prior to the Great Depression. The history of American agriculture, including the development of agricultural policy, was presented in more detail in Chapter 2.

The Government's Policy Role Prior to the Great Depression Era

Starting with the first colonial settlements in the early 1600s and for the following 300 years, government policy regarding agriculture primarily dealt with land settlement and development. After the War for Independence, the federal government eventually obtained ownership of most unsettled land in the continental United States. Maximum individual freedom and economic liberty were major tenets adhered to in developing this frontier territory in the expanding nation. Therefore, establishing private property rights and developing a land distribution system became vitally important concerns of the federal government.

After independence, the new federal government continued the colonial policy of allocating frontier land to all settlers who would work it. This policy of free land applied to families of settlers, whether poor immigrants or wealthy citizens, and helped initiate the American tradition of the small, self-sufficient family farm. It also reinforced the policy of private property rights, with the nation's agricultural land put under the complete control and ownership of individuals.

The *Homestead Act of 1862* standardized the land allocation process by giving each family of settlers 160 acres, provided the family lived on the land for five years. Eight other homestead acts followed the 1862 Act and increased the acreage allotment when it became evident that 160 acres were not adequate to support a family in most areas. The Homestead Act actually played only a small role in the disposal of public lands in the late 1800s. Most farmers and ranchers bought their land from speculators.[1]

Many other government actions and policies directly benefited agriculture and enhanced its development. Improvements in transportation had an enormous effect on agricultural trade, making it possible for farmers to sell their crops across the nation and throughout world. The federal policies that encouraged and subsidized the building of railroads tied the nation together and opened vast areas to agriculture. With railroads, livestock and commodities could be efficiently transported long distances to rapidly expanding urban areas.

Federal irrigation projects and policies made many areas west of the Mississippi River agreeable to farming. In these dry western areas, individual farmers did not have the resources to fund the large-scale irrigation projects that were necessary for farming to succeed, and the states were unwilling to fund them. The *Reclamation Act of 1902* provided funding to construct dams and other water-control facilities. Originally, each farm owner was allocated water sufficient for 160 acres of land, but over time and due to political pressure and lack of enforcement, the acreage quota was increased. Thus, federal irrigation policy allowed farmers to purchase heavily subsidized water for prices far below the government's actual production costs.

The Great Depression, not felt in the general economy until the early 1930s, actually hit the agricultural sector in the 1920s. The years before and during World War I were prosperous for agriculture. Exports were very strong, especially during the war. However, farmers' economic well-being began to deteriorate in 1920, when the United States became a creditor instead of a debtor nation. The level of exports (specifically, agricultural exports) fell sharply.

With the lower foreign demand, farm commodity prices also fell. Unfortunately, the prices farmers paid for the goods and services they needed did not fall nearly as much. This created a painful squeeze for farmers that worsened throughout the 1920s and was a major contributor to the Depression.

The Depression was the worst period of economic and social upheaval ever experienced by rural America. Throughout the 1920s farm organizations and others advised farmers to begin voluntarily cutting back production to raise prices. When these efforts failed, farmers attempted to create more formal marketing cooperatives for staple crops. After these also failed, farmers began turning to the government for help in solving the farm problem.

---

[1] Adapted from Willard W. Cochrane, *The Development of American Agriculture: A Historical Analysis,* 2nd ed. (Minneapolis: University of Minnesota Press, 1993), pp. 80–84.

Characteristics of Production Agriculture

Resources in agriculture are relatively immobile. The **fixed asset theory** explains that producers cannot respond quickly to changes in the marketplace because they are set up to produce only one good or service in the short run. In agriculture, there are often few alternative uses for land and capital goods. Land, the principal resource, is often kept in production as long as it returns at least enough to cover variable production expenses.

Labor, although more flexible than land, has tended to be "trapped" in farming, since employment off the farm has often been difficult to obtain. Historically, farm labor has typically been provided by the operator and his or her family. Often, these individuals felt they lacked necessary skills for other endeavors. Many authorities argue that labor immobility is much less an issue in agriculture today than it was in the past.

Similarly, capital goods, such as machinery, equipment, specialized facilities, etc., often have few alternative uses. For example, milking machines and mechanical cotton pickers are not easily converted to other uses. Because of the relative immobility of land, labor, and capital goods, these resources tend to remain in agriculture even though the returns on them may be very low, especially when compared to the returns on these resources in other industries.

Agricultural production is a biological process; thus, it is inherently unpredictable and highly variable. Farmers are largely at the mercy of the weather, plant and animal diseases, and insect pests. The result is sharply variable production rates and prices. Because farmers, to some extent, react to price fluctuations in their planning, they often make the variability worse. For example, if the prices of corn are relatively high, corn production will likely rise, increasing the quantity supplied. This, in turn, drives prices down.

Because of the inherent variability in agricultural production, both producers and consumers have traditionally supported government intervention to moderate the market and reduce its natural instability. The purpose has been to restrict supply so as to achieve the desired price, stabilizing farm income.

Farmers, in general, are price-takers. The agricultural sector of the U.S. economy, compared to any other sector, is closest to being a purely competitive market. What farmers do in total affects the market. However, in contrast to participants in imperfect markets with price-making firms that can influence market prices, individual farmers have no control over the market prices of their commodities. In fact, farmers have incentive to produce as much as possible, given their capital and resource constraints, as long as price exceeds average variable costs. Collectively, this strategy reduces the price they receive, because the aggregate quantity supplied will increase.

Demand for Agricultural Products

There is an unresponsive, or "price-inelastic," demand for food (Figure 24–3). Since the 1930s and especially since the 1950s, aggregate food consumption in the United States has not been very responsive to price. This means that if the price of food declines, the amount consumed does not increase significantly. Conversely, if food prices rise, consumers tend to purchase about the same amount of food by sacrificing something else rather than reducing their food intake.

The effects of shifts in supply when demand is inelastic

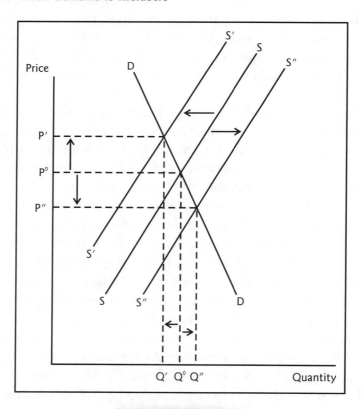

Figure 24–3

Although the overall responsiveness to price changes has been small, the degree of responsiveness to price varies among food types and commodities. Basic foods exhibit the least change (their demand is more inelastic), while foods considered luxuries change the most (their demand is more elastic).

The observed inelastic price demand for food in the United States has important implications for farm policy. For example, a reduction in the quantity supplied leads to a greater proportional increase in price and thus farm income. This helps explain why the federal government felt it could intervene in the market to artificially raise prices by reducing food and fiber supplies (which were usually in surplus, anyway). Some argue that this was simply attempting to treat a symptom of the deeper problem—too many resources in the agricultural sector.

There is slow growth in demand for food in the United States. Most modern industrial nations, like the United States, can be classified as "food-mature" economies, in which most of the population is relatively well fed and clothed. As incomes rise in such economies, only a small portion of the increase will be spent on additional food. Although there may be large shifts in the types of food purchased as prices change, the income elasticity of demand for food is low in these nations.

In modern, industrial nations, most of the additional income spent on food is not on quantity but on services and value-added factors relating to food. Food-mature nations also exhibit a slow population growth and thus a slow increase in the demand for food.

Supply of Agricultural Products

There has been rapid technological change and productivity growth in agriculture. During most of the twentieth century and especially since the 1930s, there has been a huge increase in the productivity of the resources used in farming. Total production per unit of input (land, labor, and capital) has increased dramatically. The result has been a food supply that has grown, for the most part, faster than food demand, making surpluses and depressed prices a chronic problem for U.S. farmers.

Cochrane's *Treadmill Theory*[2] explains a problem farmers encounter with rapid technological change and the accompanying productivity growth. Cochrane assumes that farmers confront an inelastic demand curve in the short

---

[2] Ibid., pp. 427–429.

Cochrane's Treadmill Theory illustrated

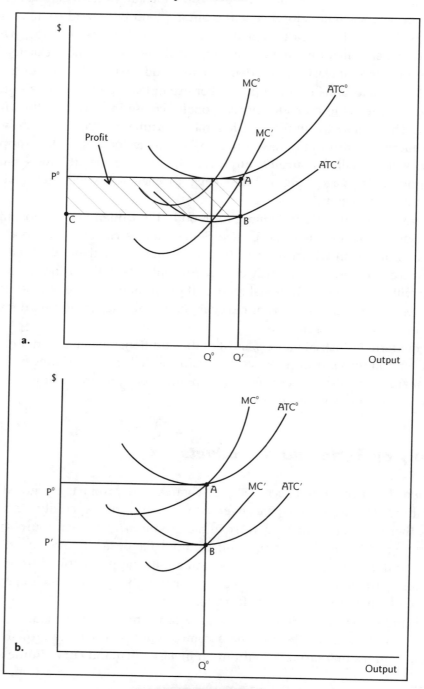

Figure 24–4

run. The adoption of new technology by an innovative farm causes its average total cost, the total cost per unit of output, to go down, from ATC^0 to ATC' in Figure 24-4a. The quantity produced increases from Q^0 to Q', since the farm can produce more at less total cost per unit. Since individual farmers are unable to influence the market, the early adopters are able to make an economic profit by selling their commodities at the market or supported price. The shaded block, P^0ABC, represents the innovator's profit. This economic profit is possible because the product price lags behind the technology-induced increase in output.

As more and more farmers adopt the new technology, supply increases, causing the price to fall. As the price begins to fall, other farmers also adopt the technology in an effort to lower their per unit costs, causing the price to fall still further, making it more difficult for those who do not implement the new technology and reducing profit for the early adopters.

Eventually the market will reach a new equilibrium, at an output price where marginal costs equal marginal revenue at minimum average total cost (P' in Figure 24-4b), given widespread adoption of the new technology. Each firm produces $0Q^0$ for sale at P'. As was the case before the new technology was introduced, no firm captures economic profits in the marketplace. Each firm just covers its total production costs, including operator labor and management.

The combined effects of these key characteristics of production agriculture and the supply of and demand for agricultural products supported the public's view of agriculture as unstable, risky, and unprofitable. Since no farmer acting alone could alter these basic characteristics or raise prices by cutting production, many felt that the federal government needed to step in to control production and alleviate these problems.

Societal Values and Beliefs

In addition to the basic characteristics of agriculture that made federal assistance sound reasonable, many in society felt that there was something unique and special about farming that warranted government intervention. This involved the emotional and philosophical aspects of farms as rural, small, family-owned and operated enterprises. Farming was viewed not merely as a profession and as the basic occupation of humankind but also as a way of life that should be preserved.

Those with this viewpoint considered farming morally superior to other work. This constitutes the basis of **"agrarian fundamentalism"** or **"Jeffersonian agrarianism." Although this philosophy has not been clearly articulated on a public scale, it has been a dominant feature of American culture since the early days of the nation. Thomas Jefferson, who popularized this concept, felt that a nation of small landowning farmers offered the best protection for upholding the ideals of a democratic society. These family farmers/landowners were seen as honest and hard-working individuals who had a vested interest in the success of the nation since they owned a piece of the nation. In a 1785 letter to John Jay, Jefferson wrote:

> Cultivators of the earth are the most valuable citizens. They are the most vigorous, the most independent, the most virtuous, and they are tied to their country and wedded to its liberty and interests by the most lasting bonds.[3]

The concept of agrarian ideology can be summed up in three basic ideas: (1) agriculture is the basic occupation of humankind; (2) rural life is morally superior to urban life; and (3) a nation of small independent farmers is the proper basis for a democratic society.

Many also felt that farming should continue to be an enterprise in which families owned the land they tilled. Ideally, according to the American paradigm of rugged individualism and personal freedom, a farmer should be his or her own boss. These feelings and the Jeffersonian ideal remained a powerful force in the 1930s, when most of the population still lived in rural areas and either directly or indirectly was connected to agriculture. They provided justification to attempt rescuing the small family farms through federal farm programs.

Social Conditions in Rural America

As we mentioned previously, export markets for U.S. grain became severely depressed after World War I and remained so and even worsened through the Great Depression of the 1930s. Because farming was the mainstay for most rural areas, most farmers were poor in this depressed environment. The problem was so large that addressing it became a high priority for the federal government during the 1930s. Intervention to provide financial assistance to farmers was seen as the most practical way to deal with the problem.

---

[3] Thomas Jefferson, *Notes on the State of Virginia*, 1785, p. 818.

Economic and Political Feasibility of Assisting Agriculture

Over time, the percentage of the total population involved in production agriculture has become smaller and smaller. Also, production agriculture has become a smaller part of the economy. As these trends progressed, providing assistance to farmers was viewed as less burdensome on nonfarm taxpayers and more feasible for the federal government. Also, many citizens were not long removed from agriculture (they either grew up on a farm or still had relatives on a farm) and thus viewed farming sympathetically. Within this political environment, legislation providing assistance was politically easy to pass, and soon such assistance became institutionalized as a permanent governmental function. This institutionalization gave rise to a new constituent base, which thus has made it politically difficult to discontinue federal assistance programs.

Social Contract

Another justification for special financial assistance to farmers was the concept of the **social contract.** The idea of a social contract goes back to early political philosophers, such as Voltaire. Relative to farm support programs dating to the Depression Era, the social contract is an implicit contract between society and agriculture. The key elements of this contract define how these two groups are to interact in the political process or marketplace.

Within the concept of a social contract, society made a commitment to protect farm incomes and prices with the *Agricultural Adjustment Act of 1933*. Agriculture, in turn, made a commitment to provide an abundant supply of high-quality and low-cost food and fiber for society. Society agreed to provide price support guarantees that would be set in the "political marketplace" instead of the "economic marketplace." Farmers, by means of farm organizations, legislators, and congressional committees, would basically exercise sovereignty over this political marketplace and would make farm policy.

The social contract continues to have much influence, as is evidenced by the existence of farm programs today. However, serious compromise with the urban majority has been necessary for several decades to ensure the continuation of this arrangement. Today, the agricultural social contract is being reevaluated as public perception of agriculture changes and as budgetary pressures force Congress to eliminate or sharply reduce many government programs.

When farm support programs were put into place prior to World War II, society viewed some outcomes of a market economy as not desirable. Accordingly, many economic, social, and political justifications for implementing agricultural price and income policies were offered. A primary justification was the assurance of an adequate food supply. With prices subsidized or set above the market equilibrium price, farmers would be encouraged to produce a greater quantity. With surplus quantities, the United States could minimize its dependence on food imports and expand agricultural exports. Additional benefits of price supports included enhancing the political stability and economic viability of rural communities and businesses, stabilizing agricultural prices, and encouraging resources to remain in agriculture that likely would have been shifted to other uses if free market mechanisms prevailed.

An Overview of the Agricultural Sector

In the previous section, we discussed several factors that influenced the development and acceptance of government intervention in the agricultural sector. In this section, we will briefly discuss selected characteristics of the agricultural and agribusiness sectors as they relate to agricultural policy today.

The Structure of the Farm Sector

There are only about 2 million farms in the United States. The number of farms has fallen steadily from its peak of 6.8 million in 1935. The greatest decline in farm numbers came in the 1950s and 1960s, due in large part to the joint effects of the rapid adoption of labor-saving technology, which depressed returns to excess labor in farming, and the rapid growth in nonfarm job employment opportunities. The number of farms is projected to be just under 2 million in the year 2000.

Total cropland, which includes land harvested, abandoned, or idled under various government programs, has remained relatively stable since 1920. The total of cropland, grassland pasture, and range has also been stable over the last 80 years.

With total land in farms remaining stable and farm numbers declining, the average size of farms in the United States has increased significantly, from 175 acres in 1940 to 473 in 1993. Most of the increase in average farm size has come from expansion by existing farmers. Typically, when land is sold, it is bought by active farmers who live within a few miles of the land purchased.

However, averages can be misleading. There are many small farms in the U.S. farming sector that account for very little of total farm production, while the relatively few large farm operations account for a large part of farm production. For example, in 1992, just under 12.1 percent of all farms marketed 75 percent of the total value of agricultural products sold. These farms, which averaged 1,804 acres, controlled about 44 percent of the land in farms. In contrast, over 59 percent of all farms in the United States had less than $20,000 in total sales of agricultural products. These smaller farms, which averaged 160 acres, accounted for only 4 percent of total sales of agricultural products.

Resource Mix

Farmers have experienced considerable substitution of capital for both land and labor over the last several decades. This trend is expected to continue, considering the structural changes the farm sector is experiencing.

Compared to the situation in the 1960s and 1970s, a greater share of farm production resources is now rented or leased rather than fully owned. Currently, the part-owner is the most common type of agricultural producer. Young persons entering farming must rent or lease resources because full ownership of resources will be difficult to attain.

Farmers depend more on purchased inputs than raised and, thus, have become more commercially oriented. Technological innovations, such as genetically engineered crop seed and precision farming technology, are becoming more widespread. Distribution channels between factory and farm have been shortened. Services to farmers, especially agricultural chemical application, have increased as farm supply centers have become more vertically integrated through either corporations or cooperatives. Farm input suppliers provide farmers with a wider range of supplies, more credit funds, and more managerial and marketing assistance.

Labor Utilization

Total labor used in agriculture today is about one-fourth of what it was in the late 1940s. As labor prices rose because of a growing economy and attractive nonfarm opportunities, farmers substituted capital in the form of machinery and chemicals for labor. This substitution, along with improved management and higher yields, has dramatically reduced the labor required per unit of farm output. For example, labor required to produce a bushel of corn in 1950–54 averaged just over 20 minutes. By 1986, only 1.5 minutes of labor were required per bushel of corn.

While total labor inputs per farm are not expected to change drastically, the total labor utilized in the agricultural industry will continue to decline as farm numbers decline. However, this will likely occur at a much slower pace, since both the increases in mechanization and the downward trend in farm numbers have slowed. The percentage of labor inputs as proportion of total inputs will likely continue to decline.

Technology

Technological developments, relating to both farm production and the farm supply and processing industries, have major implications for structural changes within the farm firm. Substitution of capital for land and labor will continue, and more reliance will be placed on new research findings regarding yields, breeds, and control of weeds, insects, and diseases.

Computer adoption has increased efficiency on many progressive farm operations. Computers are used for many routine tasks of farm management and record keeping, and they can enhance decision making and planning on most farms. A specific example of improved decision-making capabilities through computer use is the adoption of precision farming technology. Computers also have drastically improved telecommunication services.

Perhaps the greatest technological advance will be associated with biotechnology. The development of agricultural biotechnology offers the opportunity to increase crop production, lower farming costs, improve food quality and safety, and enhance environmental quality. Biotechnology techniques can be used to increase a plant's ability to resist disease and pests, to tolerate environmental stress, and to enhance food qualities, such as flavor, texture, shelf life, and nutritional content. Biotechnology can be used for animals to diagnose disease, promote growth, and develop vaccines. Other uses include increasing food processing efficiency and developing more effective diagnosing techniques for testing food safety and environmental quality.

Profit Margins

All of these changes have led to smaller profit margins per unit of production for farm producers, which increases pressure on their decision-making processes. While total profits per farm, in general, are larger as a result of volume production, the profit margin per unit of production is smaller. This means that farmers are more vulnerable to production and price risks, unless price risks are hedged or shared with other firms under contractual arrangements or supported by the government.

Farmer Decision Making

The method of making decisions on the farm continues to become more complex. The quantity and quality of information required to make good business decisions has increased with the adoption of new and better technology. For example, farmers must consider much more carefully than ever the environmental effects of their production decisions. The timeliness and accuracy of making decisions is critical—the survival of the firm depends on it.

As operation size has increased, farm firms have taken on more characteristics of nonfarm firms. As these changes have evolved, the nonfarm population has become less sympathetic to the claim that farming is a way of life and thus needs government support.

Vertical Integration

The increasing use of purchased inputs and the more stringent demands of the marketing system have encouraged more vertical integration as a means of supplying both capital and management to farm firms. Contract buying and selling will continue to have significant effects on farm marketing procedures and prices. Vertical integration and contract farming arrangements will continue to be extended to include many more crops and types of livestock. Products with which production contracts or vertical integration accounts for over half of all production include fresh and processed vegetables, citrus fruit, potatoes, sugar, seed crops, eggs, fluid milk, broilers, and turkeys.[4]

Farmer Cooperatives

While the number of farmer cooperatives may decline in the future, their memberships and business volume should at least remain steady, providing them a portion of the total business. Individual cooperatives have become much larger business units, operating over wider trading areas. Also, cooperatives that "bargain" for farmers in buying inputs and selling outputs are more important than in the past.

Bargaining Power

Associated with changes in farm numbers and farm size are changes in the bargaining power of farmers. Farmers have been traditionally price-takers for both inputs and outputs. Pure-competition theory has been appropriate for determining optimal situations from a maximum profit point of view. As farmers gain influence on price (collectively or individually), whether through volume discounts on inputs, premiums for quantity sales of outputs, or supply manipulation, some form of imperfect-competition theory of profit maximization becomes a more relevant tool. With fewer producers and a larger volume per producer (both in terms of outputs and inputs), farmers can more easily act collectively, or even individually in some cases, to influence both price and product supply. The capacity for individual farmers or groups of farmers to influence market prices varies by commodity, with specialized commodities, such as cranberries, having the most potential for success.

Changes in Processing and Marketing

Processing and marketing are undergoing significant changes as the agribusiness sector positions itself in an increasingly global marketplace. Process-

---

[4] Mark Drabenstott, "Agricultural Industrialization: Implications for Economic Development and Public Policy," *Journal of Agricultural and Applied Economics*, 27(1):13–20 (1995), p. 15.

ing outlets are fewer but larger. Marketing costs continue to increase, causing the farm-to-retail price spread to widen. This means that the farmer's share of the food dollar continues to decline. The development of convenience and synthetic foods, quality control, and product innovations have accelerated in response to consumer demand. Processors are buying more directly from producers or groups of producers, using more specification contracts. Assemblers of farm produce are declining in number and importance. Interregional competition will be keener among most farm enterprises and agribusinesses.

Markets are becoming generally more concentrated, with fewer but larger firms. Survival of smaller firms may require banding together under some type of cooperative or coordinated structure. Many of these smaller firms have the opportunity to serve niche markets too small for the large firms to service. Product differentiation and promotion by firms is increasing. International trade of farm products, both exports and imports, will increase, given trade agreements the United States has entered into, such as the North American Free Trade Agreement (NAFTA) and the General Agreement on Tariffs and Trade (GATT) (see Chapter 29 for further discussion).

Wholesaling

Concentration is also occurring at the wholesale level. More wholesalers are attempting to control or influence retail outlets in some manner. Some wholesalers are vertically integrating back into manufacturing to offset manufacturer-operated sales branches. Agents and brokers are fewer and larger, representing more product lines. In general, wholesaling is being squeezed on both sides of the distribution chain—by manufacturers and by retailers. Wholesalers tend to offset this by increased physical and economic efficiencies through larger, automated, centrally operated facilities. Specialty-line wholesalers serving other than food retailers are increasing in importance.

Retailing

Corporate chain and affiliated grocery stores will continue to grow, while unaffiliated stores will continue to decline. Convenience and food discount stores will make further gains. Eating places and institutional markets will become more important outlets as more consumers travel and eat out.

Nonfood agribusiness retailing will increase at a faster rate than food retailing because of the greater amount of discretionary income available to U.S. consumers.

In general, retailers are adopting new technology and improvements in their physical handling and distribution of goods.

Consumers

The U.S. population continues to become more concentrated in urban, and especially suburban, areas. The number of white-collar workers continues to increase relative to the number of blue-collar workers. The number of service-related jobs is increasing relative to the number of manufacturing jobs. Per capita personal income in the United States, which equaled $20,864 in 1993, has been increasing over recent decades, as has the amount of leisure time. A better-educated and more affluent society demands more sophisticated products and recreational, cultural, and service preferences. The number of women in the work force continues to increase. As incomes rise, proportionately less is spent on staple foods, but spending for convenience foods increases. The proportion of the food dollar spent on food purchased away from home has also been increasing.

Consumer desires and preferences are receiving more attention from agribusiness and government. New life styles, changing demographics, and a growing appreciation for the link between diet and health are leading to fundamental change in the way Americans eat and the foods they buy. Perhaps the most important implication will be the shift from a mass food marketing approach to more of a niche marketing approach. Consumers will increasingly demand customized products, each aimed at a separate food market niche.[5]

Consumer protection laws have been increasing in number and scope as the whole economy becomes more consumer oriented. Regulatory agencies have become more important in supervising and regulating agribusiness firms, particularly food processing, packaging, and retailing.

Changes in the Rural Economy

Living standards of commercial farm families approximate those of nonfarm families. Rural communities are becoming more attractive residential areas for both farm and nonfarm people. Rural housing, education, services,

---

[5] Ibid., p. 14.

communication, and health facilities are more similar to those of urban areas than before. Farm people tend to hold values and goals similar to nonfarm people.

Goals of Present and Future National Farm Policy

Earlier, we defined agricultural policy as "a course of action designed to achieve certain objectives in the food and fiber economy." It encompasses policies not only for producers but also for the input sector and for processors, distributors, and consumers of food and fiber.

Goals for U.S. agricultural policy, as observed in the past and anticipated for the future, include:

1. To maintain a profitable, viable, efficient, and environmentally safe agricultural production sector capable of meeting food and fiber demands while providing satisfactory incomes to farmers for use of land, labor, capital, and management.
2. To provide for an efficient, profitable, and dynamic agribusiness sector, including input suppliers and handlers of agricultural output.
3. To provide consumers with an abundant, varied, and safe supply of food and fiber at the lowest possible cost consistent with the preceding goals.
4. To conduct a food and fiber economy within the framework of a democratic society, relying on the free market system as much as possible, consistent with all preceding goals.
5. To maintain and enhance competitiveness of U.S. agricultural products in an increasingly global market.

Each of these goals is examined subsequently in greater detail.

1. Policy for the Farming Sector

Current U.S. farm policy consists largely of (1) direct payments to farmers to maintain and stabilize farm income, (2) commodity loans, (3) cropland retirement to reduce food and fiber production and remove fragile lands from production, and (4) expansion of food and fiber demand through distribution of surplus commodities and concessional sales both at home and abroad. In addition to the commodity programs, there are various other farm programs relating to research, education, marketing, environmental preservation, regulatory activities, farm credit, international food activities, and national security.

Beginning in the 1970s, farm policy generally sought to shift from artificially high price support levels to levels consistent with world market prices. In addition, direct payments from the U.S. Treasury to farmers became more widely used as a method of supporting farm incomes. Also, voluntary-type programs replaced compulsory ones. By "voluntary" it is meant that if farmers choose to comply with the programs, they receive the full measure of benefits, but that if they refuse, they obtain fewer benefits or none at all. Voluntary programs have been oriented toward payments to farmers for voluntarily diverting acreage from certain crops.

Disadvantages to a voluntary approach include the high cost to the federal Treasury of enticing participation and then paying some of the most efficient producers not to produce. Another problem is that such programs are of little benefit to small farmers and other needy rural people who have minimal impact on the total supply of a commodity and receive little of the economic benefit.

There are some major weaknesses in the many farm policies that have been used. In general, because payments are made on the basis of volume, large farms have received most of the money spent on farm programs. For example, according to the 1992 Census of Agriculture, for farms receiving government payments, the average payment received was $9,152. However, only 23 percent of the farms receiving payments received 75 percent of government payments, averaging $20,100 per farm.

Landowners have also benefited from higher land values partly as a result of capitalizing the value of program base acreages and allotments into the value of their land. Therefore, landowners rather than farm operators have been the ultimate beneficiaries of support. Also, when high price support programs were in effect, U.S. farmers lost export markets, which encouraged expanded output in other countries. The potential consequences of government involvement in the agricultural sector are discussed more in the next chapter.

2. Policy for the Agribusiness Sector

Historically, the agribusiness sector was generally ignored in the formulating and implementing of agricultural policy. The reasons for this are many and diverse but in the main are (1) the agribusiness sector commanded relatively few votes; (2) society generally held the view that this sector could take care of itself, since agribusinesses are generally price-makers, not price-takers, as are farmers; and (3) fixed costs of many agribusinesses, unlike those of most farm operations, were a smaller part of their total costs of operation, and hence their resources were relatively more mobile. In the last 25 years, however, agribusiness has become a major player.

In the future, the agribusiness sector will be increasingly involved in the policy debate and formulation process. In fact, a coalition of many large agribusiness firms played a powerful behind-the-scenes role in the shaping of farm policy in 1995–96.

A major driving force behind increased agribusiness involvement in the agricultural policy arena stems from the potential effects on agribusiness profitability. The extensive production control program of 1983, which drastically affected demand for production inputs to the detriment of many input suppliers, is an example. In this program, called "payment-in-kind (PIK)," farmers were paid in the form of commodities to remove land from production.

Agribusinesses will no doubt continue to increase their participation in the policy process to address the potential impacts on the agribusiness sector of the policy decisions made. For example, some agribusiness firms have had significant influence on the development of certain agricultural policies, such as for sugar and ethanol.

3. Policy for the Consumer Sector

The consumer sector, until recently, has not been considered seriously as an organized group in formulating and implementing agricultural policy. The main reasons are (1) consumers have few formal organizations and do not speak with one voice, (2) elected representatives of the people feel that consumers have fared relatively well in obtaining food and fiber supplies at a reasonable cost, and (3) consumers have had to spend a decreasing share of their disposable income for food and fiber, and thus the food problem was not considered critical to them.

Since the 1970s, consumers groups have become much more involved in policy formulation, better organized, more articulate, and more knowledgeable about food and fiber products. They demand more consumer protection laws, want better regulation of sanitary conditions, and have more sophistication in purchasing food and fiber products. In general, consumers care very little about traditional farm support programs. Instead, they are more concerned with such issues as nutrition labeling, food safety, and food stamp and school lunch programs.

4. Policy Determination

Traditionally, agricultural policy has been developed by the legislative and executive branches of the U.S. government through a bipartisan coalition of Democrats and Republicans from rural areas of the nation. In many instances, congressional legislators from the cotton, rice, peanut, and tobacco areas have traded support with those from the corn and wheat areas to produce a strong coalition of farm interests. Citizens of both small towns and rural areas have also formed popular coalitions to achieve agricultural policy objectives on behalf of farmers. Also, many urban members of Congress have given their support to farm policies in exchange for support of urban legislation, particularly the food stamp program, which is typically a part of federal farm legislation.

These coalitions have worked favorably for agricultural policies since the 1880s. In recent years, however, they have weakened somewhat, more so in the House of Representatives than in the Senate. Farm and other rural areas continue to lose population. In 1924, 251 out of 435 members of the U.S. House of Representatives were from rural areas; presently, the number is less than 50, with further declines possible. Farm interests are less homogeneous and cohesive than they were in the past, and agribusiness trade associations appear to be gaining political power relative to farmer groups. There has been a change from "farm politics" with producer-dominated farm policy to the politics of a comprehensive food and fiber industry.

5. International Trade

The well-being of the agricultural sector is tied closely to the ability to export agricultural products. For example, about two-thirds of total U.S. grain production is used domestically, while just under one-third is exported. The goal of agricultural policy pertaining to international trade is to maintain and enhance competitiveness of U.S. agricultural products in an increasingly global market. Being competitive has two economic requirements: (1) Domestic prices must be at or below world prices to enable U.S. producers to maintain or increase market share. (2) Domestic production costs must be low enough to allow producers to remain in business at prevailing world market prices.

The importance of international trade to the well-being of the agricultural sector was illustrated during the 1970s and 1980s. During the 1970s, exports hit record levels. They continued to surge through 1981, when they peaked at $43.3 billion. This rapid rise in agricultural exports can be attributed at least in part to the value of the dollar, which fell 28 percent from 1970 to 1979, and to extensive crop failures in many parts of the world, which resulted in increased commodity prices.

In the early 1980s, conditions changed. The value of the dollar strengthened in world markets, while world agricultural production increased. These and other factors, coupled with a global recession, caused U.S. agricultural exports to fall from $43.3 billion in 1981 to $26.2 billion in 1986. The effects of the fall in exports hit farmers where it hurts. Crop prices, in real terms, fell 37 percent from 1981 to 1987, while livestock prices stagnated after more than doubling in the 1970s.

In the next chapter, we will examine some of the approaches to and tools of agricultural policy—past, present, and future.

Topics for Discussion

1. Discuss the government's basic role in agriculture from colonization through World War I.

2. Describe the events in the agricultural sector prior to and during the Great Depression that led to direct government involvement in agriculture.

3. Discuss the trends expected to characterize farming in the future.

4. Delineate some possible goals for future farm policies.

5. Discuss the need for coordinating farm policies among the sectors of farming, agribusiness, consumption, and government.

Problem Assignment

Prepare an essay on farm policy goals that you believe will be relevant to production agriculture in the next decade.

Recommended Readings

Barkema, Alan, and Mark Drabenstott. "Agriculture Rides the Storm Out," *Economic Review,* 79(1):29–43 (1994).

Barkema, Alan, and Mark Drabenstott. "The Farm Recovery Back on Track," *Economic Review,* 78(1):41–53 (1993).

Breimyer, Harold F. "Ten Commandments for Farm Policy—And an Eleventh," *Choices,* 5(2):3 (1990).

Cochrane, Willard W., *The Development of American Agriculture: A Historical Analysis,* 2nd ed. Minneapolis: University of Minnesota Press, 1993.

Drabenstott, Mark. "Agricultural Industrialization: Implications for Economic Development and Public Policy," *Journal of Agricultural and Applied Economics,* 27(1):13–20 (1995).

Drabenstott, Mark. "The Outlook for U.S. Agriculture: Entering a New Era," *Economic Review,* 81(1):77–94 (1996).

Drache, Hiram M. *History of U.S. Agriculture and Its Relevance to Today.* Danville, IL: Interstate Publishers, Inc., 1996.

Knutson, Ronald D., J. B. Penn, and William T. Boehm. *Agricultural and Food Policy,* 3rd ed. Englewood Cliffs, NJ: Prentice-Hall, Inc., 1995. Ch. 1 and 3.

Knutson, Ronald D., et al. *Policy Tools for U.S. Agriculture.* Report B-1548, Revised. College Station: Agricultural and Food Policy Center, Texas A&M University, January 1993.

Skees, Jerry R. "Relevance of Policy Analysis: Needs for Design, Implementation, and Packaging," *Journal of Agricultural and Applied Economics,* 26(1):43–52 (1994).

Swinnen, Johan F. M. "A Positive Theory of Agricultural Protection," *American Journal of Agricultural Economics,* 76(1):1–14 (1994).

Tweeten, Luther. *Foundations of Farm Policy,* 2nd rev. ed. Lincoln: University of Nebraska Press, 1979.

Tweeten, Luther, and Lynn Forster. "Looking Forward to Choices for the 21st Century," *Choices.* 8(3):26–31 (1993).

Wright, Brian D. "Dynamic Perspectives on Agricultural Policy Issues," *American Journal of Agricultural Economics,* 75(5):1113–1125 (1993).

25

Achieving the Goals of Agricultural Policy

For more than 60 years, the federal government has directly subsidized agriculture and farmers with significant financial support. Over this time, public policy relating to agriculture has been broad and has markedly altered the structure as well as many of the traditions of American farming. The power of the government has been used to usurp market forces by fixing prices, setting price floors, subsidizing exports, restricting imports, granting favorable credit terms to farmers, and many other means.

This chapter examines the development of traditional U.S. agricultural policy. In response to the economic despair in agriculture during the 1920s, which intensified during the Great Depression, the federal government became directly involved in the agricultural sector. This chapter also discusses major program tools of agricultural policy, their goals, and the major implications and consequences of each. Responding to budgetary pressures and other influences, the government's role in the agricultural sector is changing significantly. The chapter concludes with a look ahead to some future policy issues likely to affect the agricultural sector.

Characteristics of Traditional Policy Responses to Agricultural Problems

Since the federal government began to provide direct financial assistance to farmers in the 1930s, the goals and tools of agricultural policy have continued to evolve. One significant change has been the gradual evolution of mandatory supply controls, rigid acreage allotments, and high price supports to voluntary programs, greater flexibility in land use, and low price supports, augmented by government payments to maintain farm income (deficiency payments) while allowing prices to be set in the marketplace.

Nevertheless, since the 1930s, U.S. agricultural policy has been characterized by several dominant strategies in achieving its goals. This section discusses key policy responses to the perceived needs of agriculture and some reasons for those responses.

Supply Management

Given the chronic problem of crop surpluses, supply management has been a major feature of federal farm programs. Since the 1930s, the federal government has implemented various methods of supply control, either voluntary (by means of incentives) or mandatory. As mentioned earlier, the government intervened largely because farmers operate in a price-taking environment, with no single producer able to influence market prices and with an aggregate tendency toward overproduction. Attempts by producers to control supply through voluntary cooperative actions have been notably unsuccessful.

Supply management (usually production control) has had several objectives. A primary goal has been to increase commodity prices for farmers by "shorting" the market (Figure 25–1). This market-shorting design is possible because of the inelastic demand for food. Consumers are willing to pay disproportionately more for a smaller quantity when forced by the market. Less quantity supplied by farmers translates into smaller surpluses and thereby into reduced treasury expenditures for government price support programs. Also, by taking much of the price uncertainty out of the market, consumers are assured a more stable supply of food and fiber commodities because farmers are guaranteed a price that is usually higher than the market would allow.

The two major forms of supply management in the United States have been by restricting resources used in production (e.g., land) and by limiting the amount of output that can be marketed. Traditionally, U.S. farm policy has relied on acreage control rather than on output control for implementing supply management programs. The two are certainly not equivalent, for output control would be more effective in raising farm prices and probably be less costly to the federal Treasury. Another advantage of output control is that farmers determine the input mix to use to produce a given output—for example, using fewer acres with more fertilizer or more acres with less fertilizer.

The effects of "shorting" the market through supply management

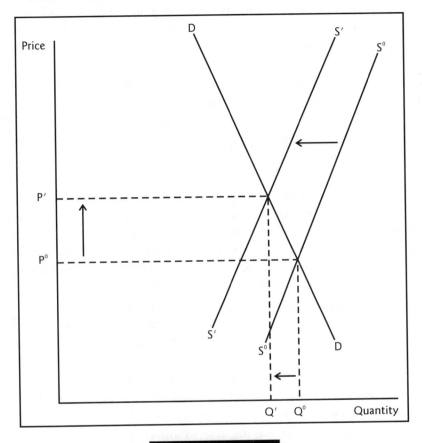

Figure 25-1

One cannot deny the efficiency of tight, comprehensive output control in raising prices and farm incomes. Consumers would bear the cost of the program through higher food and fiber prices, while the federal Treasury would pay less. Exports would likely be adversely affected by higher prices, while imports would be encouraged.

In reality, a majority of U.S. farmers would most likely oppose such a scheme of output restriction. Also, smaller farmers would not be helped greatly, because of their limited resources. And, in order to be effective, tight output restriction would need to be applied to all crops and livestock to prevent shifting of resources from a controlled commodity to an uncontrolled one. If this were done, the administrative complexity would be extremely burdensome.

Supply management policy in the United States has tended to rely on indirect methods of control. Prior to the 1960s, production controls were mandatory for all producers of program crops if they were approved by two-thirds of a majority of those producers in a national referendum. Since the 1960s, production control programs have been voluntary, with incentives provided to entice participation. Thus, producers can choose not to participate in government programs, but in so choosing, they forfeit all program benefits.

As mentioned above, the supply of program crops has been limited primarily by controlling the main agricultural input, land. There are exceptions, such as sugar, where import quotas are used to maintain high domestic prices.

Under more recent programs, acreage reduction has consisted of an annual acreage *set-aside* and/or acreage *diversion* from a farmer's historical production base. A production base is determined by the farmer's production history of the program crop (i.e., the number of acres on the farm used to produce the crop). Since 1985, the base acres have been a moving average of the acreage planted and considered planted in the last five years.

Set-aside programs require participating producers to idle and devote to a conserving use a percentage of their base acres to maintain eligibility for farm program benefits. By voluntarily participating in acreage diversion programs, producers gain access to the benefits of support programs, including nonrecourse loans and deficiency payments (discussed in the next section).

Farmers tend to satisfy the **Acreage Reduction Program (ARP)** requirements by setting aside their least productive land. However, no accompanying restrictions are placed on the use of other inputs, such as fertilizer, pesticides, water, and new technology. The result is **slippage,** which occurs when the supply management program fails to meet its objective. For example, suppose the government requires that participating producers reduce the land devoted to growing wheat by 10 percent. There is slippage if the associated decrease in wheat production is less than 10 percent. Slippage occurs because producers farm their best farm land more intensively. That not all farmers participate in farm programs also contributes to slippage, as nonparticipating farmers do not set aside any land.

Price and Income Support

Supply management programs have been used in conjunction with some type of price and income support program since the federal government initially became involved in the agricultural sector. As discussed in the preceding section, supply controls originally were designed to increase prices by reducing production. Farmers have typically viewed such controls as the political price that must be paid to receive price supports. The federal government has used several methods to support prices and incomes, the most common of which are discussed below.

Nonrecourse Loans

The **nonrecourse loan** has historically been the most common form of price support. Under this plan, the government sets the **loan rate,** a fixed price per unit of the program commodity, which serves as a price floor (Figure 25–2). The loan rate is per bushel, pound, or hundredweight (cwt.) of production. Farmers can put their commodities into government or private storage for sale at a later date and receive a loan equal to the loan rate per unit of production times the amount stored.

If the market price (P^0 in Figure 25–2) is above the loan rate (l_1 in Figure 25–2), the farmer will be better off by repaying the loan and then selling the commodity in the marketplace. However, if the market price is below the loan rate (l_2 in Figure 25–2), the farmer can simply forfeit the commodity (default on the loan) to the government and keep the loan. The government has no further claim or recourse against the farmer, no matter how far the market price may fall—hence, the name "nonrecourse loan."

Prior to the 1970s, nonrecourse loans were the major form of price support for the program crops (primarily food grains, feed grains, oilseeds, and cotton). When set well above the market price to improve farm income, the loan rate became the minimum market price. There was no need for farmers to sell their crops at prices below the loan rate because the government was there to purchase the supply at the loan rate, provided the farmers had participated in the program, which usually required some type of acreage reduction.

For a commodity of which the United States was the major world supplier, the loan rate also served as a world price floor. This provided price protection and encouraged output expansion in countries whose costs of production were below the U.S. loan rate. For a commodity of which the United States was not the dominant supplier, high loan rates effectively blocked U.S. exports of that crop.

Today, loan rates are relatively low because they are based on five-year moving averages of world market prices. Although less significant than previously, they remain in place to provide a "safety net" to protect farmers from

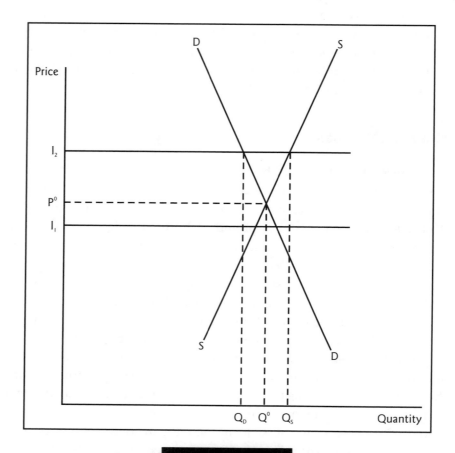

Figure 25–2

sudden decreases in world prices. Because they still establish a price floor, they potentially limit federal Treasury costs while minimizing market distortions. However, because the loan rates are usually below market prices, they do not have the same effect as they did previously, when they were set higher.

The Commodity Credit Corporation (CCC) in the U.S. Department of Agriculture administers the nonrecourse loan program. If the market price remains below the loan rate, the commodity pledged as security for the loan is transferred to CCC ownership. The CCC then sells the forfeited stocks on the market, subject to certain limitations. For example, the CCC may not sell any basic commodity at less than 115 percent of the price support loan rate or the loan repayment level, whichever is lower.

A price support program related to the nonrecourse loan program is the *Farmer-Owned Reserve (FOR)*. The 1977 Farm Bill established the FOR as a three-year extension of the CCC loan for wheat and feed grains after the regular loan has expired. The program was modified in 1980 to allow direct entry. Participating farmers keep reserve stocks in storage until the Secretary of Agriculture authorizes their release or until the extension expires. The objective of the FOR is to stabilize market prices and to stabilize grain supplies. A stable grain supply helps the United States be a more dependable source for domestic and foreign consumers.

When the entry price is set above the equilibrium market price, the FOR acts as an income support program. Stored grain stocks increase, resulting in high storage and interest costs. High loan levels and release prices (the "trigger" prices at which FOR supplies can be marketed) encourage domestic and foreign production and discourage U.S. exports. Also, the FOR supports prices only when producer participation is high and adequate storage is available.[1]

Beginning with the 1985 Farm Bill, the government set the maximum amount of wheat and feed grain that can go into the FOR. The amount for each year is based on the preceding year's ending stocks-to-use ratio, market prices, and the quantity of grain already in the FOR.

Target Prices and Deficiency Payments

High nonrecourse loan rates improved farm income, but, as we mentioned, they had some serious market-related shortcomings. **Target prices,** prices set by the government for certain commodities, were instituted in the 1970s to provide superior price and income support for farmers, but with less market distortion than nonrecourse loans. (Target prices still distort market prices by tending to pull prices down.) Target prices were initially based on costs of production, but in recent years they have been set through political negotiation (and influenced heavily by federal budget considerations) in Congress.

Target prices are designed more as income supports than price supports. Generally, when target prices are set, nonrecourse loan rates are lowered. Although the loan rate still provides a price floor, the target price determines

---

[1] Ronald D. Knutson, et al., *Policy Tools for U.S. Agriculture,* Report B-1548, Revised (College Station: Agricultural and Food Policy Center, Texas A&M University, January 1993), p. 14.

how much income will be transferred to farmers. Income transfer comes in the form of a **deficiency payment.** Deficiency payments are government payments made to farmers who participate in support programs for the "basic commodities," e.g., feed grain, wheat, rice, and upland cotton.

The deficiency payment rate is per unit of production, based on the difference between a target price (P_t in Figure 25–3) and the market price (P_m in Figure 25–3) or the loan rate (l in Figure 25–3), whichever difference is less. The total payment a farmer receives is the payment rate multiplied by the eligible production (subject to payment limitations).

The effects of a target price and deficiency payment support program

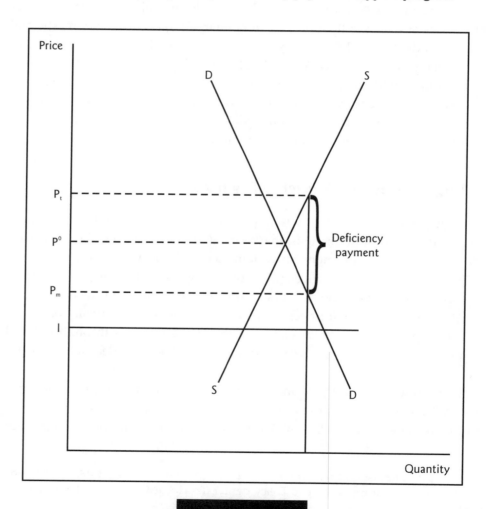

Figure 25–3

A farmer's eligible production is determined by his or her base acreage and base yield. The base acreage was defined previously. The base yield was based on a farm's yield history; however, base yields have been frozen since 1985.

In its simplest form, the deficiency payment can be calculated as the difference between the target price and the market price. For example, assume the target price for corn is $2.75 per bushel and the market price is $2.21. In this case the deficiency payment rate would be $0.54:

$$\$2.75 - 2.21 = \$0.54$$

The loan rate also plays an important role in calculating the deficiency payment rate. Thus, calculating the payment rate requires two steps—(1) determining the difference between the target price and the market price and (2) determining the difference between the loan rate and the target price. Whichever difference is smallest is the deficiency payment rate.

Target prices and deficiency payments were implemented with the signing of the Agriculture and Consumer Protection Act in 1973. At that time, loan rates were reduced to below world prices. Between 1974 and 1976, market prices were higher than target prices, which meant the treasury had no deficiency payments. In the Agricultural and Food Act of 1981, target prices were set for 1982 through 1985 without regard to inflation, crop yields, or production costs. Excess production and high government program costs resulted.[2] After world prices fell below the loan rates, exports dropped sharply, while government purchases of program crop surpluses increased as farmers forfeited their nonrecourse loans.

In the budget-tightening environment of recent years, the government has taken steps to reduce the portion of a crop for which deficiency payments are made. These steps separate, somewhat, the level of income assistance from decisions about how much to produce. This flexibility allows producers to participate in the farm programs while planting up to 25 percent (as of this writing) of their base acreage to permitted alternative crops. This acreage is known as "flex" acreage and can be divided into two types. The first 15 percent is "normal flex acres (NFA)," and the remaining 10 percent is "optional flex acres (OFA)."

---

[2] Ibid., p. 11.

Normal flex acres are not eligible for deficiency payments regardless of the crop planted. If optional flex acres are planted to program crops, they are eligible for deficiency payments. If they are "flexed" to another crop or idled, they are not eligible for deficiency payments. However, program crops or oilseeds planted on NFA or OFA are eligible for price support loans.

Other Price and Income Support Programs

Some program commodities do not have nonrecourse loans and so require alternative support strategies. For example, milk prices are supported by government purchases of enough cheese, butter, and dry milk to maintain milk prices at a predetermined target level established by Congress. Additionally, import quotas and duties are imposed to restrict imports of milk products that would undercut the support program. Likewise, the domestic sugar price is maintained well above the world market price through sugar import quotas.

Other government programs have enhanced farm income by reducing farm operating costs. Historically, these have included payments for commodity storage and subsidized credit, fertilizer, and irrigation water (in some parts of the country). Most of these programs have been phased out in recent years.

Risk Reduction and Price Stabilization

An oft-cited justification for federal farm programs is their ability to reduce risk and uncertainty for farmers while assuring domestic and export market stability. Most economists agree that, over time, price and income supports have been successful in reducing the variability of commodity prices and farm income, in contrast to the volatile, boom-and-bust cycles commonly observed in free markets. The reduction in risk made possible an increase in agricultural credit, encouraging the financing that rapidly industrialized the farm sector.

Government-sponsored commodity storage programs have attempted, with limited success, to stabilize prices by accumulating supplies during times of surplus and marketing these supplies during times of shortage. As mentioned before, the government has sometimes had to resort to import quotas and trade barriers to protect the programs from lower-priced foreign commodities.

Conservation[3]

Of all federal expenditures for resource efforts affecting agriculture, over half come from the USDA. Other key agencies involved include the U.S. Department of the Interior (USDI), the U.S. Army Corps of Engineers (Corps), and the U.S. Environmental Protection Agency. USDI and Corps programs affecting agriculture deal primarily with water conservation and management, including irrigation, flood control, and wetlands management. The EPA administers programs dealing with surface water quality, drinking water and groundwater protection, and pesticide use.

Payments to take farm land out of production and use it for conservation purposes have been made at various times in the past. The Soil Bank program, which began with the passage of the Agricultural Act of 1956, is an example. The Soil Bank program paid farmers to use conserving practices, with the joint goals of maintaining soil fertility by controlling erosion, while cutting production of program crops. More current examples include the **Conservation Reserve Program (CRP)** and the **Wetlands Reserve Program (WRP).**

Congress established the CRP as a voluntary long-term cropland retirement program in the Food Security Act of 1985. The USDA pays participants (farm owners or operators) an annual per acre rent and half the cost of establishing a permanent land cover (usually grass or trees) in exchange for retiring highly erodible or other environmentally sensitive cropland for 10 to 15 years.

The primary goal of the CRP during the late 1980s was to reduce soil erosion on highly erodible cropland. Other objectives included protecting long-run U.S. capacity to produce food and fiber, reducing sedimentation, improving water quality, enhancing wildlife habitat, reducing production of surplus farm commodities, and providing income support to farmers.

The first CRP contracts, covering 2 million acres, expired in September 1990, while contracts on more than 22 million acres expire in 1996 and 1997. Funding for the CRP was maintained in the Federal Agricultural Improvement and Reform Act of 1996.

In the WRP, authorized in 1990, the government provides easement payments to, and shares the cost of restoration with, landowners who permanently return prior converted or farmed wetlands to wetland condition. The

---

[3] Adapted from Margot Anderson (ed.), *Agricultural Resources and Environmental Indicators,* Agricultural Handbook No. 705 (Washington, DC: Economic Research Service, USDA), pp. 162–199.

goal is to have 975,000 acres enrolled in the WRP by the year 2000. Economic uses of the restored wetlands that may reduce the cost to the government are allowed. These include hunting, fishing, other recreational activities, and selective timber harvesting. Like the CRP, funding for the WRP was maintained in the FAIR Act of 1996.

The most recent approach to resource conservation the federal government has used focuses on *compliance provisions.* The compliance provisions (conservation compliance, **Sodbuster,** and **Swampbuster**) originated in the 1985 Food Security Act. These provisions require that farm operators implement specified conservation practices (also called "best management practices") and/or avoid certain land use changes to remain eligible for USDA program benefits. Specifically, any land that was farmed between 1981 and 1985 and classified as **highly erodible land (HEL)**[4] is subject to *conservation compliance provisions.* This land must be farmed using a Natural Resources Conservation Service (formerly Soil Conservation Service) approved conservation system.

The Sodbuster provision, which pertains to all HEL not cropped between 1981 and 1985, was also included in the 1985 Act. Sodbuster discourages bringing highly erosive land into production. Anyone cultivating such land must adopt a basic conservation system that reduces erosion to the T level or be subject to penalty.

Conservation compliance and Sodbuster provisions were innovative because they linked receipt of farm program benefits (including price support, loan rate, crop insurance, disaster relief, CRP, and FmHA loan programs) to conservation performance. Because of high participation rates in one or more of these programs, the government attains a higher level of compliance than just those who receive deficiency payments. A violation of either conservation compliance or Sodbuster will result in the loss of some or all USDA program benefits. While linking program payments and other benefits to conservation plans is an effective way to meet conservation goals, success still depends on program participation. If programs become unattractive to farmers for whatever reason, the compliance leverage is weakened.

Another provision enacted in the 1985 Farm Bill was the Swampbuster provision. The objective of Swampbuster is to discourage the conversion of

---

[4] *HEL* is specified as land with soils that have a natural erosion potential at least eight times their T level. *T* refers to the tolerance (T) value, the rate of soil erosion above which long-term soil productivity may be depleted.

wetlands to agricultural uses. Draining or tilling a wetland to make production possible or simply planting an agricultural crop in a wetland results in the loss of farm program benefits, including price support payments, farm storage facility loans, crop insurance, and disaster payments. These benefits cannot be resumed until the converted wetland (or another wetland converted before 1985) has been restored. Swampbuster appears to have slowed dramatically the clearing and draining of fragile lands by producers currently participating in farm support programs.[5]

Demand Expansion

We have discussed the government's attempts to reduce the quantity of food and fiber supplied to the market by means of various acreage reduction schemes. In addition, the government has adopted several programs that have strengthened demand for farm products. For example, in order to increase domestic food consumption, the government began providing food stamps for low-income people and "donated" excess commodities to needy groups and charities. The government has also funded much research into finding and developing new industrial uses for farm products. Also, commodity "check off" collections for market-expanding research and promotion programs have been authorized by the government.

A key component of the demand for agricultural products is exports. Thus, programs designed to expand exports play an important role in increasing total demand. The importance of exports to the well-being of the agricultural sector has been illustrated periodically over the years. This importance will only increase, because the rate of growth in U.S. agricultural productivity has been greater than the rate of growth in domestic demand for food and fiber products.

Food aid has been provided to foreign nations and given to industry groups working in other countries to promote U.S. agriculture. Other programs assist by granting export credits and credit guarantees, direct export subsidies, export promotion, and other incentives. The target price and deficiency payment program lowers market prices, thus encouraging exports. The Food Security Act of 1985 greatly reduced loan rates to make U.S. farm commodities more competitive on the world market. Since 1985, the marketing

---

[5] Knutson, et al., p. 36.

The effects of increasing demand through government purchases

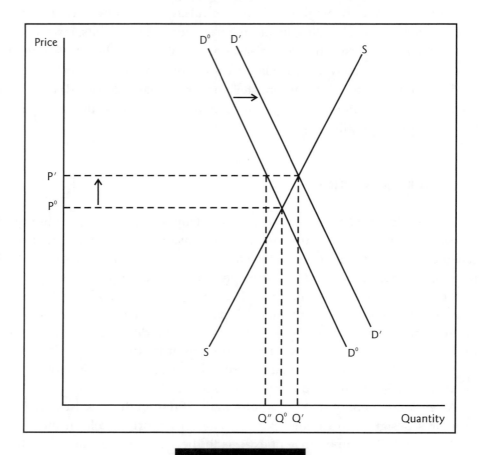

Figure 25–4

loan programs for selected program commodities have promoted sales of these crops on the world market.

The *Export Enhancement Program (EEP)*, created by the 1985 Farm Bill, is currently the primary export assistance arm of farm programs. It is designed to offset the subsidized exports of other nations and is primarily targeted at the European Community (EC). EEP also is designed to make U.S. agricultural exports more competitive on the world market by allowing the USDA to use surplus commodities or cash as export bonuses. While no particular commodity is specifically targeted for assistance in the EEP, wheat exports have received the most support.

U.S. agricultural trade policy is discussed further in Chapter 29.

Structural Consequences of Government Involvement in U.S. Agriculture

U.S. agricultural policy has been designed to achieve several stated and implied goals and objectives. As discussed above, stabilizing commodity supplies and prices and enhancing farm income have been of primary importance. Other goals have included providing an abundant supply of affordable and safe food and preserving the "family farm" structure. However, many analysts suggest that while these goals and objectives have been achieved, some serious unintended and long-term side effects have resulted. Below is a discussion of some of the past and present consequences of U.S. agricultural policy.

Government programs reduced risk, supporting a technological revolution in U.S. agriculture. The farm program price support provisions did reduce farming risks by stabilizing prices and increasing short-term profitability. Less risk translated into more willingness to borrow money for investing in technological improvements. Lenders became quite willing to extend credit to farmers, since much of their risk was eliminated (sometimes through government credit subsidies). New credit institutions were established specifically to address agricultural needs. These developments coincided with the availability of new technologies through the 1930s, '40s, and '50s. The result was a massive technological revolution in agriculture that greatly increased the productivity, production capacity, and export competitiveness of U.S. agriculture.

Agricultural policies improved farm income, but gains were difficult to maintain. While it is obvious that price supports improved farm income in some years, the improvement was undermined by increases in asset values, especially land. Indeed, much of the improvement in income was capitalized into higher asset values, which benefited only those farmers who owned the assets. These higher asset values contributed to higher long-term fixed costs of production, which, in turn, led to more pressure from farmers to increase government supports to offset these higher costs. Thus, the short-term gains for farmers led to a distortion in asset values, especially land, which contributed to an upward cost-price spiral.

Government support programs led to overinvestment in agriculture. Closely related to the previous consequence, higher farm income, due to government support, led to overinvestment in machinery and equipment, capital facilities, and new technology. Credit was often subsidized and relatively easy to obtain. The result was an increase in excess capacity to produce farm goods and thus the need for more reliance on government supply control techniques. Stronger supply control measures meant that farmers would have to increase their set-aside acreage. Farmers responded by using more capital-intensive techniques, because they had incentive to increase production on the land they farmed. This resulted in slippage, which we discussed earlier, and distortion in the mix of inputs used to produce commodities.

In spite of substantial government supply control programs, agricultural production exceeded domestic demand, and exports became a critical outlet for excess supplies, especially during the 1970s. Unfortunately, export growth slowed drastically in the 1980s, and the overinvestment in U.S. agriculture's production capacity became unsustainable. In response, for the first time since the 1920s and '30s, there was a tremendous shift of capital out of the farm sector and a major deflation of farm asset values, specifically land values. The market adjustment was very difficult for many in the agricultural and rural sectors.

Government intervention fostered major structural change in the farming sector. The combination of increased stability, greater productivity, and technological change directly led to fewer and larger farms. Individual farmers or workers could manage larger and larger operations as technology (capital) was substituted for labor. In the process, farmers became less self-sufficient as they became more dependent on nonfarm suppliers for fuel, chemicals, improved seeds, and other operating inputs, as well as the credit to buy them. More of each dollar of income from commodity sales went to pay outside input suppliers instead of being retained by farmers to cover the cost of producing inputs on the farm. Thus, the net profit margin per dollar of farm income became smaller.

To maintain at least a steady income with a smaller net profit margin, farmers had to (1) expand the size of their operations to increase total sales volume, (2) supplement farm income with off-farm income, or (3) leave farming altogether. Partly as a result of government intervention in the agricultural sector, production agriculture has been transformed from a low-technology, labor-intensive industry to a high-technology, high-capital industry with much greater concentration of production in fewer, larger farming operations. Farm size went from an average of 157 acres per farm in 1930 to an average of 473 in 1993. The number of farms peaked at 6.8 million in 1935, while today there are only about 2.0 million.

Agricultural policies have distorted resource prices and allocation. Because the amount of land allowed to be planted to program crops is often limited by farm programs, there are inherent incentives to employ more capital-intensive farming methods, such as using more fertilizer, chemicals, and other inputs to increase yields on fewer acres. Sometimes costly irrigation systems are purchased that would not otherwise be used. Program requirements have also created incentives for farmers to raise the same crops on the same land year after year. Such monocropping systems can result in the loss of benefits of crop rotation, including soil residue management and integrated pest management. Also, monocropping systems can increase the potential for nonpoint pollution, due to the nature of fertilization requirements. Because of the market-distorting effects of the government support programs, resource allocation and enterprise selection decisions have been inefficient, and costly to the economy.

Consumer effects of agricultural policies are mixed. The effects of current U.S. agricultural policy on food costs are minimal. Products made from basic commodities, like grain, probably cost less than they would if there were no farm programs. Since farm programs stimulate overproduction, lower prices are possible because the quantity supplied is in surplus. Also, consumers typically buy commodities at market prices and not at the target price (because the government pays farmers the deficiency payment directly).

While grain prices are probably lower for consumers, the prices of sugar, peanuts, and dairy products are likely higher due to the ways these commodities are supported. Import quotas or restrictions and other controls keep supplies of these commodities limited mainly to domestic production. Domestic demand is sufficiently strong to keep prices for these commodities well above world market prices. The result is fewer goods for consumers at higher prices than would otherwise be available in an unhindered free market.

The benefits and costs of agricultural policies have been distributed unevenly. Farm programs were initiated to assist an economically depressed and disadvantaged farm sector. In the past half century, farm programs have not always adapted well to changes in agriculture. Today, the typical commercial farmer has an income greater than that of the average nonfarmer and a net worth much larger. Government farm program payments are generally now regressively redistributed, from lower-income taxpayers to large commercial farmers. About 75 percent of all farm subsidy payments go to only about 7 percent of all farmers. It should be noted that these large-scale farmers produce most of our agricultural commodities—only 12 percent of all farms accounted for 75 percent of total farm product sales. Many small farmers get no benefits since they do not participate in the farm programs or do not produce program crops.

To the extent that the higher incomes of farmers, due to program payments, are being capitalized into higher land values, the main beneficiaries are current landowners. Tenants usually do not benefit from the resulting increased land values. Instead, they often face higher land rents, a result of the higher land values.

With the shift toward a more capital-intensive structure in agriculture, encouraged in large part by the existence of government support programs, the demand for farm labor has decreased significantly. Much of the burden of farm programs has been borne by displaced farm workers and their families.

American farmers produce hundreds of different commodities and products, yet only producers of the so-called basic commodities and a few specialty commodities have received subsidies. While grain and milk could be considered "basic" to the nation's food supply, it has been hard to make that argument for peanuts, sugar, tobacco, and mohair, all subsidized commodities. Beef, pork, fruits, vegetables, and many other commodities receive little or no support. This is another reason that the benefits of government farm support programs have been perceived as not fairly distributed.

While the distribution of costs and benefits of the farm commodity support programs is not well understood and documented, it is clear that the distribution is uneven. Equity issues contribute to the increasing difficulty for Congress to justify continuing supporting farm programs.

Treasury costs of agricultural policies have been high. Taxpayers, rather than consumers, have borne most of the direct costs of farm support programs in recent decades. This is because program commodity producers receive most of their benefits as direct deficiency payments, rather than through artificially high market prices. Prior to the government's shift to deficiency payments, consumers faced higher food prices, and treasury costs were low. More recently, treasury costs have been high, since most food prices have been near market levels.

Most of the treasury costs of modern U.S. farm programs have been incurred since 1980. This is a result of the shift in the 1970s from providing support through high nonrecourse loan rates to letting market prices prevail and providing support through direct income (deficiency) payments, combined with depressed exports and commodity prices in the 1980s.

Some agricultural policies have fostered protectionist trade policies. Some protectionist trade policies have been necessary to maintain domestic support programs. Maintaining prices above the world market levels requires import controls to prevent consumers from substituting lower-cost foreign goods for domestic commodities. While this problem is not as serious today because loan rates are usually below world market prices, it remains an important issue for the sugar, peanut, and dairy programs.

Some agricultural polices have been inconsistent, even contradictory. In certain respects, federal agricultural policy is disjointed and fragmented to the extent that particular programs and goals are frequently inconsistent with each other. For example, some projects encourage the expansion of agriculture (subsidized irrigation and credit, scientific research that improves production efficiency and capacity, extension educational services, etc.), while others are specifically designed to restrict the quantity produced (acreage reduction programs, set-asides, import quotas, etc.). These opposing policy goals cause added market distortions and undermine the success of U.S. agriculture.

Even with the problems and market-distorting effects of U.S. farm programs, American consumers have access to the world's best food supply at the lowest relative prices. The average U.S. consumer spends less than 15 percent of his or her income on food, the lowest percentage for any nation in the world.

U.S. Agricultural Policy Outlook

In recent years Congress has passed a new farm bill every five years. Modern farm bills have been complex and cover many areas in addition to farming, such as nutrition programs, agricultural trade, disaster relief, conservation, and research. Technically speaking, these large comprehensive pieces of legislation have been temporary amendments to the Agriculture Act of 1949. For this reason, the 1949 Act is referred to as "permanent legislation," and any subsequent amendment is called a "farm bill." The most recent farm bill is the *Federal Agricultural Improvement and Reform (FAIR) Act.*

In the coming years, Congress must grapple with several issues that influence agriculture. Some of the more important are discussed below.

Decoupling, or unlinking, is the gradual process of separating a farmer's production decisions from income supplements or price support considerations. Under this scenario, lower transfer payments are achieved by limiting the yields and acreage for which deficiency payments are made. The 1985 Farm Bill implemented a form of decoupling by freezing farm program payment yields. Federal budget pressures in 1990 resulted in a reduction in the number of base acres eligible to receive deficiency payments.

Flexibility refers to how much the farm bill restricts what crops farmers can plant. Prior to 1990, farmers generally had to plant their base crop on their base acreage to be eligible for deficiency payments. Farmers also sometimes had to idle a certain percentage of their base acreage. These planting restrictions have been criticized for distorting agricultural production as well as restricting the individual freedom of farmers. Proponents of these restrictions argue that they provide a form of crop stability and supply management. The decoupling measures mentioned above in the 1990 farm legislation also included an increase in flexibility. Farmers were allowed to plant *any* program crop on that portion of their crop base that was ineligible for deficiency payments.

Decoupling and increased flexibility have evolved into the major policy elements in farm legislation, as evidenced by the debate on the 1995 Farm Bill. The reasons for this include the influence of federal budget restrictions as well as conservative political philosophies. A notable example during the 1995 Farm Bill debate was the proposal known as *Freedom to Farm*. This proposal called for total decoupling and total planting flexibility. The proposal would have allowed farmers to plant a wide variety of crops on their land. The government payments received by participating farmers would not depend on crop prices or what the farmers planted. The fixed government payments were designed to decrease federal spending on agricultural commodity programs over a period of years. This proposal has been viewed as a possible transition to a future with no federal commodity programs.

Revenue assurance is another recent policy proposal. The concept of revenue assurance was developed in the wake of severe crop losses in the early 1990s. Devastating floods in the Midwest destroyed considerable grain crops over a wide area. Grain prices subsequently rose, so the farmers who had produced no crop also received no deficiency payments. A revenue assurance approach would therefore replace deficiency payments with an insurance-type program. Farmers would presumably receive payments when their crop revenue (from both yields and prices) fell below some targeted amount.

Means testing is another farm policy concept that has received attention. Under this scenario, the only eligible participants in federal farm programs would be farmers who "need" assistance based on certain income or size criteria. Proponents of this approach argue that limiting benefits to small family farmers would reduce federal Treasury costs as well as preserve the ideals of Jeffersonian agrarianism. There are remaining questions about whether this approach would disable the supply management and market stabilizing attributes of federal commodity programs. There have been attempts to limit the total amount paid to individual farmers, but they have not worked very well.

Green payments, sometimes called "stewardship payments," have been touted by some policy makers as potential replacements for current price and income support programs. Another thought is that a green payment approach may be used to follow up the Freedom to Farm transition period.

Instead of providing financial assistance for producing program crops, green payments would be made to farmers who practice approved soil, water, and wildlife management techniques. The rationale for such payments is to compensate farmers for the added operating expenses associated with following specific environmentally sound conventions. Green payments potentially also have the added benefit of being less susceptible to challenges under GATT and other international treaty conditions.

If green payments replace price and income supports as the primary means of government support, there will be an increase in the variability of agriculture as part of the "safety net" is removed. Since payments would be based on environmental problems, there would be a shift in the distribution of program benefits to areas where those problems are most severe. Such a move would be politically unpopular among areas now producing most of the program crops and thus benefiting most from farm programs.

The Federal Agricultural Improvement and Reform Act (FAIR) of 1996 is comprehensive or omnibus legislation. Its provisions attempt to address a broad spectrum of issues. Some of the key provisions of FAIR include the following:

- Maintains nonrecourse marketing assistance loans for rice, upland and extra-long-staple cotton, wheat, corn, and soybeans.
- Introduces planting flexibility and production flexibility contracts. Eligible producers[6] can enter into seven-year market transition contracts (i.e., support payments are decoupled from production decisions). Any commodity can be grown on contract acreage except fruits and vegetables.

---

[6] Those who had participated or had certified acreage in the wheat, feed grains, cotton, or rice programs in any of the past five years.

- Decreases payment limitation to $40,000.
- Lowers the peanut quota support rate from $678 per ton to $610 per ton through 2002; allows limited sale, lease, and transfer of peanut quota across county lines.
- Gives the Secretary of Agriculture the authority to enter into new contracts and extend expiring CRP contracts (maximum land in CRP remains at 36.4 million acres).
- Maintains the WRP.
- Provides assistance, through the Environmental Quality Incentive Program, to crop and livestock producers in implementing on-farm environmental and conservation improvements.
- Maintains Swampbuster provisions, to be overseen by the Natural Resources Conservation Service.
- Establishes the *Farms for the Future* program to preserve farm land from commercial development.
- Funds the *Market Access Program* (the Market Promotion Program, renamed) to encourage the development, maintenance, and expansion of commercial export markets for agricultural products; restricts participation to small businesses, farmer-owned cooperatives, and agricultural groups.
- Reauthorizes Food for Peace (P.L. 480); allows private sector participation for the first time.
- Reforms conservation compliance provisions, giving producers more flexibility to modify conservation practices if they can demonstrate that the new practices achieve equal or greater erosion control.
- Reauthorizes the Federal Food Stamp Program for two years while Congress continues to work on comprehensive welfare reform legislation.

Topics for Discussion

1. Define *slippage*. Discuss the factors that result in slippage.
2. Explain how nonrecourse loans function.
3. Explain how deficiency payments function as a means transferring income to farmers.

4. Briefly discuss the structural changes that have resulted because of government intervention in the agricultural sector.

Problem Assignment

Prepare an essay discussing possible effects on agricultural producers of decoupling farm support payments from production decisions.

Recommended Readings

Barkema, Alan, Mark Drabenstott, and Kelly Welch. "The Quiet Revolution in the U.S. Food Market," *Economic Review,* 76(3):25–41 (1991).

Boehlje, Michael D., and Vernon R. Eidman. *Farm Management.* New York: John Wiley & Sons, Inc., 1984. Ch. 3.

Browne, William P., et al. *Sacred Cows and Hot Potatoes: Agrarian Myths in Agricultural Policy.* Boulder, CO: Westview Press, 1992.

Colander, David C. *Economics,* 2nd ed. Chicago: Richard D. Irwin, Inc., 1995. Ch. 34.

Cramer, Gail L., and Clarence W. Jensen. *Agricultural Economics and Agribusiness,* 6th ed. New York: John Wiley & Sons, Inc., 1994. Ch. 11.

Drabenstott, Mark. "Agricultural Industrialization: Implications for Economic Development and Public Policy," *Journal of Agricultural and Applied Economics,* 27(1):13–20 (1995).

Drabenstott, Mark. "The Outlook for U.S. Agriculture: Entering a New Era," *Economic Review,* 81(1):77–94 (1996).

Drache, Hiram M. *History of U.S. Agriculture and Its Relevance to Today.* Danville, IL: Interstate Publishers, Inc., 1996.

Halcrow, Harold G., Robert G. F. Spitze, and Joyce Allen-Smith. *Food and Agricultural Policy: Economics and Politics.* New York: McGraw-Hill, Inc., 1994.

Helmberger, Peter G. *Economic Analysis of Farm Programs.* New York: McGraw-Hill, Inc., 1991.

Knutson, Ronald D., J. B. Penn, and William T. Boehm. *Agricultural and Food Policy,* 3nd ed. Englewood Cliffs, NJ: Prentice-Hall, Inc., 1995.

Knutson, Ronald D., et al. *Policy Tools for U.S. Agriculture.* Report B-1548, Revised. College Station: Agricultural and Food Policy Center, Texas A&M University, January 1993.

Manchester, Alden C. *Rearranging the Landscape: The Food Marketing Revolution, 1950–91.* Agricultural Economics Report No. 660. Washington, DC: Economic Research Service, USDA, September 1992.

Orden, David, Robert Paarlberg, and Terry Roe. "Can Farm Policy Be Reformed? Challenge of the Freedom to Farm Act, *Choices,* 11(1):4–7, 39–40 (1995).

Rossmiller, George E. "Six Problems That Affect Farm Ag. Policy," *Choices,* 7(1):14–17 (1992).

Seitz, Wesley D., Gerald C. Nelson, and Harold G. Halcrow. *Economics of Resources, Agriculture, and Food.* New York: McGraw-Hill, Inc., 1994. Ch. 18.

Stovall, John G., and Dale E. Hathaway. "The 1995 Farm Bill: Issues and Options," *Choices,* 10(2):8–14 (1995).

Tweeten, Luther. *Foundations of Farm Policy,* 2nd rev. ed. Lincoln: University of Nebraska Press, 1979.

Tweeten, Luther. "The Twelve Best Reasons for Commodity Programs: Why None Stands Scrutiny," *Choices,* 10(2):4–7, 44, 45 (1995).

Zulauf, Carl, and Luther Tweeten. "The Post-Commodity-Program World: Production Adjustments of Major U.S. Field Crops," *Choices,* 11(1):8–10 (1995).

PART FIVE

26 Federal Reserve System: History, Organization, and Functions

27 Input-Output Functions

28 Selected Principles of Agribusiness Management

29 International Trade and Agriculture

30 Economic Development in the Developing Areas of the World

26

The Federal Reserve System: History, Organization, and Functions

The Federal Reserve System was organized to deal with the disarray that periodically occurred in the nation's monetary system. This disarray frequently culminated in panic on a large scale. Such was the case in 1907, when after a period of unusually good times at the turn of the century, what was called a "rich man's panic" hit the stock market.

The panic quickly spread to the banking system. Ten New York banks failed within a single week. Runs on other New York banks by their local depositors, as well as by banks all over the country whose deposits they held, forced these banks to restrict cash payments. Banks in other places, unable to obtain cash from their New York accounts, also limited their payments on claims presented to them. This demoralization of the monetary system forced many businesses into liquidation, and repercussions from the failures racked the whole economy well into 1908.

The Aldrich-Vreeland Act, passed by Congress in 1908, created the National Monetary Commission. Composed of nine members from each House, it was to "inquire into and report to Congress at the earliest date practicable, what changes were necessary or desirable in the monetary system of the United States or in the laws relating to banking and currency." The Commission spent much time thoroughly studying the history of banking in the United States, as well as studying the development and operation of banking in major foreign countries. The final report was submitted in January 1912. It began with an enumeration of the defects of the U.S. banking system of that period. The major defects could be grouped under the headings of *currency, clearing, reserves, credit,* and *control.*

Inelastic Currency Issue

The Commission felt that the nation's currency supply was too inelastic to meet both seasonal and cyclical changes in business conditions. It was, in fact, the inability of banks to pay out currency during the panic of 1907 that caused a great deal of the economic distress. Currency needs during the agricultural harvest season in the fall created the biggest seasonal influence on currency demand. Farmers needed currency to pay labor and to repay crop loans; at the same time, the nonfarm economy needed currency to purchase the harvested crops. Cycles in the level of general business activity also caused changes in the demand for currency from year to year.

Gold and silver coin, as well as gold and silver certificates that were 100 percent backed by metal, were limited by the production of the precious metals from mines. The other major forms of currency were U.S. notes and national bank notes. U.S. notes were held constant by a congressional act. National bank notes were required to be 100 percent backed by government bonds. Consequently, the quantity of notes issued was dependent on how many bonds banks were able and willing to hold. This system led the government to a policy of issuing bonds, not because it needed revenue, but because the public needed the currency that the bonds would back at national banks. Rural areas had great difficulty in obtaining funds under this system.

Inefficient Check Clearing

The collection of out-of-town checks involved the cumbersome, expensive, and inefficient process of collecting through correspondents. Checks were often sent by circuitous routes to avoid high fees many banks charged for collection. Sellers often would refuse payment by checks drawn on out-of-town banks. Out-of-town customers had to use "New York funds" rather than their own checks to pay for goods. This meant they had to buy bank drafts from banks that had a correspondent in New York. In addition to paying a normal purchasing fee for a draft, customers often found that during times of financial panic, New York dollars were more expensive than their own.

The machinery for foreign payments worked no better. Almost all foreign trade was financed through London banks rather than through the U.S. financial system.

Inadequate Reserve Structure

At the turn of the twentieth century, the United States was the only country that required banks to maintain reserves against deposit liabilities. The reserves could be held either in cash or on deposit with other banks. The very existence of that requirement, however, meant that the cash could not be used to meet depositors' demands during runs on a bank. Thus, banks would sometimes have to close their doors, even with sufficient cash in the vault to make payment, because it was illegal to draw down these reserves below a minimum level. If a New York bank was in trouble and attempted to draw on deposits held at correspondent banks in other locations, these banks, in turn, would attempt to reduce their deposits with New York banks. Thus, the reserve structure of the period tended to channel a major disturbance anywhere in the country to New York, where it often exploded.

Outdated Credit System

Because of the rigid reserve structure and the lack of means to acquire additional reserves, a bank simply could not respond to its increasing credit demands when business was growing. Thus, the supply of bank credit was almost as inelastic as the quantity of currency.

The primary problem of a bank needing additional reserves for whatever reason was that its earning assets were almost impossible to turn into cash on short notice. In Europe, a highly developed market existed for bills of exchange so that any bank wishing to obtain funds quickly could sell its bills on that market. In the United States, banking authorities frowned on such measures. Commercial loans were typically tied up in promissory notes that were not readily marketable because they represented a very personal relationship between borrower and banker. Growing out of this circumstance of illiquidity was an attitude among bankers that the soundness of their loans was less important than the quantity of the cash they kept on hand. Only a few clearing houses of the time provided some help to members during such emergencies. Help was often slow in coming, however, and it was limited to banks located in larger cities.

Coordinated Control Lacking

The early banking system had no semblance of coordinated policy. Less than one-third of the 24,514 banks in existence in 1910 had a national charter placing them under the single authority of the U.S. Comptroller of the Currency. The rest operated under 46 different sets of state regulations. There was no agency to provide direction to the monetary system or to assist banks in coping with the other major problems already discussed. The banking system of the United States consisted of thousands of banks operating in their own separate ways. This lack of a uniform goal for achieving profit maximization often led to the conditions of panic experienced in 1907. Intense dissatisfaction with this system brought about the organization of the Federal Reserve System.

The Birth of the Federal Reserve System

In spite of the exhaustive analysis of banking problems by the National Monetary Commission, its proposals to deal with these problems never were brought to a vote in Congress. Instead, a bill sponsored by Carter Glass, Chairman of the House Committee on Banking and Currency, eventually emerged on December 23, 1913, as the Federal Reserve Act. This legislation differed from the Commission's proposal in that it reduced the influence of bankers in the administration of the system by the designation of a completely government-appointed board at the top level. In this way the Federal Reserve approached the concept of a central bank; however, its formulators were careful not to refer to it as such because of the intense feeling against government interference with private enterprise during that era. Thus, the Federal Reserve had some attributes of a central bank, but it was unlike any of the European central banks that the National Monetary Commission had studied in detail.

The Federal Reserve's Remedies for Banking Problems

The Federal Reserve Act attacked the rigid currency supply problem by adding Federal Reserve notes to currency already in circulation. These notes were issued against a 40 percent backing by gold plus a 100 percent backing by commercial paper. As businesses borrowed from commercial banks, commercial paper was created that banks could use as collateral in borrowing from the Reserve Banks. The Reserve Banks could issue more notes as they received more commercial paper in collateral for loans. Thus, the currency supply could expand with rising business needs. It could also contract as business loans were repaid. In that way the currency supply became automatically elastic.

A clearing mechanism was set up within the Federal Reserve System to facilitate check clearing from one region to another. It practically eliminated circuitous routing, expensive collection, and exchange charges. Currently, the Fed clears about 45 percent of the checks drawn on banks within the United States. The implementation of a Regional Check Processing Centers Program further reduced the time required to clear checks. The proportion of checks handled by the System increased rapidly after the program began.

The major impact of the Federal Reserve on the condition of unusable reserves was to form a pool of reserve assets of member banks from which a bank in trouble might borrow for a limited period. As a result of this mobilization, reserves could be put to use in an emergency.

A standardized procedure by which the Federal Reserve would make loans gave member banks additional flexibility in the management of their assets. The Federal Reserve made loans by discounting acceptable commercial paper. Thus, any bank could remain liquid as long as it held paper representing loans that satisfied Federal Reserve regulations.

Although the Federal Reserve System was intended to exercise some powers of coordination over the banking system, the centralized control was rather limited at first because the individual Federal Reserve Banks were more powerful than the Federal Reserve Board. Centralized authority has increased over the years, but the need for the Federal Reserve System to have more direct control over the entire banking system, not just member banks, remains a major issue.

Economic Crises and the Federal Reserve

The Federal Reserve began to operate in November 1914. World War I had already begun, and several other crises were to follow that would reveal the need for corrections in the original blueprint of the System. No crisis was so devastating as the crash of 1929, which shattered all illusions that as long as the Federal Reserve existed, another major financial crisis was impossible. It was painfully apparent after the banking holidays in March 1933 that the System that had been devised through such careful study to prevent precisely the existing situation was impotent.

Monetary architects began the slow process of picking up the pieces; they worked for two years to build a better structure. Their efforts provided first the tentative Banking Act of 1933, followed by the more far-reaching Banking Act of 1935. In the process, four of the Federal Reserve's original cornerstones were removed: (1) Banks were permitted to substitute government securities for commercial paper as collateral behind Federal Reserve notes, thus cutting the tie between the volume of business and the quantity of currency. (2) Gold was completely withdrawn from circulation, and the amount required for backing Federal Reserve notes and deposits was reduced, thus loosening the links between gold, money, and prices considerably. (3) The philosophy of an automatic money supply was abandoned, and the Federal Reserve System was given the power to change reserve requirements of banks within limits set by law, thus attempting to regulate credit rather than to protect banks' depositors from the consequences of a run. (4) The policy of decentralized control gave way to centralization through an elevation of the status of the Federal Reserve Board, thus giving the Board (a) complete power to change reserve requirements, (b) majority representation on a new "open market committee," and (c) greater power over changes in discount rates.

Aspects of Federal Reserve Operations

The basic organization of the Federal Reserve System has undergone relatively little change since the origin of the System. The principal components of the System, in addition to the almost 39,500 member banks, are the Board of Governors, the Federal Open Market Committee (FOMC), and the 12 Federal Reserve Banks. All but the FOMC were described in Chapter 5.

The FOMC is responsible for determining what transactions the Fed conducts in the open money markets of the nation. In effect, the FOMC is responsible for implementing monetary policy.

The Committee is made up of the seven members of the Board of Governors and five district Reserve Bank presidents. The president of the Federal Reserve Bank of New York is a permanent member. The other 11 Bank presidents serve one-year terms on a rotating basis. By tradition, the chair of the

Board of Governors is the chair of the FOMC, and the president of the Federal Reserve Bank of New York is its vice-chair. The group meets in the Board's offices in Washington usually once every four weeks throughout the year. Because of the profound impact that the decisions of the FOMC have on the economy, a summary of the minutes of each meeting is not released to the public until 90 days have passed. The minutes in greater detail are kept secret for extended periods.

The Federal Reserve Act provides for the Federal Advisory Council, which meets in Washington at least four times a year and makes recommendations regarding the affairs of the Federal Reserve System. The Council is made up of 12 members, one selected by the board of directors of each Federal Reserve Bank.

The Fed uses several other committees and groups to address many operating and policy problems. Some of the more prominent of these groups are the conferences of (1) the chairs, (2) the presidents, and (3) the first vice-presidents of the Federal Reserve Banks. These groups usually meet several times during each year. Additionally, the Advisory Committee on Truth-in-Lending meets from time to time at the call of the Board. Various staff-level committees draw upon specialists from within the System.

About 62 percent of the 64,000 commercial banks in the United States are members of the Federal Reserve System. A national bank is required to be a member of the System, whereas banks chartered by any of the 50 states may elect to become members. The number of member banks increased over 31 percent through the 1980s, while the number of nonmember banks increased only 3.7 percent.

There are several benefits available to member banks that become eligible to use all of the System's facilities. Member banks may (1) borrow from the Federal Reserve Banks when temporarily in need of additional funds; (2) collect checks, settle clearing balances, and transfer funds to other cities through the Federal Reserve's facilities; (3) obtain currency as needed; (4) share the System's informational facilities; and (5) participate in the election of directors from their district. Several of these services are provided to nonmember banks through their correspondents on a fee basis. A bank holding company is often able to perform these services at reduced cost to its members.

In their role as fiscal agents for the government, the Federal Reserve Banks hold Treasury checking accounts, process public purchases of U.S. Treasury securities, allot securities among bidders, accept payments from buyers, deliver securities, redeem securities, transfer securities to other cities, pay interest coupons, and conduct transactions in the market for various Treasury accounts. The Treasury reimburses the Reserve Banks for most fiscal agency functions.

Instruments of Monetary Policy

The Fed's efforts to regulate the nation's supply of money and credit encompass both attempts to ensure a sufficient growth in the money supply to accommodate an expanding economy and actions in the shorter run to regulate such growth in order to combat inflationary or deflationary pressures.

The effects of these actions are registered first on the cost and availability of bank reserves. Soon afterward, the effects spread to the supply of money and credit, to interest rates in financial markets, and to the liquidity of financial institutions. Ultimately, changes in financial conditions affect business and consumer expenditures, output, employment, and prices. Changes in these end results comprise what is usually termed the "business cycle."

The primary tools used by the Federal Reserve to exercise monetary policy are (1) the coordinated use of open market operations, (2) the regulation of member bank discounting with the Federal Reserve Banks, and (3) changes in member bank reserve requirements. Earlier discussions in Chapter 5 indicated the potential impact of each of these measures. They are not generally utilized alone but in concert to achieve a desired effect. In practice, two of the measures, the discount rate and reserve requirements, have less importance in and of themselves. The discount rate tends merely to signal monetary policy changes through certain phases of the business cycle rather than to serve as an effective vehicle for policy change. The discount rate typically is not allowed to rise to the heights of other money market rates during periods of credit restraint, and it often fails to decline in step with those rates during periods of monetary ease. Thus, it is not usually a regulator of member bank borrowing, and, therefore, it does not play a solid role in monetary policy that is usually attributed to it.

Changes in reserve requirements are a powerful tool of monetary policy, but because of its extreme force, this tool is used cautiously and infrequently. A change in requirements of even a fraction of a percent can have a major impact on the credit-creating capacity of the banking system. Furthermore, in periods of monetary restraint, banks tend to draw down their reserve base to a bare minimum. If the Federal Reserve increases requirements during such a period, the strain on banks to meet higher reserve requirements may be severe. Under such conditions, banks are likely to withdraw from membership in the Federal Reserve System in increasing numbers, limiting the Fed's direct control of bank credit. Altering reserve requirements, then, is not the handy tool of monetary policy that in theory it might seem to be.

This leaves open market operations as the primary tool for implementing monetary policy. Major advantages of this tool over other measures are its flexibility and its relative obscurity. It can be employed almost instantaneously to alter the course of money market developments, and the majority of the public is unaware of the direction or degree of movements in policy at the time such actions are taken.

In the past, the record of policy actions has typically not been released for about 90 days following meetings of the Federal Open Market Committee. Secrecy is maintained to keep the information from those segments of the public in a position to benefit from its early release. However, some critics argue that decisive moves by the FOMC can hardly escape detection by knowledgeable participants in money markets and that the major purpose of secrecy is to shield the FOMC from pressures that could result from immediate public awareness of its actions.

How Open Market Operations Regulate

The Federal Reserve purchases and sells securities, primarily U.S. government securities, in the open money market. Although federal agency securities and bankers' acceptances are also eligible instruments, only limited quantities of these are traded.

When the Fed buys securities, checks written in payment are deposited in commercial banks, and the supply of bank reserve funds is expanded. When the Fed sells securities, the payments of purchasers contract the supply of banks' reserves when the purchasers' checks reach the central banks. Thus, the initial impact of open market operations is on the availability and cost of bank reserves. Afterward, the effects spread to the supply of money and credit; interest rates in financial markets; and liquidity of financial institutions, businesses, and the general public. Eventually these changes in financial conditions affect current expenditures, output, employment, prices, and personal income.

For the open market instrument to work effectively, the Fed must be able either to buy or to sell a sufficient volume of assets on a timely basis to keep bank reserves in line with prevailing policy objectives. Most types of assets cannot be traded readily enough to be well suited to open market operations. The U.S. government securities market generally averages several billion dollars a day and can accommodate whatever volume of transactions the Federal Reserve needs to make.

The bulk of the Fed's buy and sell orders are handled "over the counter" by specialized dealer firms that make regular markets. The majority of these firms are located in New York City, but they have communication links with customers and dealers throughout the country, providing broad boundaries to the market. U.S. government securities are held in some quantities by economic units of virtually all types, including many individual investors, who, to the extent that they are connected with New York securities dealers, can also be affected by the Federal Reserve's open market operations.

The Federal Open Market Committee designates the New York Federal Reserve Bank to serve as its agent in executing the open market transactions the Committee authorizes. The manager of the System Open Market Account, a senior officer of the New York bank, has immediate responsibility for carrying out open market operations. All 12 Federal Reserve Banks participate in the System Account. A daily conference call involving the System Account management, the senior staff at the Board of Governors, and a Reserve Bank president who is currently serving as a voting member of the Committee coordinates day-to-day operations of the Account. The call is followed by a memorandum sent to all Committee members informing them of action that the System Account Manager expects to take during the day in light of developing conditions.

The discussion of policy at the monthly committee meetings centers on three general areas. First, members express their persuasions as to the state of the economy and the prospects for its future performance. Second, members make explicit policy recommendations for both the short and long run. Finally, through give-and-take discussions, the Committee attempts to reach a policy consensus that can be expressed in terms that will provide meaningful guidance to the Manager of the Account in the conduct of day-to-day open market operations.

The targets of open market operations include rates of growth in bank reserves and monetary aggregates and ranges of tolerable changes in money market conditions. The federal funds rate has become a principal focal indicator of money market conditions. This is the rate at which banks are willing to lend or borrow immediately available reserves on an overnight basis. It is an extremely sensitive indicator of the availability of bank reserves. When reserves become tight or scarce in relation to demand, the funds rate rises. When reserves become more plentiful, the rate falls. These changes are sometimes abrupt, and immediate action is required to keep the funds rate between predesignated limits.

Policy makers need an extensive volume of background information in order to make objective judgments concerning the economy's status and its likely future course. To provide this information, intensive and systematic staff preparation precedes each meeting of the FOMC. The results of this preparation are presented to the Committee both in written reports prior to the meetings and in oral presentations at the meetings themselves. In addition, the Account Manager provides detailed written and oral reports of his or her transactions since the Committee's last meeting.

The Account Manager's Operating Techniques

The Federal Reserve engages in open market transactions nearly every day, mostly to offset the effects of technical market factors that would be inconsistent with the aims of monetary policy. Market factors that can cause large day-to-day changes in bank reserves include currency in public circulation, Federal Reserve float-checks credited to the reserve account of one bank but not yet debited to the account of the bank on which they are drawn, and the Treasury's balance at the Federal Reserve. The Manager's operating techniques must be sufficiently adaptable to allow compensation for abrupt changes in bank reserves (which might result from the Manager's own previous actions). If the projected need for reserve adjustment extends for more than a bank-statement week, the Manager engages in outright purchases or sales of securities for prompt delivery. When the market situation requires temporary adjustments in bank reserves, the Manager makes "repurchase agreements" (RP's) with security dealers to add reserves or "matched sale-purchase transactions" to withdraw reserves.

Outright sales and purchases of the System are typically made through an auction process in which dealers submit bids or offers for securities. These tenders are arranged according to price, and the Manager accepts amounts bid or offered in sequence until the order is covered. Most of the System's outright transactions occur in Treasury bills, the most active sector of the securities market.

A repurchase agreement consists of the sale of a short-term security, with the condition that, after a period of time, the original seller will buy the security back at a predetermined price. The Fed uses repurchase transactions with dealers or banks when temporary adjustments are needed. The maturity of RP's used by the Fed is always less than 15 days. This arrangement allows the System to inject reserves temporarily, because they are automatically drawn back when the RP's mature.

Matched sale-purchase transactions are utilized to effect the opposite temporary adjustment in reserves. To absorb reserves, the Account Manager engages in a contract for immediate sale to a securities dealer and a matching contract for subsequent purchase from the dealer. These arrangements are usually set to mature within seven days. When sales are made, surplus reserves flow from banks through the dealers to the System; the flow is reversed when the System purchase is implemented.

Other Policy Instruments

The Federal Reserve manages two additional monetary policy instruments, each of which has more selective impact on deposits and credit flows than the three general controls already discussed. One selective tool is the authority to set maximums, or ceilings, on the interest rates that member banks may pay on savings and time deposits. These ceilings are effective only when interest rates are high relative to the ceilings. Then, because of the more stringent controls applicable to deposits in small denominations, these interest rate ceilings impact more heavily on smaller banks that cannot offer the large certificates of deposits (CD's) for which interest controls are either more liberal or nonexistent. Larger banks, by means of large-denomination CD's, can continue to attract funds and make new loans for a period after smaller banks have exhausted their funds.

In some cases, smaller banks tend to be eager purchasers of large CD's during tight-money periods. They can pool the funds of many small deposits and buy the larger CD's that their individual customers cannot purchase. There is a strong temptation for smaller banks to curtail loans in local communities even more than necessary during tight-money periods in order to take advantage of higher rates available in other locations. Some argue that interest rate ceilings should be revoked, since their primary function is to discriminate against the interest earnings of small savers without safeguarding the supply of local funds for use in the community during periods of tight money. Proponents of interest rate ceilings argue that their removal would only increase the cost of credit without increasing the availability of funds for borrowing.

The second selective tool of monetary policy is the authority to regulate margin requirements (down payments) on stocks and bonds purchased on credit. This tool is effective in limiting the expansion in stock market credit at all times, not just during periods of extremely tight credit. Margin requirements are more limiting, however, when they reach high levels (they typically are raised to their peaks during periods of monetary restraint). This tool may also encourage the movement of funds out of stock and bond markets and into larger financial institutions during periods when the highest rates of return on deposits are available.

Emergency Credit

In addition to the privilege available to member banks to borrow from the Federal Reserve for normal adjustment purposes, banks may obtain emergency credit when facing financial stringency caused by adverse financial developments. In such cases the Federal Reserve serves as the "lender of last resort." Machinery also exists to allow nonmember financial institutions to borrow from the Federal Reserve under conditions of emergency.

Seasonal Borrowing Privilege

In 1973 the Board of Governors instituted a seasonal borrowing privilege for certain member banks that lack effective access to national money markets. This privilege is designed to help smaller banks meet demands for credit that normally peak at the same time that deposits are being drawn down. This is a normal occurrence for banks located in farming or resort areas.

The following are specific conditions that a member bank must meet to obtain eligibility for seasonal credit:

1. Lack reasonable reliable access to national money markets.

2. Have a seasonal need arising from a recurring pattern of movement in deposits and loans that persists for at least eight weeks.

3. Meet from its own resources that part of the seasonal need equal to at least 5 percent of its average deposits over the preceding calendar year.

4. Arrange with the Reserve Bank for seasonal credit in advance of the actual need for funds.

Unlike borrowing for adjustment purposes, seasonal credit, for some banks, may remain outstanding for a number of months.

Topics for Discussion

1. Discuss the historical background of the Federal Reserve System.
2. Discuss the Fed's remedies for banking problems.
3. Discuss the Fed's open market operations. Why is this tool more widely used than other Federal Reserve economic tools?

Problem Assignment

Obtain a public copy of the minutes of past Federal Open Market Committee meetings, and analyze them for trends in monetary policy.

Recommended Readings

Board of Governors, Federal Reserve System. *The Federal Reserve System: Purposes and Functions.* Washington, DC: Federal Reserve Board, 1984.

Colander, David C. *Economics,* 2nd ed. Chicago: Richard D. Irwin, Inc., 1995. Ch. 14.

Degen, Robert A. *The American Monetary System: A Concise Survey of Its Evolution Since 1896.* New York: Free Press, 1987.

Gwartney, James D., and Richard L. Stroup. *Economics: Private and Public Choices,* 7th ed. Orlando, FL: The Dryden Press, 1995. Ch. 12.

Hall, Robert E., and John B. Taylor. *Macroeconomics: Theory, Performance, and Practice,* 3rd ed. New York: W. W. Norton & Co., Inc., 1991.

Kidwell, David S., Richard L. Peterson, and David W. Blackwell. *Financial Institutions, Markets, and Money,* 5th ed. Orlando, FL: The Dryden Press, 1993. Ch. 25 and 26.

Timberlake, Richard H. *Monetary Policy in the United States: An Intellectual and Institutional History.* Chicago: The University of Chicago Press. 1993.

27

Input-Output Functions

In Chapter 12, we discussed the factor-product model, in which one input was used to produce one output. Although using a production function with only one variable input may be easier to understand, considering several inputs in the production of one output or in the production of several outputs is closer to reality. For example, farmers generally have a number of different crops they can produce from a variety of different inputs. This chapter presents the basic economic theory involved in choosing the inputs and outputs that will result in the most profitable production.

The Factor-Factor Model

The **factor-factor model** expands on the factor-product model by considering two variable inputs instead of only one. The production function for the factor-factor model can be expressed as

$$y = f(x_1, x_2 / x_3, \ldots, x_n)$$

where y is output, x_1 and x_2 are variable inputs, and $x_3, \ldots, x_n$ are fixed inputs. In this case, the producer must decide how much output to produce and what combination of the two variable inputs to use to produce that output.

Given the production function, the economic rationale for determining the most profitable combination of two inputs to employ in producing one output is based on the following:

$$\frac{MPP_{x_1}}{r_1} = \frac{MPP_{x_2}}{r_2}$$

where MPP_{x_1} and MPP_{x_2} are the marginal physical products of using x_1 and x_2 in the production of y and r_1 and r_2 are the per unit costs of inputs x_1 and x_2.

If we assume that output is held constant at some level, then we can determine the various input combinations that will produce that output level. Since output is fixed, gross dollar returns are the same at all combination levels of input use (pure competition). Profit maximization, then, simply calls for producing output at the least-cost or least-outlay input combination.

Least-cost input combinations exist where the ratio of substitution of Input 2 (x_2) for Input 1 (x_1) is equal to the ratio of input prices, r_1 to r_2. The ratio of substitution is referred to as the "marginal rate of technical substitution" (MRTS) and can be expressed as follows:

$$\text{Substituting } x_2 \text{ for } x_1: \text{MRTS} = \frac{\Delta x_1}{\Delta x_2}$$

or

$$\text{Substituting } x_1 \text{ for } x_2: \text{MRTS} = \frac{\Delta x_2}{\Delta x_1}$$

If we add the input price ratio, either r_1 to r_2 or r_2 to r_1, we develop the following decision rule:

$$\text{MRTS} = \frac{\Delta x_1}{\Delta x_2} = \frac{r_2}{r_1}$$

or

$$\text{MRTS} = \frac{\Delta x_2}{\Delta x_1} = \frac{r_1}{r_2}$$

Assuming it is profitable to produce at all, it is possible to determine the best combination of resources for a given level of production by finding the rates at which resources substitute for one another and their cost ratios.

The slope of the input substitution curve, also called an "isoquant," which is discussed in the next section, must be equal to the inverse of the slope of the cost line for a least-cost input combination. This relationship represents the optimum combination of two resources for a particular level of production.

This least-cost combination criterion does not tell how profitable the enterprise is. It merely suggests a combination of resources. Profit will be determined by all costs, total product, and price of the product.

We will now show the factor-factor problem graphically using isoquants and isocost lines. Isocost lines, or budget lines, indicate possible combinations of the two inputs that can be purchased with a given amount of money.

Nature of Isoquants

An **isoquant** normally refers to a line representing a given quantity of total output that can be produced with various combinations of two inputs, x_1 and x_2. "Iso" means equal, and "quant" is short for quantity. Thus, isoquant literally means "equal quantity." Each isoquant refers to a different but constant level of production, and isoquants farther from the origin indicate higher levels of production.

The shape of the isoquants shows the substitutability of the two inputs in the production process. If two inputs are perfect substitutes, the isoquants are straight lines, indicating that inputs substitute at a constant ratio. The gain from additional units of one input exactly offsets the loss from reduced usage of the other input (Figure 27–1–a).

Isoquants that are slightly convex to the origin of the graph illustrate input relationships in which inputs are good, but not perfect, substitutes for each other in the production process. This is the relationship generally encountered. Functionally, this relationship means that two inputs do not exchange at the same ratio at all levels of substitution (Figure 27–1–b).

If two inputs are poor substitutes, the isoquants become highly convex and ultimately curl away from the axes, meaning, for example, that more units of x_1 would be needed along with more units of x_2 to hold output at a given level (at high levels of use of x_1) (Figure 27–1–c).

Another way of determining the relationship between two inputs is by noting the slope of the isoquants. The slope of an isoquant at a particular point is the marginal rate of technical substitution (MRTS) at that point. Consider a small movement down an isoquant where a small amount of x_1 is substituted for x_2. By definition, output is constant as the gain in output from use of more x_1 is offset exactly by the loss in output from use of less x_2. The gain in output is the marginal physical product of the additional units of x_1 ($MPP_{x_1} * \Delta x_1$). The loss of output is foregone marginal physical product of the reduced units of x_2 ($MPP_{x_2} * \Delta x_2$). Thus, the slope of an isoquant at any point is equal to the ratio of the marginal physical products of the two inputs. The MRTS is diminishing; namely, isoquants are convex to the origin. Movement down an isoquant means that less and less x_2 can be traded off for given

Substitutional possibilities between two inputs

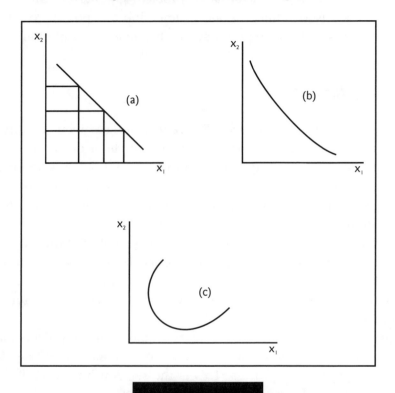

Figure 27-1

increments of x_1. Movement up an isoquant means less and less x_1 can be traded off for given increments of x_2.

Convexity to the origin is only one of three special characteristics of isoquants. Convexity indicates diminishing rates of technical substitution between inputs. The second characteristic is that isoquants do not intersect. If isoquants were permitted to intersect, it would mean that two levels of output would be possible with the same combination of inputs (a technical impossibility).

The third characteristic is that isoquants slope downward to the right. This essentially means that decreased usage of one input must be compensated by increased usage of the other input if output is to remain constant. This implies a substitute relationship between the two inputs.

Isoquants can be used to show returns to scale. Returns to scale are shown if the two inputs are the only inputs in the production process. If a graph represents a production process using only x_1 and x_2, then it can be used to determine returns to scale. A straight line drawn from the origin and intersecting the isoquants will show returns to scale. The spacing of isoquants being intersected reveals the response of output to varying the two inputs in the ratio shown by the slope of the line from the origin. The straight line from the origin is first crossed by isoquants at smaller, then at equal, and finally at increasing intervals. This corresponds with increasing, constant, and decreasing returns to scale (Figure 27–2).

An isoquant map showing returns to scale

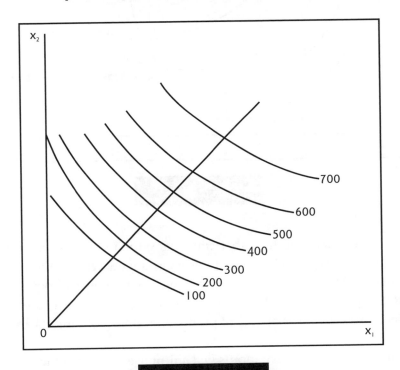

Figure 27–2

Just as we had stages of production in the one-variable-input case (the factor-product model), we have them in the two-variable-input case (the factor-factor model). However, the boundaries between Stages I and II are less easily explained.

To show the stages in the factor-factor case, we make use of ridgelines. Ridgelines connect isoquants where the isoquants are parallel to the axes. Hence, there are two ridgelines, one connecting the isoquants where they are

Ridgelines and stages of production

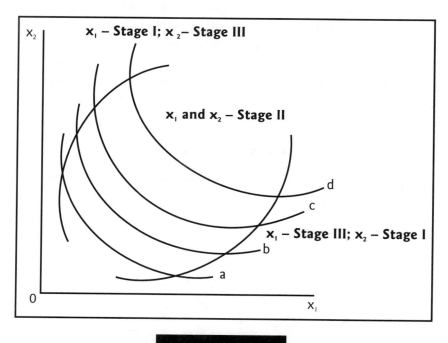

Figure 27-3

parallel to the x axis and the other connecting the isoquants where they are parallel to the y axis. The former connects isoquants where the slope of the isoquants, the MRTS, equals zero, while the other connects isoquants where the MRTS equals infinity. These are shown in Figure 27-3. The area between the ridgelines is the rational area of production, since both inputs are in Stage II.

What combination of two inputs will a firm use in the production process? An isoquant merely shows the combination of the two inputs that will produce a given constant level of output. We need more information if we are to determine the least-cost combination to use. That information is the prices of the two inputs. With knowledge of production response (isoquants) and input costs, we can determine the best combination to use to produce a given level of output.

Since the productive response is shown by the isoquant, we need to compare it with input prices to develop the best combination. But quantities, not prices, are plotted on the axes of the factor-factor graph. Fortunately, prices or costs can be represented on the same graph by developing a ratio between the two input prices. This ratio, when plotted on the graph, produces an isocost (equal cost) line. How can this ratio be established? It can be set up in either of two ways:

1. Establish a hypothetical outlay to be spent on the two inputs. Divide this outlay by the two input costs. Plot the resultant two quantities on the x_1 and x_2 axes. Finally, connect these two points with a straight line (the isocost line) (line a in Figure 27–4).

   For example, assume that we have \$16 to spend on inputs x_1 and x_2 and that the cost per unit of x_1 and x_2 (denoted r_1 and r_2) is \$2 and \$4. Then the isocost line can be derived.

   $$x_1 = \frac{\text{outlay}}{r_1} = \frac{\$16}{\$2} = 8 \quad \text{and} \quad x_2 = \frac{\text{outlay}}{r_2} = \frac{\$16}{\$4} = 4$$

Isocost lines

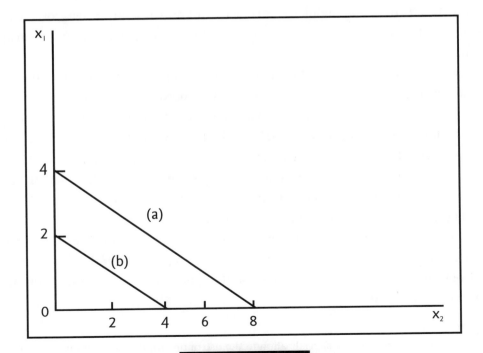

Figure 27–4

2. Plot the price per unit of x_2 as a quantity on the x_1 axis. Then plot the price per unit of x_1 as a quantity on the x_2 axis. Connect these two points with a straight line. The slope of this line (the isocost line) is equal to the inverse of the price ratio (line b in Figure 27–4).

Since the firm wants to produce any given level of output at the least cost, it will use that combination of inputs that will produce the given output with the smallest total outlay. This occurs where the isocost line is tangent to (i.e., just touches) the isoquant indicating the desired production.[1] To find the tangent point graphically, simply use a straightedge to represent the slope of the isocost line and move the straightedge parallel to itself until it is tangent to the desired isoquant. The tangent point shows the optimum combination of the two inputs to use for that output, given the production function and the input costs. The tangent point represents the least-cost input combination. Any other combination is more costly or will not produce the desired production level.

An illustration of the optimum combination of two inputs to produce a given output is shown in Table 27–1. The optimum combination consists of about 2 units of x_1 and about 22 units of x_2. Not only are input costs lowest at $35, but at that point the substitution ratios between the two inputs are also equal.

Note that as x_1 is increased from 0 to 1 unit, $5 in cost is added but that cost is also reduced by $30, since there is less of x_2 used. Next, with the increase from 1 to 2 units of x_1, we add another $5, but again we reduce the units of x_2 used, this time cutting cost by $15. With the next unit increase of x_1, we increase cost by $5 but also reduce cost by $5 with the reduction in x_2 (25 to 20). We have reached an equivalent of $5 in gain and loss (Table 27–1). At the next stage, we spend $5 more but save only $3; thus, we lose $2.

Figure 27–5 illustrates data presented in Table 27–1. Note the 100-unit isoquant and the isocost line drawn tangent to that isoquant. The least-cost combination ($35) is indicated at 2.5 units of x_1 ($5 per unit) and 22 units of x_2 ($1 per unit). The slope of the isoquant at that point is equal to the slope of the outlay or isocost line.

---

[1] In the case of complete specialization in the use of one input, the optimum combination occurs where the isocost line first touches the isoquant.

Table 27-1

Least-Cost Combination of Two Inputs to Produce a Given Output[1]

Combinations of Two Factors or Inputs		Total Output (y)	Total Cost of Inputs Used			Marginal Rate of Technical Substitution	
x_1	x_2		x_1	x_2	Both Inputs	$\left(\dfrac{\Delta x_2}{\Delta x_1}\right)$	$\left(\dfrac{r_1}{r_2}\right)$
------- Units used -------		Units	-------- Dollars --------			-------- Number --------	
0	70	100	0	70	70	30	5
1	40	100	5	40	45	15	5
2	25	100	10	25	35	5	5
3	20	100	15	20	35	3	5
4	17	100	20	17	37	2	5
5	15	100	25	15	40	2	5
6	13	100	30	13	43	3	5
7	10	100	35	10	45	—	—

[1] Assume that x_1 costs \$5 per unit and that x_2 costs \$1 per unit.

Point of tangency of isoquant and isocost lines, illustrating the least-cost output

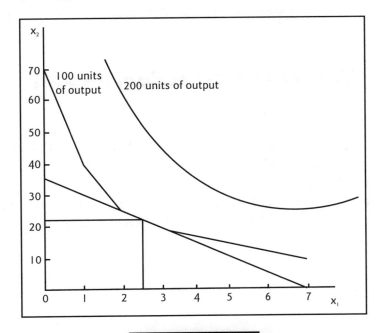

Figure 27-5

Three or More Inputs–One Output

To this point, we have discussed producing one output with one variable input and with two variable inputs. What can we say about resource allocation, i.e., the input combination, when producing one output with more than two inputs? The decision rule is basically the same as with two inputs.

$$\frac{MPP_{x_1}}{r_1} = \frac{MPP_{x_2}}{r_2} = \frac{MPP_{x_3}}{r_3} = \ldots = \frac{MPP_{x_n}}{r_n}$$

A relatively convenient method to determine the optimum combination of inputs to produce an output is using least-cost input combinations. While least-cost input combinations are not necessarily profit maximizing—for example, when capital is limited—the profit-maximizing input combination is always a least-cost input combination.

For situations involving several inputs, the necessary computations can become quite involved. In such cases, linear programming is suggested as a more efficient method of determining optimum input and output levels.

The Product-Product Model

We have discussed situations in which one output is produced using one or more inputs. We now consider the **product-product model**, where two or more outputs are produced using one or more inputs. In our discussion, we will consider a firm producing two outputs using one input.

Suppose we wish to produce two products, or outputs, with one input, such as labor. We will start by defining the production functions for each output:

$$y_1 = f(x)$$
$$y_2 = g(x)$$

where y_1 and y_2 are the outputs and x is the input that can be used to produce either output, but the firm has a fixed amount to use. Assume r is the per unit price of the input, x, and P_1 and P_2 are the per unit prices of the outputs, y_1 and y_2.

Given these assumptions, we can define the marginal value product functions for each output.

$$MVP_{x-y_1} = P_1 * MPP_{x-y_1}$$

and

$$MVP_{x-y_1} = P_2 * MPP_{x-y_1}$$

By combining the marginal value product functions with the input price, we can develop the decision rule to use when considering how much of each product to produce.

$$\frac{MVP_{x-y_1}}{r} = \frac{MVP_{x-y_2}}{r}$$

Table 27–2 shows an optimum combination of two hypothetical outputs or enterprises (y_1 and y_2) with one input, labor (x). The objective is to maximize revenue, since the input cost is constant for the firm. Should all six units of the input be used to produce y_1? If this were done, total revenue would be $48. If all six units of the input were used to produce y_2, the firm would generate $44 of revenue. Is there a more profitable combination? There is. Using about 4.5 units of x to produce y_1 and 1.5 units to produce y_2 yields $64 of revenue. At the point of optimum product combination, the marginal physical product of x used to produce y_1 times the price of y_1 equals the marginal physical product of x used to produce y_2 times the price of y_2.

From the data in Table 27–2, the optimum product combination from the use of the input x would be as follows:

$$MPP_{x-y_1} * P_1 = 4 * \$2 = \$8$$

and

$$MPP_{x-y_2} * P_2 = 4 * \$2 = \$8$$

Table 27–2

Determining Optimum Two-Enterprise Combinations with One Input Factor

Units of Labor Input Used		Total Output		Marginal Physical Product (MPP) of Labor		Marginal Rate of Product Substitution	Price Ratio[1]	Total Revenue ($)
Enterprise		Enterprise		Enterprise				
y_1	y_2	y_1	y_2	y_1	y_2		$P_2 \div P_1$	
0	6	0	22			4.00	1.00	44
				4	1			
1	5	4	21			4.00	1.00	50
				4	1			
2	4	8	20			2.00	1.00	56
				4	2			
3	3	12	18			2.00	1.00	60
				4	2			
4	2	16	16			1.00	1.00	64
				4	4			
5	1	20	12			0.33	1.00	64
				4	12			
6	0	24	0	–	–	–	–	48

[1] Assume that P_1, the price per unit of y_1, equals $2 and that P_2, the price per unit of y_2, equals $2.

Note in Table 27–2 that when the use of the input to produce y_1 is cut by 1 unit from 6 to 5, revenue is reduced by $8. Using that unit of input to produce y_2 increases revenue by $24. Reallocating another unit of the input from the production of y_1 to the production of y_2 reduces revenue by $8, but also increases revenue by $8. However, shifting the next unit of input from producing y_1 (from 4 to 3 units) to y_2 reduces revenue by another $8, while the additional production of y_2 increases revenue by only $4. Revenue is maximized when about 4 units of x are used to produce y_1 and about 1.5 units to produce y_2.

The data in Table 27–2 are presented graphically in Figure 27–6. The optimum output combination is where 18 units of y_1 are produced using 4.5 units of the input and 14 units of y_2 are produced using 1.5 units of the input, x. Total revenue, TR, equals

$$TR = P_1 * y_1 + P_2 * y_2 = \$2 * 18 + \$2 * 14 = \$64$$

Maximum profit combination using one input

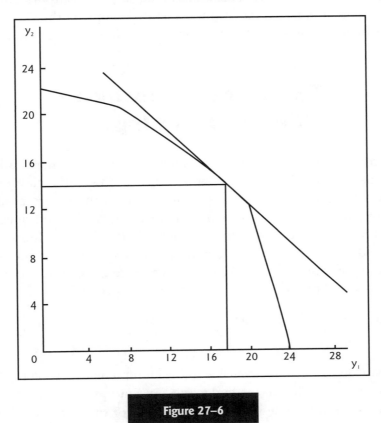

Figure 27–6

Three or More Outputs–One Input

To this point, we have considered situations involving the production of one output and two outputs using one input. What can be said about producing three or more outputs using one input?

Given the input, x, which can be used to produce any of the three outputs, y_1, y_2, and y_3, and the output prices, P_1, P_2, and P_3, the decision rule guiding how the variable input is used is

$$MPP_{x-y_1} * P_1 = MPP_{x-y_2} * P_2 = MPP_{x-y_3} * P_3$$

The additional value from the last unit of input used to produce y_1 must equal the additional value from the last unit of input used to produce y_2, and so on. If marginal value products are not equal, then the firm can increase its returns by reallocating the variable input to the output or outputs with higher marginal value products. If the firm has a limited supply of the variable input available, then it should use the input only where it is the most profitable.

Two or More Outputs– Two or More Inputs

Let's consider the more realistic situation where two or more variable inputs are used to produce two or more outputs. This is the case on many farms, where land, labor, capital, and management inputs are available in varying or fixed quantities and two or more enterprises, or outputs, can be produced from them. Consider the situation where a firm produces two outputs, y_1 and y_2, using two variable inputs, x_1 and x_2. Given output prices, P_1 and P_2, and input prices, r_1 and r_2, the decision rule guiding optimum output combinations and resource allocation can be expressed as follows:

$$\frac{P_1 * MPP_{x_1-y_2}}{r_1} = \frac{P_2 * MPP_{x_1-y_2}}{r_1} = \frac{P_1 * MPP_{x_2-y_1}}{r_2} = \frac{P_2 * MPP_{x_2-y_1}}{r_2}$$

Two basic techniques are available to the entrepreneur in reaching an optimum combination. The first is budgeting. An entrepreneur can construct a budget for each enterprise under consideration for the operation. The decision maker projects yields and prices for the products, estimates variable costs of producing these outputs, and allocates an equitable portion of fixed costs to each enterprise.

The next step is to calculate the net revenue per unit produced for each enterprise and project what total net revenue would be if all resources were allocated to a specific enterprise, such as soybeans, corn, or hogs. The effect of limiting resources must also be considered. For example, producing only soybeans may quickly exhaust available spring labor, while producing only hogs may quickly exhaust available cash. Thus, whole-farm budgeting must be considered in light of resource limitations.

The second technique is linear programming (LP). Business firms, including farm businesses, as we have already discussed, are concerned with the allocation of limited resources to achieve certain objectives. The objective of the decision maker is to achieve the best outcome or outcomes with the available resources. Linear programming is a popular mathematical programming tool that provides guidance in making decisions related to this objective.

The clearly defined objective of linear programming must be to optimize. Subject to appropriate constraints, this can be done by maximizing (profits, for example) or by minimizing (production costs, for example), but never by both.

The constraints reflect the limiting effects of the scarce resources the firm has to use in production. The limited resources could be money, production capacity (including land), personnel, time, and technology. The constraints impose restrictions on the activities—for example, on the enterprises of a farm plan, such as those considered in the objective function.

Relationships Between Enterprises

This chapter has focused primarily on competitive relationships between enterprises, but as discussed in Chapter 11, there are four types of relationships between enterprises: (1) competitive, (2) complementary, (3) supplementary, and (4) joint. As shown in Figure 27–7, from A to B are complementary relationships, where an increase in production of y_1 is accompanied by an increase in production of y_2. From B to C are supplementary relationships,

Various types of enterprise relationships

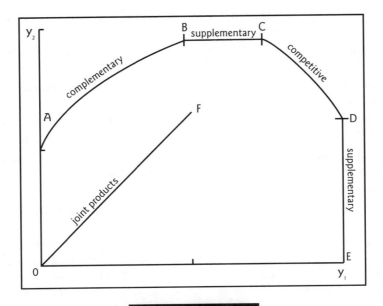

Figure 27–7

where an increase in y_1 does not affect production of y_2. From C to D are competitive relationships, where an increase in y_1 leads to a reduction in y_2 or where an increase in y_2 leads to a reduction in y_1. From D to E are supplementary relationships, where an increase in y_2 does not affect y_1. From 0 to F the relationships are joint, where an increase in y_2 is accompanied by a direct proportional increase in y_1. For example, increasing cotton lint production also results in close to a direct proportional increase in cotton seed production.

Topics for Discussion

1. Discuss the economic decision rule guiding input usage decisions in the factor-factor model, using two variable inputs to produce one output.
2. Discuss the economic rationale for the product-product model, using one input to produce two outputs.

Problem Assignment

Prepare hypothetical data tables, and illustrate the least-cost combination for (1) two inputs and one output and (2) one input and two outputs.

Recommended Readings

Boehlje, Michael D., and Vernon R. Eidman. *Farm Management.* New York: John Wiley & Sons, Inc., 1984. Ch. 5.

Cramer, Gail L., and Clarence W. Jensen. *Agricultural Economics and Agribusiness,* 6th ed. New York: John Wiley & Sons, Inc., 1994. Ch. 5.

Debertin, David L. *Agricultural Production Economics.* Macmillan Publishing Co., 1986. Ch. 5, 6, 15, 16, 17.

Dillon, John L., and Jock R. Anderson. *The Analysis of Response in Crop and Livestock Production,* 3rd ed. Elmsford, NY: Pergamon Press, Inc., 1990. Ch. 2 and 3.

Doll, John P., and Frank Orazem. *Production Economics: Theory with Applications,* 2nd ed. Melbourne, FL: Krieger Publishing Co., 1992 reprint of 1984 ed. Ch. 3 and 4.

Kay, Ronald D., and William M. Edwards. *Farm Management,* 3rd ed. New York: McGraw-Hill, Inc., 1994. Ch. 6, 8, 10.

Luening, Robert A., Richard M. Klemme, and William P. Mortenson. *The Farm Management Handbook,* 7th ed. Danville, IL: Interstate Publishers, Inc., 1991.

Mjelde, J. W., et al. "Integrating Data from Various Field Experiments: The Case of Corn in Texas," *Journal of Production Agriculture,* 4(1):139–147 (1991).

Seitz, Wesley D., Gerald C. Nelson, and Harold G. Halcrow. *Economics of Resources, Agriculture, and Food.* New York: McGraw-Hill, Inc., 1994. Ch. 6.

28

Selected Principles of Agribusiness Management

In any business organization, there is a person or group of persons with the responsibility for making sure the work gets done—the manager or managers. In this chapter, we will explore some basic management principles pertaining to agribusiness firms. We will concentrate primarily on principles involved in monitoring business performance.

The Functions of Management

The manager of a business has the responsibility of ensuring that a variety of tasks are carried out. These tasks can be grouped into general categories, the *functions* of management. The functions describe what managers do, regardless of the part of the firm under consideration—production, processing, distribution, sales, etc.

The functions of management include planning, organizing, implementing, and controlling. Of course, these functions overlap, at least to some extent. Decisions fulfilling one function should not be made without considering their effects on the other functions and on potential outcomes.

Planning involves considering all the activities that determine the future direction or course of action for the business. Planning begins with the firm's identifying and defining its purpose and objectives. Some questions to consider include: Why do we exist? What are our goals? How can we achieve them? Answering these questions carefully is a major step in establishing an identity for the business.

Time frame—short term, intermediate term, or long term—is important in making planning decisions. Short-term planning should lead into and mesh with intermediate-term planning, which, in turn, should lead into and mesh with long-term planning. All three types of planning are involved in moving the business toward achieving its overall goals and objectives.

Organizing involves combining the people, resources, and business activities necessary to bring the goals and objectives identified in the planning stage into reality. A key aspect of this function centers on business organization and structure, which influence how the efforts of individuals within the business are coordinated.

Implementing the plans set for the business involves overseeing and directing how those plans are executed, including delegating the work appropriately and communicating expectations clearly. Thus, implementing is concerned with employee relationships. Another important aspect of the implementing process is coordinating the activities to avoid "bottlenecks." A good example of this is inventory management in a feed mill. There it is essential to make sure enough grain and supplements are available so the mill can fill mixed-feed orders on a timely basis, without any shortages. At the same time, though, inventory levels must not be too large, because that would involve commitment of much capital that could be used elsewhere in the business.

Controlling is the management function that involves the activities necessary to ensure that the policies of the business are, in fact, being carried out. Managers must be able to measure the progress toward the goals set in the planning process.

A business needs a system of internal controls that keeps managers aware of individual, departmental, and overall performance. Such a system helps a manager determine whether policies are being carried out and whether policy changes are needed. It also provides important performance criteria, both physical (e.g., output per employee) and financial (e.g., profits, net worth, and cash flow). The complexity of the control system is usually dictated by the size and complexity of the organization.

The Business Decision-Making Process

In carrying out the functions of management described above, managers spend much of their time making decisions that affect the success of the business. Management decision making is described in this section as an orderly process involving six interrelated steps:

1. *Identifying the problem or opportunity.* Identifying and defining the problem or opportunity is usually the most important step in the decision-

making process, since the perception of the problem or opportunity influences the actions that will be taken in response.

2. *Determining alternative courses of action.* Once a problem or opportunity has been clearly identified and defined, the next step is developing a list of all pertinent courses of action to solve the problem or capitalize on the opportunity. After the alternatives identified in this step have been reflected upon, the several that appear most promising should be selected for more careful consideration.

3. *Analyzing the alternatives.* The possible courses of action have been narrowed to those that appear the best for the situation faced. Analyzing these alternative courses of action involves evaluating them in much more detail to determine how each would accomplish the goals and objectives of the firm. The evaluation involves selecting the best measurable criterion (e.g., sales, costs, units produced, etc.) and gathering appropriate data, necessary facts, and other information. It also involves selection of an appropriate and objective method of analysis.

4. *Selecting the best alternative.* Given the information provided in the previous steps, the next step is deciding which course of action is "best." The firm's long-and short-run objectives must be considered as this selection is made.

5. *Implementing the decision.* Now that the "best" action has been selected, the manager must ensure that the decision is executed in as timely a manner and as efficiently as possible. Good communication is a must for successful implementation.

6. *Following up.* The final step is evaluating the performance of the decision. Such a review helps a manager decide if the course of action chosen needs to be changed or modified, given the additional information available. Steps 1 through 5 should never be viewed as final, but in need of constant evaluation and corrective adjustment. This is especially important in a business environment characterized by changing technology, laws, regulations, market structure, and personal conditions.

Economic Activities of a Firm

Managers execute the functions of management in three types of economic activities: production, marketing, and financing. Production activities require decisions concerning what products will be produced or what services will be provided, how and when they will be produced or provided, and in what amounts. Marketing activities involve decisions related to obtaining inputs for the production process, as well as pricing and distribution of products and services. Financing activities require management decisions regarding the firm's capital needs, including how to gain control of additional capital and how to use available capital.

The economic activities clearly overlap. When making decisions in one area, managers should carefully consider how the other areas will be affected. A firm will not want to produce a new product without carefully considering the market potential of that product. Decisions on what products to produce and in what quantities also determine, in part, the firm's capital requirements and their timing. Production decisions should be made in light of seasonal trends in marketing. Marketing decisions, such as the selection of input suppliers and marketing outlets for products and the timing of purchase and sales transactions, are usually influenced by financing activities, such as credit availability.

Monitoring Business Performance

Managing a business requires a wide range of information on physical and financial performance. Physical performance measures include such things as total plant production, output per employee, time involved in a production process, etc. Our focus here, though, is on monitoring financial performance.

Managers need to keep accurate records. Too often the information is recorded in the mind of a manager, where it is subject to forgetfulness, or worse yet, the manager's biases. In other cases, managers keep sporadic and informal records, with notes jotted here and there. In either case, chances are good that a manager will be using inaccurate and incomplete information when making decisions that influence the future of the operation.

Financial analysis involves maintaining and using records and other information needed to measure the financial performance of the business. There are two broad reasons why a manager should keep a good set of financial records.

The first is to report business performance to others, including owners, lenders, and taxing bodies. By using information contained in the financial statements, the manager can analyze the financial performance and strength of the business. By keeping a set of records over time, the manager can monitor the financial progress of the firm. The manager can use the records to justify the need for a loan and to show the ability to repay a loan.

The second broad reason for keeping financial records is to be able to use the data for managerial decision making. A complete set of financial records allows the manager to evaluate the performance of particular enterprises or departments within the business. The manager can use projected financial statements to help evaluate alternative enterprises and investments.

Sources of Information for Financial Control

Key sources of information for financial analysis are accurate financial statements prepared from well-maintained records. In the discussion here we will briefly introduce two important financial statements, the balance sheet and the income statement. Other useful financial statements include the statement of cash flows and the statement of owner equity.

It is critical that financial statements be prepared on a consistent basis. This means that the income statement, the statement of cash flows, and the statement of owner equity cover an identical period and that there is a balance sheet for the beginning of that period and one for the end.

We should note that income tax returns, while useful for verification purposes, are not considered a part of the basic set of financial statements or a substitute for the income statement.

The Balance Sheet

We will start with the balance sheet, which is also called the "net worth statement," the "statement of financial condition," or the "statement of financial position." The balance sheet is a snapshot of the firm's financial condition at a point in time. It is formally defined as a statement that provides information about an entity's assets (what it owns), liabilities (what it owes), and equity (what it is worth) and about the relationships of these elements to each other at a particular time. The balance sheet delineates the entity's resource structure (the major classes and amounts of assets) and its financing structure (the major classes and amounts of liabilities and equity).

The classic accounting identity underlying the balance sheet is

$$\text{Assets} = \text{Liabilities} + \text{Equity}$$

Liabilities represent claims on the firm's assets by lenders and other creditors. Equity or net worth represents the claims of owners on those assets. The total claims of creditors and owners cannot exceed the total value of the assets.

Before we get into more specific detail about the balance sheet, we should note one thing the balance sheet is not designed to do. The balance sheet, by itself, is not intended to show the value of the business, although some people may disagree with this statement. The balance sheet is useful, however, in estimating the value of a business when it is considered along with other financial statements and other information. But even a market-value–based balance sheet represents only an estimate of tangible asset market value. True value can be determined only from an actual sale of the assets to a third-party purchaser.

Assets are classified according to their liquidity. **Liquidity** refers to the firm's capacity to generate cash quickly and efficiently to meet its financial commitments as they come due without disrupting normal business operations. Highly liquid assets are those that can be quickly converted to cash without affecting the operation of the business or the net value of the business. Highly illiquid assets are those that are difficult to convert to cash, and doing so results in a sharp loss in value to the business.

Current assets include cash and near-cash inventory items that will be converted to cash during the year in the normal course of business. They also include assets whose conversion to cash will only minimally disrupt the normal business operation. Besides actual cash, examples of current assets include checking accounts, accounts receivable to be paid during the year, and inventory on hand that will be sold during the year.

Noncurrent assets yield services to a business over several years. Many noncurrent assets eventually are fully depreciated and are replaced, or they are liquidated. Depreciable assets include machinery, equipment, and buildings. Real estate is also a noncurrent asset. Other noncurrent assets include investments in capital leases and investments in other entities.

Current liabilities are the existing obligations payable within the operating year. Examples include accounts payable to the local farm supply store, that portion of notes payable to lending institutions that will fall due within the next 12 months, accrued interest, taxes, and rent.

Noncurrent liabilities are those debt obligations with a maturity of more than a year. They primarily involve the noncurrent portion of notes payable on non–real estate and real estate assets. A subclassification of noncurrent liabilities by loan maturities allows the manager to evaluate the firm's liquidity and evaluate the debt structure relative to the firm's assets.

The Income Statement

Most firms are in business to earn a profit. Profit, or net income, is the single most important factor that determines whether a firm stays in business.

Accurately measuring net income generated from business operations is the key to accurately evaluating current financial performance, the progress made, and future potential. A properly prepared income statement provides a way to measure net income.

The income statement, also called the "profit-and-loss statement" or the "operating statement," is a summary or measure of the revenue (receipts or income) and expenditures (costs) of the business over a specified period (usually one year) to determine its profit position. The income statement provides information about the sources for changes in equity that occur from one balance sheet to the next. The objective of the income statement is to generate a net income that reconciles the change in equity during the year.

The income statement usually has four parts:

1. Revenue

2. Expenses

3. Taxes

4. Other adjustments

Revenue is generated by cash sales and changes in inventory. To properly identify the value of everything produced during the year, certain items must be included as revenue. Among these are all cash received from the sale of products and services, all other cash receipts, all changes in the value of inventories of products on hand from the beginning to the end of the year, and all changes in accounts receivable from the beginning to the end of the year. The changes in inventories and accounts receivable are calculated using the beginning and ending balance sheets.

Expenses are grouped into three categories: cash operating expenses (except interest), noncash adjustments to expenses (e.g., depreciation), and interest on borrowed capital. Separating operating and financing (interest) expenses makes the income statement more useful for analysis purposes because it allows the manager to compare operating expenses efficiency without different debt loads and interest payments clouding the picture. The noncash adjustments are also separated out. They ensure that the expense section accurately reflects the costs incurred in this year's operation.

Expense is not synonymous with expenditure. *Expense* refers to the cost of goods and services used in the process of earning revenue for a given period. *Expenditure* refers to an actual outlay. For example, assume a manager purchases for $20,000 a capital asset that will be used every year for the next 10 years. Assume also that there will be no salvage value. The expenditure in this case is $20,000, while the expense is the annual depreciation, $2,000 (assuming straight-line depreciation, which allocates the original cost on a year-by-year basis in relation to the proportion of total services yielded annually by the capital asset).

Depreciation is an accounting procedure by which the original or purchase cost of an asset is prorated over its projected useful life to reflect the approximate rate at which the asset's total service flow is consumed. In other words, depreciation is the allocation of the expense that reflects the "using up" of assets. Typical depreciable assets in which agribusinesses invest include machinery and equipment, buildings, and real estate improvements.

To calculate the amount of depreciation that should be charged as an expense during any period of an asset's useful life, the following information is needed:

1. The original cost (usually the purchase cost) of the asset
2. The depreciation method that will be used to determine how much of the asset was used up during the period
3. The salvage value of the asset, which is an estimate of what the asset will be worth at the end of its useful life

Managers should keep a separate account for each depreciable asset, in which the original cost, accumulated depreciation, and depreciated value are shown.

Three methods are commonly used to calculate economic depreciation: straight-line, declining balance, and sum-of-the-years' digits.

Straight-line depreciation provides the same amount of depreciation during each year of an asset's useful life. The formula for calculating depreciation using the straight-line method is

$$D_{sl} = \frac{OC - SV}{N}$$

where

OC = original cost of the asset
SV = projected salvage value
N = asset's expected useful life in years

For example, let's say a feed mill bought a delivery truck for $50,000. Assume it has a useful life of 10 years, with a salvage value of $5,000. The depreciation is calculated as follows:

$$D_{sl} = \frac{50{,}000 - 5{,}000}{10} = \$4{,}500 \text{ per year}$$

Straight-line depreciation is the easiest to calculate and probably the most widely used.

The declining balance method provides an accelerated charge-off for depreciation. The formula for the declining balance method is

$$D_{db} = \frac{X}{N} R$$

where

N = the years of useful life
R = the remaining book value of the asset at the beginning of the year

The (X/N) part of the equation gives the depreciation rate. The double declining balance, where x = 2, is commonly used, so

$$D_{ddb} = \frac{2}{N} R$$

The declining balance method uses a fixed rate of depreciation each year and applies that rate to the asset's remaining value at the beginning of the year, until the salvage value is reached. In the example, the depreciation in the first year would be

$$\frac{2}{10} (50{,}000) = 10{,}000;$$

in the second year,

$$\frac{2}{10} (50{,}000 - 10{,}000) = 8{,}000;$$

in the third year,

$$\frac{2}{10} (32{,}000) = 6{,}400;$$

and so forth.

The sum-of-the-years' digits method calculates depreciation in a given year using this formula:

$$D_{sy} = \frac{n}{\sum \text{ of the years' digits}} (OC - SV)$$

where n is the number of years of useful life remaining and OC and SV are as defined before.

For the example, the denominator is

$$10 + 9 + 8 + 7 + 6 + 5 + 4 + 3 + 2 + 1 = 55$$

An easier way to calculate that is to use the formula

$$\frac{N(N+1)}{2} \rightarrow \frac{10(10+1)}{2} = \frac{110}{2} = 55$$

So, for the example, depreciation for the first year is

$$\frac{10}{55} (50{,}000 - 5{,}000) = 8{,}182$$

In the second year, the numerator is decreased by 1:

$$\frac{9}{55} (45{,}000) = 7{,}364$$

and so on.

Financial Analysis and Control

We briefly discussed the balance sheet and income statement. Now we will discuss how to use them. Financial statements provide information needed to evaluate the profitability, solvency, and liquidity of the business. Measuring and monitoring business performance over time is a critical part of the financial control process. Without measures of business performance, there is no way for managers to evaluate their decisions and no way to tell whether they need to change plans because of poor performance. Our goal in this section is to introduce key measures of financial position or financial performance.

Financial position refers to the total resources controlled by a business and the total claims against those resources at a single point in time. Measures of financial position provide an indication of the capacity of the business to withstand risk from future operations and provide a benchmark against which to measure the results of future business decisions.

Financial performance refers to the results of production and financial decisions over one or more periods of time. Measures of financial performance include the impact of external forces beyond anyone's control (such as drought, grain embargoes, etc.) and the results of operating and financing decisions made in the ordinary course of business.

The need to measure financial position and financial performance increases as managers rely more on capital, either borrowed or invested. With borrowed capital (debt), the manager must be able to show the lender that the capital will be used in a way that will assure the lender's being paid the agreed-upon rate of interest (for "rent" of the money) and being repaid the entire amount of principal advanced. Since the lender is going to receive repayment of the principal with interest only if the business is successful, he or she will want assurance that the risk of failure is reasonably low.

Invested capital (equity) is usually viewed from two perspectives: capital invested by the owners and capital invested by others not involved in the operation (which could include nonactive stockholders or partners in the business). Both active managers and passive investors need measures of the return on the capital invested, because both have the option of leaving their capital invested in the business or investing it in something else.

The financial measures we are will discuss can be grouped into several broad categories: liquidity, solvency, profitability, and activity ratios. They measure either financial position or financial performance.

Liquidity Measures

Liquidity refers to the ability to meet financial obligations as they come due in the ordinary course of business, without disrupting the normal business operations. The common liquidity measures are the current ratio, net working capital, and the quick ratio.

The current ratio measures the amount of current assets per dollar of current liabilities:

$$\text{Current ratio} = \frac{\text{Current assets}}{\text{Current liabilities}}$$

For example, if the current ratio is 2, then the firm has $2 of current assets for every dollar of current liabilities.

Net working capital is the difference between current assets and current liabilities.

$$\text{Net working capital} = \text{Current assets} - \text{Current liabilities}$$

It reflects the amount of current assets being financed using noncurrent debt. If the current ratio is less than 1, net working capital will be negative, indicating potential liquidity problems.

The quick ratio, also called the "acid test ratio," is calculated as follows:

$$\text{Quick ratio} = \frac{\text{Current assets} - \text{Inventories}}{\text{Current liabilities}}$$

By subtracting inventories, the quick ratio recognizes that a firm's inventories are often one of its least liquid current assets. Inventories, especially work-in-progress and carry-over inventories near the end of a season, are often very difficult to liquidate quickly without discounting prices significantly.

Another useful liquidity measure is the current asset turnover:

$$\text{Current asset turnover} = \frac{\text{Net sales}}{\text{Average current assets}}$$

The current asset turnover, which shows sales revenue per dollar of average current assets, measures the extent to which a firm works its current assets.

Solvency Measures

Solvency refers to the amount of borrowed capital (debt), leasing commitments, and other expense obligations a business has relative to the amount of owner equity invested in the business. Debt capital is interest bearing and has a date by which it must be repaid. Solvency measures provide an indication of the firm's ability to repay all financial obligations if all assets were sold (at the prices used when completing the balance sheet). Also, it provides an indication of the ability to continue operations as a viable business after a financial adversity, which usually results in increased debt or reduced equity.

Common solvency measures include the debt ratio, the debt-to-equity ratio, and the times interest earned ratio.

The debt ratio, which is calculated

$$\text{Debt ratio} = \frac{\text{Total debt}}{\text{Total assets}},$$

measures the proportion of a firm's total assets financed with borrowed funds. As the debt ratio increases, so do the firm's fixed interest charges. If the debt ratio becomes too high, the cash flows the business generates during slow periods may not be sufficient to meet interest payments.

The debt-to-equity ratio, also called the "leverage ratio," is calculated this way:

$$\text{Debt-to-equity ratio} = \frac{\text{Total debt}}{\text{Total equity}}$$

It relates the amount of a firm's debt financing to the amount of equity financing. The higher the leverage ratio, the greater the percentage of a firm's financing that comes from outside sources.

The times interest earned ratio, which is calculated

$$\text{Times interest earned ratio} = \frac{\text{Earnings before interest and taxes}}{\text{Interest charges}},$$

uses income statement data to measure a firm's use of financial leverage. It tells the manager how well the firm's current earnings are able to meet current interest payments. Earnings before interest and taxes are used because the interest charges are paid out of operating income.

Profitability Measures

Profitability refers to the extent to which a business generates a financial gain from the use of its resources, i.e., its land, labor, management, and capital. Profitability ratios measure how effectively a firm is generating profits on sales, total assets, and owner equity.

The gross profit margin ratio, which is calculated

$$\text{Gross profit margin ratio} = \frac{\text{Sales} - \text{Cost of sales}}{\text{Sales}},$$

measures the relative profitability of a firm's sales after the cost of sales has been subtracted. It shows how well the management is making decisions about pricing and the control of production costs.

The net profit margin ratio, which is calculated

$$\text{Net profit margin ratio} = \frac{\text{Earnings after taxes}}{\text{Sales}},$$

measures how profitable a firm's sales are after all expenses, including taxes and interest, have been deducted.

The return on investment (ROI) is calculated as

$$\text{ROI} = \frac{\text{Earnings after taxes}}{\text{Total assets}}.$$

It measures a firm's net income in relation to its total asset investment.

The return on owner equity (ROE) is calculated as

$$\text{ROE} = \frac{\text{Earnings after taxes}}{\text{Owner equity}}.$$

It measures the rate of return the firm earns on its owner equity. The ROE may be a more appropriate measure of return if a firm leases or borrows a large portion of its assets, such as land, equipment, etc. For businesses that are not incorporated, the earnings after taxes must reflect an adjustment taking out the opportunity value of the owner's time and management.

Activity Ratios

The activity ratios measure the activity of a business relative to its inventory levels, its customer credit payments, and its own bill paying.

The inventory turnover ratio, which is calculated

$$\text{Inventory turnover ratio} = \frac{\text{Cost of sales}}{\text{Average inventory}},$$

measures the activity of inventory by the number of times inventory is used up or turned over during the year.

The accounts receivable turnover, which is calculated

$$\text{Accounts receivable turnover} = \frac{\text{Credit sales}}{\text{Accounts receivable}},$$

identifies how long a firm must wait to get its money from credit sales. It is determined by dividing the credit sales in the accounting period by the value of accounts receivable.

The accounts payable turnover, which is calculated

$$\text{Accounts payable turnover} = \frac{\text{Credit purchases}}{\text{Accounts payable}},$$

identifies how fast a firm pays its own bills. It is determined by dividing the firm's credit purchases in the accounting period by the value of accounts payable.

Budgets

An effective financial manager must have a thorough knowledge of the sources and uses of cash in the business. Even if a firm has a strong balance sheet and income statement, it may be difficult to generate cash to meet cash commitments. Budgets provide much of the information necessary to anticipate and avoid problems. Two types of budgets are discussed: (1) operating budgets and (2) cash flow budgets.

Operating Budgets

The first phase of an operating budget is to estimate sales volume in terms of both units and selling price per unit. Then, both variable costs and fixed costs are projected, based on that sales volume. Net margins are also projected. Ratio analysis can be used, based on the projected operating statement and balance sheet.

As the business year progresses, the operating budget can be evaluated against actual results. This technique calls for specific forecasts of how much will be spent for each item during a given month, quarter, or year. Actual bills are compared with estimates. The reasons for the differences, if any, are then sought. Where actual results are worse than the forecasts, corrective actions can be taken.

Cash Flow Budgets

Cash flow budgets, also called the "sources and uses of funds," estimate cash flows, both outflows and inflows, per specified period, i.e., monthly, quarterly, semiannually, or annually. They include only anticipated cash transactions, either revenues or expenditures; they do not include noncash items, such as depreciation.

A cash flow budget has several important uses:

1. The cash flow budget formalizes the business planning process, since it requires managers to think through the operating year, anticipating business transactions and their timing.

2. Because the cash flow budget helps in identifying the expected timing of cash flows, managers can use it to develop better borrowing and repayment plans. Cash flow budgets are especially important for seasonal-type businesses.

3. The cash flow budget serves as a financial control tool, allowing managers to track actual cash flows and compare them to the projected amounts. Managers can monitor the performance and progress of business plans and modify those plans as necessary throughout the year.

4. The cash flow budget allows coordination among business plans of all the firm's divisions by bringing them together into a "big picture."

5. The cash flow budget assists with the management of cash reserves, which combined with financing arrangements, provide the way for firms to meet cash obligations as they come due.

6. The cash flow budget provides important information necessary to complete pro forma financial statements.

Pro Forma Financial Statements

In addition to preparing operating and cash flow budgets, the financial manager should also develop a pro forma income statement and a pro forma balance sheet for each budget period. *Pro forma* refers to setting up accounting information in advance, with projections for a future period. These statements will reflect expected trends in the profit prospects and general financial condition of the firm.

Pro forma financial statements indicate where the firm will be financially at the end of the period if the period unfolds as anticipated. The tendency may be to think "Why bother?" since things seldom turn out as expected. However, it is this business uncertainty that makes completing the pro forma financial statements that much more important.

Marketing and Sales Management[1]

In this section, we will briefly introduce some key marketing and sales policies as they relate to a firm's marketing management. First, we must explain some underlying concepts.

Sales Management Concepts

Market potential is an estimate of the maximum possible sales opportunities present in a particular market segment *open to all sellers of a product or service* during a specific future period. For example, an estimate of the number of gallons of herbicide that might be sold in the Mississippi Delta in a given year represents the market potential for herbicide in the Mississippi Delta during that year. Market potential indicates how much of a particular product can be sold to a particular market segment during some future period, assuming appropriate marketing methods.

Analyzing market potential involves a three-step process. First is identifying the market, which requires answering such questions as: Who buys the product? Who uses it? Who else might be prospective buyers and/or users? Market identification studies reveal demographic and economic characteristics that differentiate the market segments that make up the product's market potential.

---

[1] Adapted from Richard R. Still, Edward W. Cundiff, and Norman A. P. Govoni, *Sales Management: Decisions, Strategies, and Cases,* 5th ed. (Englewood Cliffs, NJ: Prentice-Hall, Inc., 1988).

The second step in analyzing market potential is determining why customers buy the product and why potential customers might buy it. This involves answering these questions: Why do people buy? Why don't people buy?

Given the information gained in the first two steps, the third step is selecting and using market factors, which are market features or characteristics related to the product's demand, to identify market segments with potential buyers.

Sales potential is an estimate of the maximum possible sales opportunities present in a particular market segment *open to a particular firm selling a product or service* during a specific future period. For example, an estimate of the number of gallons of herbicide that the XYZ Farm Supply Store might sell in a given year in the Mississippi Delta is the herbicide sales potential of the XYZ Farm Supply Store for that year.

Sales potential represents sales opportunities available to a particular firm, while market potential indicates sales opportunities available to an entire industry. Sales potentials are derived from market potentials based on historical market share relationships and on the firm's and its competitors' selling strategies and practices.

Sales forecast is an estimate of sales, in dollars or physical units, in a future period under a particular marketing program and an assumed set of economic and other factors outside the unit for which the forecast is made.

The principal factors in forecasting sales volume are (1) defining the market area, (2) evaluating the firm's past and present sales data, (3) projecting general and specific economic conditions for the market area, (4) projecting the firm's expected sales volume, and (5) evaluating sales projections against actual results.

A sales forecasting method is a procedure for estimating how much of a given product or product line can be sold if a given marketing program is implemented. Forecasting methods range from relatively simple (basing forecasts on compiled results of interviews with industry experts) to quite sophisticated (using regression analysis and econometric modeling).

Depending on the method, some of the more important sales forecasting variables include:

1. Reliable economic data (current and historical)
2. Product price and its effect on sales
3. Product and service quality
4. Substitute products available and their prices
5. Advertising and promotion programs

The sales forecast itself consists of the following:

1. *Sales by product lines.* Sales are broken down by products and services to aid in planning.
2. *Sales figures,* preferably in units rather than dollars, because prices change, but units remain stable. Units also lend themselves more suitably to inventory control.
3. *Sales by territories.* This categorization is for the purpose of controlling the sales effort. The facts also aid in determining business conditions in the respective areas.
4. *Sales by month, week, or some other period* to identify seasonal fluctuations.
5. *Sales according to accounts.* Products sold at different prices to wholesalers, retailers, etc., require forecast of sales by customer classes to determine expected income and cash receipts. This categorization also helps in evaluating the advisability of sales to each customer class.

Sales Policies

There are three major types of sales policies: product policies (what to sell), distribution policies (to whom to sell), and pricing policies (how much to charge). These identify the guidelines within which the firm attempts to meet its sales objectives.

Product policies serve as a guide for making product decisions. The products a firm sells determine its basic nature. As entrepreneurs pursue opportunities to make and/or market certain products, a business comes into being. As it grows, the management makes product decisions that affect the future of the business: Do we keep or drop old products? Do we add new products? Do we expand existing product lines and/or add new ones? These decisions, others concerning product design and quality, and product-related matters such as guarantees and service should be addressed in the firm's product policies.

Distribution policies determine the functions of a firm's sales force or department. The choice of one or more particular marketing channels sets the pattern for sales force operations, both as to geographic area and as to the customers from whom sales personnel solicit orders. The decision on the number of outlets at each distribution level affects the size and nature of the firm's sales force and the scope of its activities.

A firm's pricing policies influence the ease with which its sales force makes sales. Product or service pricing policy is influenced by several factors. The economic structure of the market, including the degree of competition, helps determine pricing policy relative to the competition. If competition is price-based, the firm sells its product at the same price as its competitors. If competition is nonprice-based, the firm generally meets the competition's price in an attempt to avoid using price as a competitive weapon. Instead, it will focus on nonprice attributes, such as product and/or service quality.

A firm also has pricing policy influenced by the cost of its product. Long-run sales revenues must cover all long-run costs for the firm to survive, but short-run sales revenues need not cover all short-run costs. Pricing methods and strategies, which are guided by pricing policies, are discussed in the next sections.

Pricing Methods

Full-Cost Pricing

Many businesses take the view that the price of each product must cover all the costs of that product. This approach, called "full-cost pricing," means that the price will cover labor, materials, and overhead costs, plus a predetermined percentage for profit or margin.

However, few businesses adhere rigidly to a full-cost pricing policy. Many managers express a preference for such an approach, but when it comes actually to establishing prices, their decisions are influenced by demand and other factors that have little relationship to the seller's costs.

Flexible Markups

A common practice is to use full cost, not as an inflexible point at which the price is to be set, but as a floor below which the price will not be allowed to fall—a reservation price or a reference to which flexible markups are added. Two kinds of flexibility are found in actual pricing: (1) adjustments over the course of time to changes in demand or in competition and (2) variations among different products due to differences in the market for individual products.

Pricing with Product Costs Unknown

Some firms cannot use a full-cost formula for pricing because their costs cannot be estimated precisely enough. Garden and landscape nurseries illustrate this situation. Their products develop over a period of years and are subject to uncertain damage from weather and disease and to shifts in buyers' requirements. It is difficult to allocate labor costs, overhead, and funds tied up in products in process. Therefore, pricing decisions are strongly influenced by demand and competition.

Gross-Margin Pricing

Another widely used pattern of pricing is that of adding a markup to the cost. This method, gross-margin pricing, is customary in both retailing and wholesaling. Some firms compute the markup as a percentage of cost; others, as a percentage of selling price.

As shown below, it is simple to convert markups based on the retail price to markups based on cost, and vice versa:

Retail Price Markup of:	equals	Product Cost Markup of:
(percent)		(percent)
50.0		100.0
33.3		50.0
25.0		33.3
20.0		25.0
10.0		11.0

The applicable formula is

$$\text{Cost basis markup (\%)} = \frac{\text{Gross margin}}{\text{Cost of goods}} * 100$$

where gross margin is the difference between the sales price and the product cost.

One may also use a simple formula for figuring selling price when markup is based on sales:

$$\text{Selling price} = \frac{\text{Cost of item}}{100\% - \% \text{ desired markup}} * 100$$

As in full-cost pricing, firms using the gross-margin method usually do not apply the same markup to all items or at all times. It is more profitable to take into account the effect of different prices on sales volume and then decide which products will bear high markups and which will require low ones. Price elasticity of demand is an important concept here.

Suggested and "Going Rate" Prices

Another approach to pricing is to follow an external guide. Some managers prefer not to make their own pricing decisions but to rely on prices suggested by manufacturers or wholesalers. Still other managers simply follow the prices set by similar firms.

Profit-Margin Formula

The profit-margin (p-m) formula of pricing takes into account cost, volume, and profit relationships. This formula distinguishes between variable and fixed costs. It uses only the variable element as the starting point in setting prices; the fixed element is accounted for separately as a part of general plant overhead. The profit margin is the difference between net sales revenue and the total variable costs of the products sold. The first pricing objective is to cover all the variable costs charged directly to the item. The second objective is to produce, in addition, the largest possible number of profit-margin dollars that can be applied to overall fixed costs.

The profit-margin formula is a highly practical pricing tool because it gives the basis for sound management decisions. It helps, for instance, in deciding whether to make or to buy a part, to purchase additional machinery or to replace old equipment, to expand or contract capacity, and to increase or decrease advertising outlays.

Pricing Strategies[2]

Firms may use a variety of pricing strategies to influence sales and profits or net margins. Only a few are discussed here.

---

[2] Adapted from Stephen L. Montgomery, *Profitable Pricing Strategies* (New York: McGraw-Hill, Inc., 1988).

Skim Pricing

Skim pricing is pricing at very high levels, disregarding costs, to skim the "cream" off the market. This might be an appropriate strategy when there is little danger of short-term competition, when product uniqueness creates relatively price-insensitive demand, or when organization policy demands recovering start-up costs rapidly.

Slide-Down Pricing

The slide-down pricing strategy, a modification of the skim pricing strategy, involves moving prices down over time. A skim pricing strategy is followed initially, and then the price is lowered periodically to penetrate more of the market.

Penetration Pricing

Penetration pricing involves setting the price sufficiently low that even marginal customers are attracted to the product and a mass market is developed. This might be considered in order to establish an initial market position rapidly and discourage new competitors from entering the market. Penetration pricing can also be used to maximize market share, especially at the expense of competitors who do not react to the lower prices.

Segment Pricing

Segment pricing is pricing the same products and services differently in different markets, perhaps using slightly modified product or service strategies. This pricing strategy is important when various types of buyers are willing to buy at different prices for slight product or service differences.

Loss Leader Pricing

Managers use loss leader pricing when they use low prices for certain products to attract buyers for other products and services. This strategy, which is often used to build customer traffic, works best when complementary products are available to be sold at higher prices with the loss leader items.

Promotion[3]

Promoting a firm's products and services is a key aspect of marketing and sales management. Promotion programs are designed to communicate to those consumers who make up a firm's sales potential that the product or service they need is available at the right price and in the right place and form, with the specific goal of convincing them to purchase it.

Three promotional methods—mass selling, personal selling, and sales promotion—are used to accomplish the goal of convincing potential consumers to become buying customers. Most agribusiness firms use a combination of the three. The mix is determined by a firm's sales policies.

Mass Selling

Mass selling, selling to a large number of potential consumers at one time, is a form of communication. For mass selling to work, the firm must have a clear and simple promotional message to send to its potential consumers. That message requires a well-defined and measurable marketing objective. If the marketing objective is to introduce a new product, the promotional message might be designed to inform the potential consumers. If the marketing objective is to increase sales a certain amount within a certain period, the promotional message should be designed to convince potential consumers to buy now. Promotional messages can also be designed to remind consumers in the sales potential that a product is really good and that it will meet their needs.

The two basic forms of mass selling are advertising and publicity. Advertising, which includes all forms of paid mass selling, can be classified into three broad categories—institutional, pioneering, and competitive. The objective of institutional advertising is to develop good will for a firm or a general product category. The objective of pioneering advertising is to develop the demand for a general category of products rather than a specific brand. The objective of competitive advertising, of which there are several kinds, is to develop the demand for a specific product. Publicity refers to forms of unpaid mass selling, including writing newspaper articles, putting on seminars, sponsoring events, and supporting charitable activities.

---

[3] Adapted from James G. Beierlein and Michael W. Woolverton, *Agribusiness Marketing: The Management Perspective* (Englewood Cliffs, NJ: Prentice-Hall, Inc., 1991).

Personal Selling

Personal selling, together with other marketing elements such as pricing, advertising, product development and research, market channels, and product distribution, is a means for implementing marketing programs. At its simplest, personal selling is direct face-to-face contact with potential buyers by a person representing the firm. This person must be capable of successfully persuading prospects or customers to buy the products or services from which they derive utility or satisfaction.

In today's marketplace, selling is a profession. A *profession* is an "occupation requiring extensive education and training." Sales personnel, including those involved with agribusiness sales, are professionals. They must know their product and its application thoroughly. But they must also understand what people of different personality types expect and how they react. Sales personnel must be able to interact in different ways with different types of people.

There are four general types of selling situations, each requiring its own approach for successful personal selling. These situations, which we will only list here, are defined by the type of customer being served: (1) wholesalers and retailers, who resell the product; (2) purchasing agents, who buy the good or service for their company to use as an input in a production process; (3) product users, who are similar to purchasing agents in that they buy the product for use in a production process but are too small to have a designated purchasing agent; and (4) consumers, who actually consume the good or service.

Sales Promotion

Sales promotion includes all activities that complement personal and mass selling. Contests, coupons, in-store signs and banners, and giveaways are all part of promotion. Sales promotion is often difficult to separate from mass selling and personal selling because it is usually most effective when done together with other forms of promotion.

Break-Even Point Analysis

Earlier we discussed the decision rule derived from economic theory that states that to maximize profit, we produce at the level where marginal revenue equals marginal cost. In most agribusinesses, determining the most profitable point of production by using this decision rule is difficult, if not impossible. Instead, break-even analysis, also called "cost-volume-profit (CVP) analysis," is often used.

Break-even analysis allows a manager to evaluate how revenue, fixed costs, and variable costs affect levels of profit and to determine the break-even point. The break-even point is the sales volume at which total revenue equals total costs, with neither a profit nor a loss realized. To find the break-even point, the manager needs the total fixed and variable costs at any given sales volume. Recall that whether costs are fixed or variable depends on whether they vary with the number of units of output.

The contribution margin is that portion of revenue left after variable costs (also called "direct costs") have been covered. That amount can then be applied to fixed costs for the period. Total direct costs per unit are subtracted from the product price to derive contribution margin:

$$\text{Contribution margin} = \text{Price} - \text{Direct costs per unit}$$

or

$$\text{Unit contribution} = \text{Total revenue} - \text{Total direct costs}$$

Normally, it does not make sense to sell a product at a price less than the direct or variable costs to produce that product. If a firm did so, it would be better off to shut down. Fixed costs will be incurred no matter what, so a unit contribution covering at least part of the fixed costs will be helpful. If the product price exceeds direct costs per unit, the excess is used to cover fixed costs. Any revenue remaining after covering fixed costs is profit.

This formula can be used to determine the break-even point:

$$\text{Break-even quantity} = \frac{\text{Total fixed costs}}{\text{Price per unit} - \text{Direct costs per unit}}$$

Changing any element of this equation affects the results; increasing or decreasing fixed costs, selling price, and/or direct costs per unit changes the quantity needed to break even.

For example, suppose a business with annual fixed costs of $100,000 produces its product at a direct cost of $60 per unit and sells its product for $100 per unit. The break-even sales point is 2,500 units:

$$\text{Break-even quantity} = \frac{100{,}000}{100 - 60} = 2{,}500$$

The advantage of figuring the break-even point by means of this formula is simplicity. The break-even chart, however, gives a broader, "moving" picture of business activity. It shows not only the specific break-even point but also the amount of profit or loss at any point of volume.

On a break-even chart, the measure of sales volume is on the horizontal axis of a graph, and total costs and total revenue are on the vertical axis. Total costs and total revenue are then plotted (Figure 28–1).

Break-even sales quantity

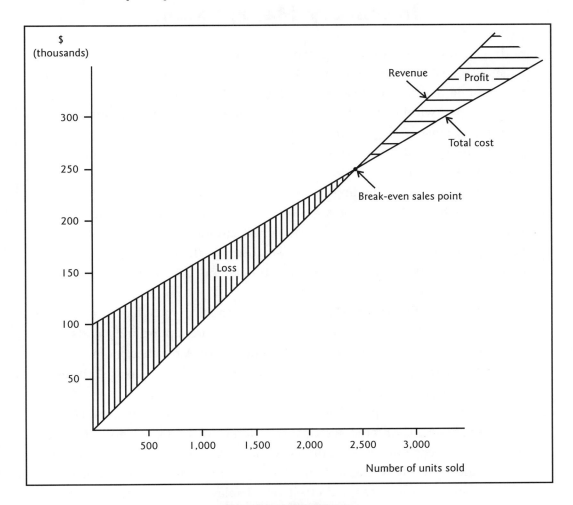

Figure 28–1

The point where the two lines cross represents the break-even point. The area between the two lines to the left of the break-even point represents the loss at those sales levels. The area between the two lines to the right of the break-even point represents the profit at those sales levels.

A slight variation in the break-even analysis is to include a profit target. The profit target, the minimum total dollar amount of profit the firm expects to earn, is added to the total fixed costs in the numerator of the break-even point equation above.

Inventory Management

In general, inventory management is a critical aspect of agribusiness management. Agribusinesses may hold inventories to meet year-round demand for commodities that are produced (or harvested) only once a year. For example, someone within the agribusiness system, generally either producers or processors, must hold grain inventories. Other agribusinesses, such as farm supply stores, hold inventories of goods that are produced year-round but used only seasonally.

Another reason for holding inventories is to avoid *stockouts*, i.e., running out of something the firm or a customer needs. Depending on what is involved, stockouts can result in costly plant shutdowns or lost customers. Holding inventories can also lower costs, because there are generally price discounts associated with large-quantity orders and transportation costs per unit are also lower.

However, holding inventories is also costly. The firm must have a physical location allocated to inventory storage, which means opportunity costs are incurred since that area cannot be put to another use. A number of other costs associated with storage—insurance, special equipment, taxes, interest, etc.—must also be considered.

There are several important guidelines for inventory management: (1) Inventories should be ordered systematically to avoid shortages and lost sales, but excessive accumulations should be avoided. (2) Upon delivery, merchandise should be carefully checked and verified. (3) Inventory stocks should be both efficiently warehoused and effectively displayed. (4) Inventory turnover ratios should be carefully monitored. (5) Carrying cost of inventories should be accurately determined.

Economic Order Quantity (EOQ)

Business managers are always faced with the problem of determining the right quantity of inventory to maintain. One basic approach is to use the *economic order quantity* model:

$$EOQ = \sqrt{\frac{2 * A * C}{P * I}}$$

where

EOQ	=	economic order quantity
A	=	units of item needed during period
C	=	cost of processing an order
P	=	price per unit of goods being purchased
I	=	annual inventory carrying charge as a percentage of annual inventory value

For example, assume a firm sells 10,200 units of a product per year, the purchase price is $4 per unit, the cost of processing an order is $15, and the annual inventory carrying charge as a percentage of annual inventory value is 0.20. Given these values, the economic order quantity for this firm is

$$EOQ = \sqrt{\frac{2 * 10{,}200 * 15}{4 * 0.20}} = 618.5$$

Therefore, if 10,200 units are sold annually, the EOQ is 619, and about 17 orders should be placed during the year (10,200/619 = 16.5). An amount for safety stocks may be added to insure a minimum level of inventory.

The economic order quantity model is merely a basic inventory model. Its usefulness lies in its simplicity and its approach to evaluating alternative inventory management decisions. However, the EOQ model does have limitations that cannot be overlooked. For example, its underlying assumptions are difficult to uphold in actual practice. Also, it fails to address the seasonality of most agricultural products.

Topics for Discussion

1. Discuss the functions of management.
2. Why is financial analysis important?

3. Distinguish between market potential and sales potential.
4. Discuss a firm's sales policies.
5. List the various pricing methods.
6. Discuss the promotional methods a firm may use.
7. What is inventory management?

Problem Assignment

Select one of the topics presented in this chapter, and discuss its implementation with a business manager.

Recommended Readings

Beierlein, James G., Kenneth C. Schneeberger, and Donald D. Osburn. *Principles of Agribusiness Management,* 2nd ed. Prospect Heights, IL: Waveland Press, Inc., 1995. Ch. 2, 4, 10, 11, 12, and 17.

Beierlein, James G., and Michael W. Woolverton. *Agribusiness Marketing: The Management Perspective.* Englewood Cliffs, NJ: Prentice-Hall, Inc., 1991. Ch. 15–20.

Devino, Gary T. *Agricultural Finance.* Danville, IL: Interstate Publishers, Inc., 1981.

German, Carl L., U. C. Toensmeyer, Jarvis L. Cain, and Robert J. Rouse. "Planning for the Retail Farm Market," *Journal of Food Distribution Research,* 26(1):82–88 (1995).

Lee, Jasper S., James G. Leising, and David E. Lawver. *AgriMarketing Technology: Selling and Distribution in the Agricultural Industry.* Danville, IL: Interstate Publishers, Inc., 1994.

Montgomery, Stephen L. *Profitable Pricing Strategies.* New York: McGraw-Hill, Inc., 1988.

Moyer, R. Charles, James R. McGuigan, and William J. Kretlow. *Contemporary Financial Management,* 5th ed. St. Paul, MN: West Publishing Co., 1992. Ch. 19.

Newman, Michael E., and Walter J. Wills. *Agribusiness Management and Entrepreneurship,* 3rd ed. Danville, IL: Interstate Publishers, Inc., 1994. Ch. 1, 2, 12, 13, and 15.

Powers, Nicholas J. "Sticky Short-Run Prices and Vertical Pricing: Evidence from the Market for Iceberg Lettuce," *Agribusiness: An International Journal,* 11(1):57–75 (1995).

Roden, Foster. *Foundations of Business Finance,* 3rd ed. Houston, TX: Dame Publications, Inc., 1994. Ch. 16.

Still, Richard R., Edward W. Cundiff, and Norman A. P. Govoni. *Sales Management: Decisions, Strategies, and Cases,* 5th ed. Englewood Cliffs, NJ: Prentice-Hall, Inc., 1988. Ch. 1–5.

29

International Trade and Agriculture

The purpose of this chapter is to provide a basic overview and understanding of international trade—what it is, why it is important, and how it functions—with particular emphasis on agricultural trade.

Nature of International Trade

International trade, the exchange of goods and services between countries, is merely an extension of domestic market concepts into a global arena. Instead of buyers and sellers across the street or across states, the world market implies buyers and sellers across national borders.

It is important to remember that each country has its own domestic issues and problems and its own approach to addressing those issues and problems. This influences how the country views international trade. Hence, international trade is more than an exchange of material goods—it has far-reaching policy implications affecting such areas as national security, mutual assistance, monetary balances, economic growth and development, technological advancement, and altruistic concerns.

Production and marketing decisions, which in our discussion to this point have considered mainly domestic resources and domestic demand, are extended by foreign trade to include all trading nations in the world. In the words of John Stuart Mill, "The benefit of international trade is a more efficient employment of the productive forces of the world."

The broader market has encouraged specialization, just as the move from subsistence to commercial farming has led to specialization. Specialization is the outcome of the *principle of comparative advantage,* which we discussed in

Chapter 11. Even if two countries can produce the same commodities, they generally find it beneficial for each to concentrate production on a few goods and trade them for others. The United States, for example, exports soybeans to Japan, while Japan exports cameras and stereo equipment to the United States. Both countries can produce those commodities, but they are better off economically because of specialization and international trade.

In local U.S. grocery stores, we can buy bread, meat, eggs, milk, and so on—products that were most likely produced domestically. However, the coffee, tea, cocoa, bananas, and spices were probably imported. Many of these imported goods are regarded as essential by consumers (like coffee first thing in the morning). Add our dependency on automobiles and electronic equipment that use imported petroleum products, rubber, tungsten, zinc, and other imported materials, and the importance of international trade becomes obvious.

As a result of trade, consumers in trading nations are better off in at least four respects:

1. A wider distribution of scarce resources is made possible, allowing the increase of total output.
2. Specialization is encouraged because each country will concentrate on those commodities and services for which it has a comparative advantage, lowering production costs.
3. Communication and social interaction are promoted and serve as a basis for peaceful coexistence.
4. Improved technology is exchanged and applied.

International trade is important to the United States, in general, and to the agricultural sector, in particular. Trade is probably even more important to countries more limited in size and productive resources. Without export markets, some nations would find it extremely difficult to maintain domestic employment and income and to accumulate the foreign exchange necessary to import goods needed for economic and social growth and development. As developed economies have grown and less developed economies have industrialized, trading opportunities have increased considerably.

It has been said that international trade is a substitute for mobility of resources, such as labor and capital. Labor and capital move more freely within a country that allows resource mobility than between countries. For example, workers in the United States can move from Mississippi to Michigan without territorial restraints. However, crossing from China to the United States is far more difficult and at times strictly forbidden. Also, in democratic nations, workers have the freedom to change type of employment—for example, from the farm sector to manufacturing—while in some counties, such change is difficult.

The international movement of capital (in this case, negotiable currency) is limited by governmental restrictions, tax laws, political bargaining, and a host of other institutional restraints. Since nations issue their own currency, there are many banking complications involved in foreign transactions. "Foreign exchange" refers to the currency (or credits) from other countries that is acceptable to exporters. As far as U.S. exports are concerned, dollar holdings by foreign countries are very important. A country with a liberal supply of dollars or dollar credits will be more likely to import commodities from the United States.

Basis for Trade Relations

We have seen in the material of other chapters that a firm's marketing decisions are integrated with its production and financing decisions. In Chapters 20 and 21, we discussed types of market outlets and how local, state, and federal regulations influence commodity marketing. Many such regulations exist, increasing the complexity of marketing goods and services domestically across state boundaries. However, these are almost insignificant when compared to the challenges of international trade.

The level of world trade has fluctuated considerably over the years, largely in response to industrial growth and technological changes in production, transportation, and communication. When output increases, trade increases. When output decreases, trade decreases. Political developments also influence the level of trade. This was observed during the Great Depression, when the decrease in world trade due to falling incomes was made worse by trade restrictions imposed by countries around the world.

At the time of the Declaration of Independence in the United States, Adam Smith stressed the absurdities of economic isolation and advocated free trade as a means for each nation to increase its wealth. He recognized the principle of comparative advantage and shifted the emphasis from the mercantile attitude of self-sufficiency to that of increased satisfaction and wealth realized from the division of labor. His passage from *The Wealth of Nations* deserves repetition:

> It is the maxim of every prudent master of a family, never to attempt to make at home, what it will cost him more to make than buy. The tailor does not attempt to make his own shoes, but buys them of the shoemaker. The shoemaker does not attempt to make his own clothes but employs a tailor. The farmer attempts to make neither the one nor the other, but employs those different artificers. All of them find it for their interest to employ their whole industry in a way in which they have some advantage over their neighbors, and to purchase with the price of a part of it, whatever else they have occasion for.

The above statement implies that an absolute advantage is needed to justify trade. This philosophy persisted until David Ricardo explained in his 1817 book, *Principles of Economics,* that a comparative labor cost advantage makes it economically feasible to carry on trade.

A further refinement was provided by John Stuart Mill in 1848, when he examined the matter of international values. Instead of taking as given the level of output, with labor costs different in each country, he assumed different productivities for given amounts of labor; the outputs differed, and the value of output per unit of labor naturally differed. Mill maintained that the ratio of values between two exchangeable commodities really determined whether trade would occur. In essence, the actual ratio at which trade occurs depends upon the exchange value of these commodities.

International trade also depends upon costs of production, which are the combined costs of land, labor, and capital, sometimes referred to as rents, wages, and interest. These costs, of course, vary from country to country because of differing supply and demand conditions. Labor, for example, may be inexpensive in Hong Kong because of its abundance, while land, on the other hand, is relatively expensive. In the United States the reverse situation occurs. Countries with abundant and cheap land will tend to be low-cost producers of such agricultural commodities as wheat, corn, and soybeans, while those with abundant and cheap labor will concentrate on labor-intensive goods. Likewise, capital-intensive industries will seek out areas where interest rates are low and technology and skilled labor are available.

When discussing our simple explanation of specialized production in various countries, we must also consider that in a dynamic system, changes can occur quite rapidly. As supply and demand conditions change, the prices of goods and services change and, accordingly, the relative trade positions of countries also change. Add to this, transportation costs, import duties, and other barriers to trade, and the complexity of international trade increases quickly.

Transportation costs are an extremely important element in international trade. In recognition of this, subsidies are often provided by the exporting country so the delivered cost of the commodity will still be competitive with the cost of the commodity from other suppliers in the market. The cost of transporting products internationally, added to the handling cost at port of embarkation, further boosts the cost of products and weakens their competitive position in the world market. For the United States, transportation costs are substantial, as they reflect the higher wage rates paid to U.S. workers. The difference in shipping rates between foreign and U.S. flagships further accents the difference in transportation costs. Because of lower rates, the use of foreign ships has almost completely replaced the use of U.S. freighters.

Although not mentioned by the early economists as a factor for serious consideration, social costs receive attention among trading nations. In this context, social costs refer to the impact of imports and exports on employment, communication, education, environment, cultures, migration, tradition, religion, health, land tenure, and, more generally, the way of life of any given society. In recognition of these social costs, nations are increasingly involved in negotiating trade agreements and imposing trade regulations that give governing bodies authority to control prices, require licenses, and impose duties, quotas, and taxes to protect their societies against indiscriminate trading by individual business firms with short-term profit motives rather than long-range social goals.

Trade Barriers[1]

Countries use a variety of methods to restrict trade, including tariffs, import quotas, embargoes, voluntary export restraints, and regulatory trade restrictions. We will discuss these briefly in this section.

---

[1] Much of the material in this section is adapted from Ronald D. Knutson, et al., *Policy Tools for U.S. Agriculture,* Report B-1548, Revised (College Station: Agricultural and Food Policy Center, Texas A&M University, January 1993).

Tariffs, or customs duties, are government-imposed taxes on goods when they cross national boundaries. Tariffs are the most commonly used trade restriction tool. Import tariffs make goods more expensive than they would otherwise be. As a result, consumption of domestically produced goods is encouraged. In the United States, import tariff rates may be relatively low or even nonexistent on products not produced domestically, such as bananas, coffee, and rubber. They are generally much higher on products that compete with domestic goods, such as apparel, footwear, and textiles.

Generally, an import tariff may be either a **specific tariff,** a fixed charge per unit of the good being imported, or an **ad valorem tariff,** a percentage of the value of the good being imported. Another type of import tariff is the **countervailing duty,** which is a tariff designed to offset an export subsidy of a competing country.

The effect of an import tariff on production and consumption in a domestic market is illustrated in Figure 29–1. Assume that the world price and the domestic price (P^0) are the same before imposing the import tariff. In this case, domestic production equals Q_1, while domestic demand equals Q_2. Domestic demand exceeds domestic supply, and the excess demand is satisfied by importing. The quantity imported equals ($Q_2 - Q_1$).

Imposing the import tariff (t) raises the domestic price of the imported good from P^0 to ($P^0 + t$). The quantity demanded of domestically produced goods increases because their price is lower than the price of imports with the tariff. However, this increase in the quantity demanded drives the price up, until the price of domestic goods equals the price of imports with the tariff. Thus, domestic production increases to Q_3 while domestic demand decreases to Q_4; the quantity imported falls to ($Q_3 - Q_4$). The government revenue resulting from the import tariff is represented in the shaded area.

Export tariffs are essentially taxes on goods exported from a country. They are used by many developing countries to provide hard currency to the government. Export tariffs discourage exports, often with the goal of maintaining larger domestic supplies.

An **import quota** imposes a limit on the amount of a good that may be imported in a given period. Once the quota is filled, no more of that good may be imported, regardless of the price. Quotas and tariffs have similar effects on restricting trade, but they differ in who receives the revenue. As we mentioned, with a tariff, the government gets the revenue; with a quota, it accrues to firms with the right to import the protected good.

The effect of an import quota is illustrated in Figure 29–2. As before, the domestic price and the world price (P^0) are the same before imposing the

Effect of a tariff

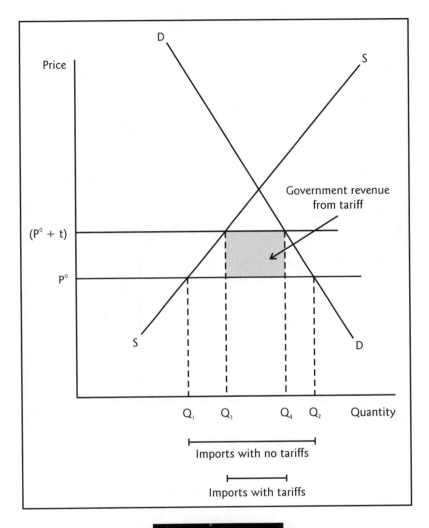

Figure 29–1

quota. Again, domestic production equals Q_1, domestic demand equals Q_2, and the quantity imported equals $(Q_2 - Q_1)$. Putting the quota into effect reduces the quantity supplied, which, in turn, results in higher domestic prices. These higher prices stimulate increased domestic production, often by less efficient producers who would not be competitive without the quota in place. Eventually, prices reach equilibrium at P', where domestic consumption

Effect of an import quota

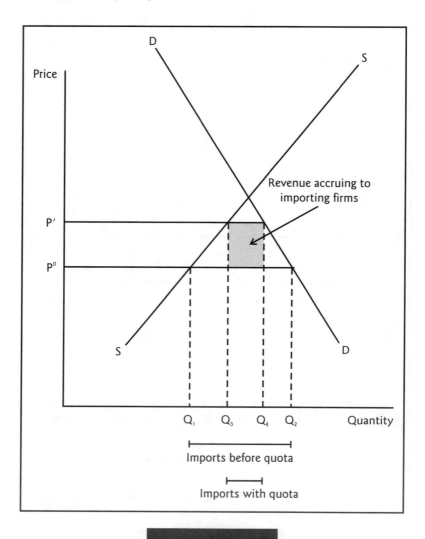

Figure 29–2

just equals domestic supply plus the import quota. The shaded area represents the revenue that accrues to the firms importing the quota amount.

An **embargo** is a government order restricting imports or exports of a good. Embargoes are usually instituted as a political response to an international dispute, rather than for economic reasons. Export embargoes, which set absolute limits on the quantity that can be exported, are used to keep commodity prices lower in the exporting country, prevent domestic shortages, and/or achieve foreign policy objectives. Instituting export embargoes may have resulted in loss of confidence in the United States as a dependable supplier, which contributed to the loss of this nation's share of world agricultural trade in the early 1980s.

A **voluntary export restraint** is an agreement in which foreign governments are asked to limit their exports of specific commodities to a given quantity. This controls the importation of those commodities, protecting domestic producers. Voluntary export restraints tend to cause wider price fluctuations in world markets than would occur in a free market environment. By protecting the domestic industry, they also decrease production efficiency in the importing country.

Tariffs, quotas, and embargoes are direct methods to restrict trade. Indirect methods, called **regulatory trade restrictions**, are also used to restrict trade, but in less obvious ways.

Regulatory trade restrictions are government-imposed rules and regulations that limit imports. For example, to protect its citizens, a country may have restrictions in place that forbid importing fruits and vegetables sprayed with certain insecticides. Another type of regulatory trade restriction involves complicated customs procedures that discourage importers from even attempting the process.

Regulatory trade restrictions are types of *nontariff trade barriers*. Nontariff trade barriers cover all forms of import restrictions, including quotas, other than tariffs. The use of nontariff trade barriers has increased, in part, because of the General Agreement on Tariffs and Trade (GATT) and its emphasis on reducing tariff trade barriers. Common U.S. nontariff trade barriers relate to health and sanitation restrictions on animal and plant products. For example, meat imports are prohibited from countries that have foot and mouth disease. Some nontariff barriers are justified, but others are simply protectionist.

There are a variety of nontariff trade barriers, including:

1. The **variable levy,** an import tax imposed when the world price falls below a government-set minimum import price, yielding a domestic price at the predetermined level. The variable levy is an effective barrier to trade because it eliminates the price advantage (e.g., from production efficiency) of the imported good. The variable levy is the principal mechanism used by the European Union (EU) to restrict agricultural imports.

2. Labeling and marketing regulations.

3. Monopolies, including state agencies, quasi-government agencies, or private institutions operating under governmental authority to determine the imports to be allowed and the conditions of entry.
4. Advance security deposits on imports.
5. Material or industrial standards.
6. Import licensing that discriminates among exporting countries.
7. Bilateral agreements, which restrict trade with countries not party to the agreements.

Reasons for Restricting Trade

A variety of arguments for protection against imports are offered. Some are more valid than others, yet they can all become discussions that determine economic and political policies. Some of the key arguments are listed here with a brief explanation to identify the nature of the protection sought.

Protection of infant industries. Infant industries are generally regarded as high-cost industries because they have not achieved the efficiencies associated with economies of size. Such producers argue that only if the government will erect trade barriers will they be able to lower average costs and become competitive. Cheaper imports may provide a serious competitive burden and dominate the market to the extent that infant industries are forced out of business. The textile industry was a good example of a protected infant industry in the early history of the United States.

Economists recognize the infant industry argument but suggest that highly industrialized nations have less need for protecting infant industries than do developing nations. Also, once protective measures are established, they are difficult to rescind. This results in production inefficiency. The protected industries have no incentive to "grow up" and find ways to compete internationally. More economists would support the infant industry argument if the trade restrictions came with a definite time frame and conditions when they would end.

Self-sufficiency for national security. In the event of war, either of a military or economic nature, each country would like to be self-sufficient; accordingly, trade restrictions are needed to strengthen industries producing essentials for survival—for example, food and defense-related materials. In this instance one must weigh the merits of nationalism and military security against the efficient allocation of world resources. Often, the national security argument is extended to goods that have little to do with national security. Economists would argue that subsidies to strategic industries would be less burdensome than economic isolationism, which imposes a greater cost burden on every consumer.

A major drawback to the national security argument is that restrictions on the trade of military goods can be evaded. Countries can have other countries purchase the goods for them. These third-party sales, called *transshipments*, are common in the world market. They limit the effectiveness of absolute trade restrictions.

Protection of domestic employment. On the surface, this trade philosophy sounds practical and appears increasingly fashionable in times of economic recession. The fallacy lies in the fact that while restricting imports may have immediate short-term benefits for domestic employment, in the long-run, it becomes increasingly obvious that international trade is a two-way street: the exports of one country are the imports of another. Without imports by the United States, a foreign country is unable to accumulate dollars with which to purchase U.S. exports. Moreover, achieving full domestic employment by having an excess of exports over imports is akin to exporting unemployment to the importing nation.

It is also worth noting that nations affected adversely by the quotas and tariffs of other nations will usually retaliate with quotas and tariffs of their own. This was painfully evident when the Smoot-Hawley Tariff Act of 1930, which imposed the highest tariffs ever enacted, drew almost immediate retaliatory measures against the United States by at least 15 countries. Two of these, Italy and Switzerland, declared a complete embargo against American goods. Imports into and exports out of the United States fell sharply by 1932, with the depressing effect felt worldwide. Imposing the high tariffs undoubtedly helped perpetuate the massive economic depression of the 1930s.

Diversification for stability. Protection to encourage diversification is closely related to the ideas of encouraging self-sufficiency and independence. Some countries, due to size, geographic location, and resource endowments, are essentially economically dependent on exports of only a few commodities. Barriers against imports of other goods, it is alleged, would induce domestic producers to diversify and produce those items formerly imported. The benefits would be realized in the form of greater economic independence and domestic stability. However, again, the gains from specialization would be foregone.

Protection of wages and standards of living. The benefits of trade are spread over the entire population, while the costs of free trade are usually borne by specific small groups of people. While the benefits exceed the costs, those hurt by free trade advocate political intervention. For example, some advocates for trade restriction contend that cheap foreign labor permits foreign countries to impose damaging competition by exporting low-cost products. Consequently, American goods lose their market, ultimately resulting in unemployment, lower wages, and lower living standards. There is some merit to this reasoning, although the argument ignores the fact that wage rates and cost of production per unit are related through productivity. Low wages may reflect the low productivity of the worker. On the other hand, high wages may result in lower unit costs because productivity may be relatively high.

Though the arguments for protection against foreign imports are many and have varying degrees of economic validity, they still do not cancel out the strong case for free trade. As we mentioned earlier, international trade is a two-way street. The principle of comparative advantage explains the potential for both the importing and the exporting nations to benefit from trade. Aside from the economic benefits, there are important political and social benefits as well. Communication between nations, each with its own distinct culture, is an important step in establishing and maintaining world peace. In the long-run, a nation must import in order to export. More importantly, the United States is almost totally dependent upon trade for certain commodities not available domestically.

Subsidizing Exports

The government has tried a number of methods to increase the competitiveness of excess domestic production in the world market. Providing **export subsidies**, payments to exporters by the government, is an example. Like the specific and ad valorem tariffs discussed previously, there are specific and ad valorem export subsidies. With a **specific export subsidy**, a fixed amount is paid per unit of the good being exported, while with an **ad valorem export**

subsidy, the amount paid is a percentage of the value of the good being exported.

The effect of an export subsidy is illustrated in Figure 29–3. Assume that, before implementing the export subsidy, the world price is P^0, with exports equal to $(Q_1 - Q_2)$. When the export subsidy (s) is first implemented, producers receive more for their good by exporting rather than by selling domestically. As a result, they export as much as possible, creating a domestic

Effect of an export subsidy

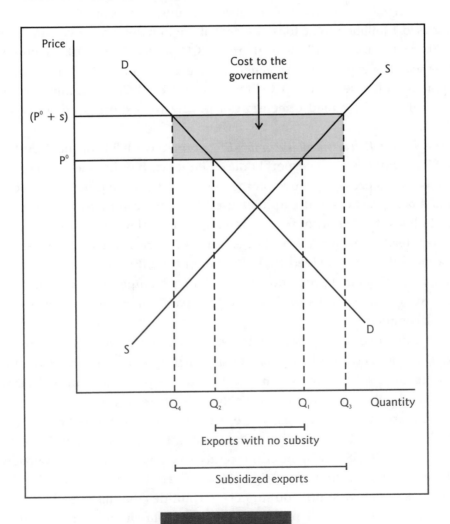

Figure 29–3

shortage, which, in turn, drives the domestic price up. Eventually the domestic price will increase to (P^0 + s), the world price plus the export subsidy. Producers increase output to Q_3, while domestic demand falls to Q_4, and the quantity exported, ($Q_3 - Q_4$), increases. The shaded area represents the cost to the government of providing the export subsidy.

The *export Payment-in-Kind (PIK)* and the *Bonus Incentive Commodity Export Program (BICEP)* were two export subsidy programs the U.S. government used to expand export markets for selected U.S. commodities by making the commodities more price competitive in the world market. Under export PIK, the government provided an in-kind export commodity bonus for each regular commercial purchase of a specified amount. For example, if a country purchased 1 million metric tons of wheat, it might have received an additional 100,000 metric tons of PIK wheat from CCC stocks. The 100,000-metric-ton bonus was the export PIK. In general, the use of export PIK was limited to surplus commodities held in CCC inventories. The BICEP program was similar, but it had the added dimension of paying the export bonus in cash or in-kind.

The *Export Enhancement Program (EEP)* replaced BICEP in the 1990 Farm Bill. EEP continues to use export bonuses, either cash or in-kind, to make U.S. commodities more competitive on the world market or to offset the effects of unfair trade practices and subsidies used by other countries. EEP has been used to boost the export of wheat, poultry, flour, barley, rice, sorghum, cattle, livestock feed, vegetable oil, and eggs. Wheat has been the major beneficiary of export subsidies provided by EEP. In the future, EEP is likely to focus on market expansion and promotion, rather than on competing with other countries' export subsidies. Its goal will be to expand all exports, not just target specific commodities.

The U.S. government uses a number of other programs, such as *Food for Peace (P.L. 480)* and *Food for Progress* to subsidize exports. The objective of P.L. 480 is to dispose of surplus commodities, develop markets, provide emergency food aid, and assist friendly nations in their economic development. Countries are selected for assistance based on economic and political criteria, as well as need. The Secretary of State makes the final decisions regarding who receives P.L. 480 assistance. In recent years, Egypt, Morocco, Zimbabwe, and Jamaica have been major recipients of U.S. aid through P.L. 480. Food for Progress is designed to provide support to countries that have made commitments to introduce or expand free enterprise elements in their agricultural economy.

Trade Agreements

Multilateral Trade Agreements

Multilateralism refers to cooperation among a number of countries in trade negotiations and in other means to serve the interests of the participating countries.

The *General Agreement on Tariffs and Trade (GATT)*, established in 1947, is an example of a multilateral trade agreement. It is a multilateral United Nations treaty that has been signed by more than 100 countries, including the United States. The objective of GATT is to increase trade among nations through negotiated reductions in tariffs and other trade barriers. These actions are designed to prevent the development of rounds of retaliatory trade barriers. In fact, GATT was largely the product of efforts to prevent the recurrence of the severe trade restrictions that contributed to the world depression of the 1930s.

A foundational aspect of GATT is the **most-favored nation** principle, which provides equal tariff treatment for imports from all GATT members. Thus, a most-favored nation will be charged as low a tariff on exports it purchases as any other country.

There have been eight "Rounds" of GATT negotiations. The first seven Rounds were Geneva (1947), Annecy (1949), Torquay (1950), Geneva (1956), Dillon (1960–61), Kennedy (1964–67), and Tokyo (1973–79). The eighth was the Uruguay Round (1986–93). Agricultural trade was a central focus of negotiations in the Uruguay Round, as were trade-related issues that are increasing in importance, including trade in services, trade-related investment measures, and trade-related intellectual property matters.

Regarding agricultural trade, the Uruguay Round of GATT requires member countries to convert all nontariff measures to tariffs (a process called *tariffication*) prior to implementing the trade agreement. Industrial countries must, over six years, cut tariffs from a 1986–88 base by 36 percent, on average, with a minimum cut of 15 percent on any given tariff line.

Domestic support for agriculture must be reduced by 20 percent from a 1986–88 base as the agreement is implemented. So-called green-box subsidies—certain government service programs, decoupled income support, social safety net programs, structural adjustment assistance, and environmental programs, among others—are exempt from reduction under GATT. Export subsidies must be reduced from a 1986–90 base by 36 percent in value for each product during the implementation period.

Several provisions are in place to provide greater flexibility for developing countries. For example, reductions in tariffs, domestic support, and export subsidies are set at two-thirds of the levels for industrialized countries and spread over 10 years, instead of 6. Least-developed countries are exempted from all reduction commitments.

The Uruguay Round lasted for just over seven years. It resulted in agreement to reform all the negotiating areas and to establish a new intergovernmental organization, the **World Trade Organization (WTO).** The WTO, which officially began functioning January 1, 1995, essentially succeeds GATT. Nations that ratified GATT are members of the WTO. The structure of the WTO is intended to provide for more effective decision making and greater involvement in trade relations. The WTO has been given authority to oversee adherence to trade agreements on goods and services, as well as intellectual property.

The *World Trade Organization Agreement* lists these five functions of the WTO:

1. Facilitate the implementation, administration, and operation of the GATT terms and further the objectives established

2. Provide a forum for negotiation among member countries concerning their multilateral trade relations, and then assist in implementing the results of those negotiations

3. Oversee dispute settlement between member countries

4. Through the Trade Policy Review Mechanism, improve adherence by member countries to rules, disciplines, and commitments made under the multilateral trade agreements

5. Cooperate, as appropriate, with the International Monetary Fund and the International Bank for Reconstruction and Development (the World Bank) to achieve greater coherence in global economic policy making.[2]

---

[2] Philip Raworth and Linda C. Reif, *The Practitioner's Deskbook Series: The Law of the WTO* (New York: Oceana Publications, Inc., 1995).

Regional Trade Agreements[3]

Much attention has been focused on economic integration on a regional basis. **Regional trade agreements** involve trade negotiations among countries within a given region.

A *free trade area (FTA)* is an area in which member nations lower or eliminate tariffs and perhaps other trade barriers among themselves on broad categories of goods, while each maintains its own independent trade policy toward nonmember nations. No other economic integration among members occurs.

In 1993, Congress approved the *North American Free Trade Agreement (NAFTA)*, which establishes a free trade area among the United States, Canada, and Mexico. The agreement is technically three agreements: between the United States and Canada, between the United States and Mexico, and between Canada and Mexico. Canada and Mexico are the second and third largest export markets for U.S. agricultural products (Japan is the largest). Under NAFTA, tariffs and other trade barriers among these countries will be reduced through the 1990s. U.S. agriculture has much potential gain in NAFTA, especially through improved trade opportunities with Mexico. The outlook for increasing exports of corn, wheat, and oilseeds to Mexico is very good, while meat exports have already expanded. The importance of Mexico as a U.S. trading partner will only increase.

Like a free-trade area, a *customs union* involves lowering or eliminating trade barriers among participating countries. A customs union also involves establishing a common trade policy toward nonmembers. No deliberate integration of factor markets or other economic policy occurs.

A *common market* is essentially a customs union with some integration of input markets, such as for labor and capital. Some partial alignment of tax, fiscal, monetary, and agricultural policy may occur in a common market.

An *economic federation* or *economic union* is somewhat more integrated than a common market. Input markets and product markets are closely integrated. There is substantial unification in government policy in an economic federation, which blurs the original distinctions between member countries. Full political integration, as far as relationships with the rest of the world are concerned, denotes an advanced stage of economic federation.

---

[3] Adapted from James P. Houck, *Elements of Agricultural Trade Policies* (Prospect Heights, IL: Waveland Press, Inc., 1986; reissued, 1992), Ch. 14.

The European Union (EU) is an economic federation that has evolved from the European Community (EC) and the European Economic Community (EEC). The EEC was organized in 1957 by the *Treaty of Rome* to form an economic integration of six western European countries: France, West Germany, Italy, Belgium, the Netherlands, and Luxembourg. By 1995, it had expanded to include nine more countries: The United Kingdom, Ireland, and Denmark joined in 1973; Greece in 1981; Portugal and Spain in 1986; and Sweden, Austria, and Finland in 1995. It is intended to be a unified area in which commerce is carried on freely, much as it is among the states of the United States. The area has common policies with respect to agriculture, transportation, taxes, and international trade.

The EU operates under the *Common Agricultural Policy (CAP),* which defines the policy the EU uses to support agricultural prices and subsidize exports. The EU is regarded as a "major villain" in world agricultural trade because it uses *export restitutions* (subsidies equal to the difference between the EU's high domestic support price and the world price) to export its agricultural surpluses to the rest of the world. Variable levies used by the EU, which are also specified in the CAP, insulate EU producers and consumers from world price signals of shortage and surplus. The CAP, while playing an important historical role in holding the EU together, has been a cause of major internal conflict because it has absorbed a large share of the EU budget. That high cost has been a principal impetus for CAP reform.[4]

Bilateral Trade Agreements

Bilateral trade refers to trade between any two nations. Bilateral trade agreements are contracts between two countries specifying the quantity of a commodity to be traded during a certain period. The objective of bilateral trade agreements is to normalize trade, develop markets, and retain markets for a country's products. A trade agreement is a means of opening a new market and maintaining a competitive position. It assures the importing country a minimum supply and the exporting country a market for its product. A bilateral trade agreement is, in essence, a barrier to trade, because it ties up markets over the life of the agreement.

---

[4] Luther Tweeten, *Agricultural Trade: Principles and Policies* (Boulder, CO: Westview Press, Inc., 1992), pp. 234–235.

Unilateralism

Unilateralism means that each nation designs its own international trade policies without consultation or collaboration in wider trade forums. With each nation acting in its own interest, confrontation rather than cooperation is likely to be all too frequent. One benefit of unilateralism is that it can reduce collusion that distorts prices.

Unilateralism could benefit large-country traders, like the United States, because of their clout in the world market. While global free trade is preferred, it is not always feasible. Sometimes the United States and other countries must use confrontation with non-free traders to break down barriers to trade country by country. The confrontational tactics could predominate if multilateral and regional trade agreements fail.

Trade and the Balance of Payments

Balance of payments is a measure of a nation's total payments to and receipts from the rest of the world. In our case, it compares the dollar value of the goods and services we buy from foreigners with the value of the goods and services they buy from us. The main purpose of the balance of payments is to inform monetary authorities of the international position of the nation and to aid banks, firms, and individuals engaged in international trade in their business decisions.[5]

The balance of payments can be likened to a checking account. The account is replenished by the earnings of merchandise exports, returns on foreign investments, travel by foreigners in the United States, and services provided by the United States to foreign countries. Depletion of the account arises from the importation of goods produced abroad, government grants and loans, private long-term investments abroad, foreign travel (including military spending), services provided by foreign countries, and other dollar outflows.

When the outflow of dollars exceeds the inflow of dollars, payments exceed receipts, and we have a balance of payments *deficit* in our international trade account. Balance of payments deficits can be corrected over the long run

---

[5] Dominick Salvatore, *International Economics*, 5th ed. (New York: Macmillan Publishing Co., 1995), p. 362.

U.S. ag trade surplus to be record wide in 1996

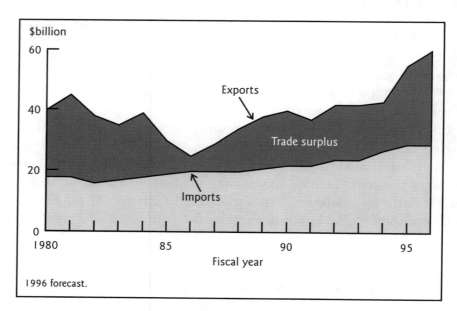

Source: U.S. Department of Agriculture

Figure 29–4

by decreasing the volume of payments to foreign countries (i.e., by reducing imports or purchases) and by increasing receipts (i.e., by expanding exports or sales). In this respect, the deficits prevailing in the past were often eased by increasing exports of agricultural and industrial products.

The United States maintained a surplus of exports over imports in most years until the early 1970s. The trade deficit grew rapidly through the 1970s and into the 1980s, when it peaked at nearly $160 billion. The trade deficit was $75.7 billion in 1993, with a deficit on the trade of goods of $132.6 billion. With a trade surplus of $56.9 billion, the export of services, an area expected to see rapid growth in the future, offset some of the goods trade deficit.

Agricultural exports have done their part to lower the trade deficit. Through 1995, there have been 36 consecutive years of agricultural trade surpluses, which contrasts with the persistent trade deficits in nonagricultural trade. The U.S. agricultural trade surplus was $24.6 billion in 1995, the second-largest trade surplus ever and a major increase from 1986, when the surplus approached zero.

International Trade and the Agricultural Sector

The United States is a major player in world agricultural trade. With the emergence of NAFTA and the WTO, the importance of international trade will increase. In this section, we briefly discuss U.S. agricultural exports and imports.

U.S. Exports

Agricultural exports make up about 10 percent of the total value of all commodities exported from the United States (Figure 29–5). At 81.5 percent of the total value of agricultural exports, crop products dominate the value of U.S. agricultural commodities exported (Table 29–1).

Total value of U.S. exports, 1993

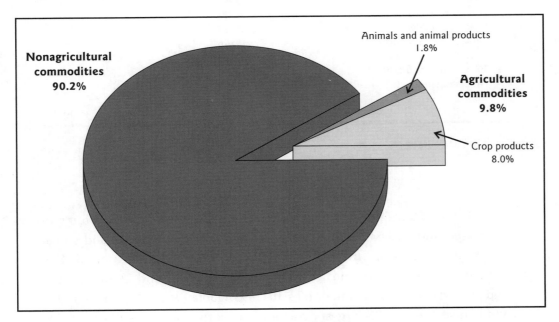

Source: U.S. Department of Agriculture

Figure 29–5

Table 29-1

Value of Exports of Agricultural Products, 1993

Products	Value	Percent
	- - - - - - $1,000 - - - - - -	
Animals and animal products	7,880,805	18.5
Live animals	357,544	0.8
Dairy products	761,613	1.8
Meat products	3,349,396	7.9
Poultry and poultry products	1,313,100	3.1
Hides and skins	1,287,681	3.0
Other animal products	811,471	1.9
Crop products	34,708,996	81.5
Cotton	1,537,660	3.6
Fruits and nuts	3,831,711	9.0
Feed grains	5,260,549	12.4
Rice	766,408	1.8
Wheat and wheat products	4,994,297	11.7
Feeds and fodders	2,146,787	5.0
Oil seeds and products	7,210,676	16.9
Vegetables	3,220,106	7.6
Tobacco, unmanufactured	1,442,763	3.4
Other crop products	4,298,039	10.1
Totals	42,589,801	100.0

Source: U.S. Department of Agriculture.

The United States remains the world's largest supplier of bulk commodities. Oilseed and oilseed products (mainly soybeans), feed grains, and wheat and wheat products are the primary U.S. crop commodities exported. U.S. exports of corn, soybeans, and wheat made up 61.8, 59.7, and 31.8 percent of world exports of those commodities in 1993 (Table 29–2). Not only is the United States the world's largest agricultural exporter, but since 1985, the U.S. share of agricultural trade has grown more than the share of any other exporter, including the EU, the United States' largest competitor. In 1995, the U.S. share was 23 percent, the highest level in over a decade.

Table 29-2

**Production and Exports of Selected Farm Products,
United States and the World, 1993
(in metric tons, except as indicated)**

Product	Unit	Production		Exports		U.S. as a Percentage of World	
		U.S.	World	U.S.	World	Production	Exports
Wheat	Million	66.0	560.0	31.5	99.1	11.8	31.8
Corn for grain	Million	165.0	454.0	34.5	55.8	36.3	61.8
Soybeans	Million	50.0	112.0	17.0	28.5	44.8	59.7
Rice, milled	Million	5.1	344.0	2.5	14.4	1.5	17.4
Tobacco, unmanufactured	Thousand	734.0	8,630.0	230.0	1,812.0	8.5	12.7
Cotton	Million bales[1]	16.2	82.5	5.2	24.8	19.6	21.2

[1]Net weight of a bale is 480 lbs.
Source: U.S. Department of Agriculture.

Significant changes in U.S. agricultural policy are in store with the signing of the Federal Agricultural Improvement and Reform (FAIR) Act of 1996. With the farm policy changes, U.S. agricultural producers will make their production decisions more in response to market forces than to government policy. Feed grain trade is expected to increase markedly, with the acreage of feed grains increasing in response to this demand strength. The same is expected for oilseeds.

Consumers around the world have a large and growing appetite for U.S. agricultural products, particularly high-value consumer foods and beverages. Per capita income in many countries that import U.S. agricultural products is rising. This, coupled with more interest in Western-style foods, greater demand for convenience, and growing interest in a safe and healthy diet, has helped make consumer foods the fastest-growing segment of U.S. agricultural products sold overseas. The U.S. share of the global market expanded from about 10 percent in the mid-1980s to about 16 percent by the mid-1990s. High-value-product exports will continue to increase (Figure 29–6).

Japan was the largest buyer of U.S. agricultural products in 1993, followed by Canada, Mexico, and the Republic of Korea (Table 29–3). Considered as a unit, the EU is the largest market for U.S. agricultural exports.

Exports of high-value products to rise faster than bulk

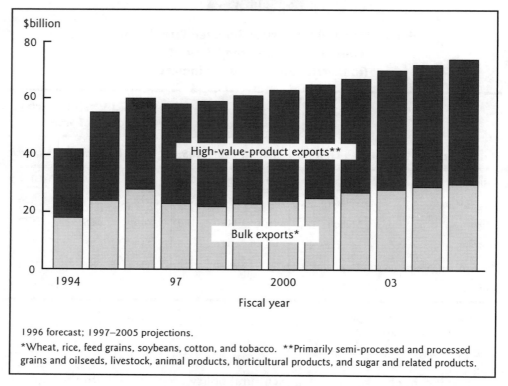

Source: U.S. Department of Agriculture

Figure 29-6

The entire Pacific Rim is a rapidly growing region for U.S. agricultural exports. The trend is expected to intensify, especially as China becomes more dependent on imported food items over the next several years. Asia is the fastest growing economic region in the world and the largest regional market for U.S. agricultural exports. The value of agricultural exports to Asia amounts to over 40 percent of total U.S. farm exports.

Economic growth, with corresponding growth in demand for U.S. farm products, in Central and South America has been strong. While the rate of growth in export sales to Central and South America will likely be less than the rate of growth in export sales to the Pacific Rim, these slower-growing markets will become more important.

Table 29-3

Total Value of U.S. Agricultural Exports, Top 10 Countries of Destination, Fiscal Years 1991 Through 1993

Country	1991	1992	1993
	------------------ Millions of dollars ------------------		
Japan	7,735.8	8,382.8	8,461.5
Canada	4,408.7	4,812.2	5,219.6
Mexico	2,884.7	3,676.0	3,660.0
Republic of Korea	2,158.8	2,200.1	2,040.9
Taiwan	1,738.6	1,915.9	1,999.1
The Netherlands	1,561.0	1,812.3	1,801.2
Former USSR	1,758.4	2,704.2	1,560.7
Germany	1,134.9	1,090.8	1,146.2
United Kingdom	883.3	882.0	916.0
Hong Kong	744.7	817.0	880.1

Source: U.S. Department of Agriculture.

U.S. Imports

The United States is one of the largest importers of agricultural products in the world market. In 1994, the United States imported agricultural products worth over $26 billion (Table 29–4). While the United States is a major import market for agricultural products, agricultural imports make up under 5 percent of total U.S. imports (Figure 29–7).

Imports provide consumers with agricultural products that are either not produced domestically or not available in sufficient quantities in the United States. Major imports not produced in the United States include bananas, some spices, coffee, tea, cocoa, and rubber. Domestic production of other goods, such as certain cheeses and tobaccos, is insufficient to satisfy domestic demand, so imports meet the excess demand.

Canada, Mexico, and Brazil are the top sources of agricultural products imported into the United States. However, if the EU countries are considered jointly, then the EU is the top exporter of agricultural goods into the United States.

Table 29-4

Value of Agricultural Imports, by Selected Major Commodity Group, with Leading Countries of Origin, 1994

Commodity Group	Value	Leading Countries of Origin
	$1,000	
Competitive products		
Meat and products	2,657,548	Canada, Australia, New Zealand
Dairy products	963,376	New Zealand, Italy, Ireland
Grains and feeds	2,339,092	Canada, Italy, Thailand
Fruits, nuts, and vegetables	4,725,954	Mexico, Chile, Spain
Sugar and related products	1,128,712	Canada, Dominican Republic, Guatemala
Wine and malt beverages	2,084,650	France, The Netherlands, Italy
Oilseeds and products	1,562,955	Canada, Italy, Philippines
Tobacco, unmanufactured	613,182	Turkey, Brazil, Greece
Noncompetitive products		
Bananas and plantains	1,071,834	Costa Rica, Colombia, Ecuador
Coffee and products	2,485,433	Brazil, Colombia, Mexico
Cocoa and products	1,033,906	Canada, Ivory Coast, Indonesia
Rubber and gums	965,390	Indonesia, Thailand, Malaysia
Tea	185,515	China, Argentina, Germany
Spices	326,890	Indonesia, India, Madagascar
Total imports	26,818,015	Canada, Mexico, Brazil

Source: U.S. Department of Agriculture.

Topics for Discussion

1. Explain why international trade is important to the United States. Cite at least five possible advantages of trade.

2. Discuss the application of the principle of comparative advantage as it applies to international trade.

3. List and briefly discuss the arguments for trade barriers against cheap foreign imports.

4. Identify at least six nontariff barriers to international trade, and briefly explain how they function.

5. Discuss the export subsidy programs the United States has used.

6. Why is international trade important to U.S. farmers?

7. What are the likely prospects for agricultural exports in the future?

Total value of U.S. imports, 1993

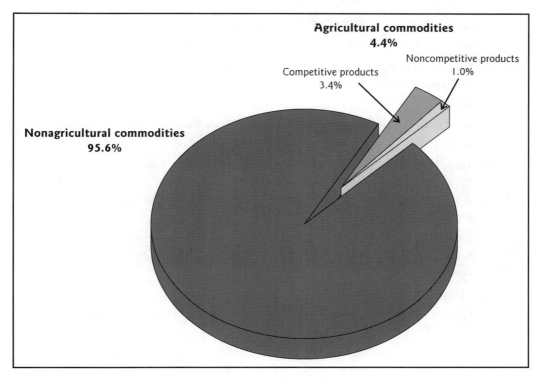

Source: U.S. Department of Agriculture

Figure 29–7

Problem Assignment

Select a specific agricultural commodity, and determine how trade of that commodity with Canada and Mexico has been affected by U.S. entry into the North American Free Trade Agreement.

Recommended Readings

Colander, David C. *Economics*, 2nd ed. Chicago: Richard D. Irwin, Inc., 1995. Ch. 37.

Cramer, Gail L., and Clarence W. Jensen. *Agricultural Economics and Agribusiness*, 6th ed. New York: John Wiley & Sons, Inc., 1994. Ch. 16.

Goldberg, Ray A. "New International Linkages Shaping the U.S. Food System," *Choices*, 8(4):15–17 (1993).

Gwartney, James D., and Richard L. Stroup. *Economics: Private and Public Choices*, 7th ed. Orlando, FL: The Dryden Press, 1995. Ch. 31.

Halliburton, Karen, and Shida Rastegari Henneberry. "A Comparative Analysis of Export Promotion Programs for U.S. Wheat and Red Meats," *Agribusiness*, 11(3):207–221 (1995).

Houck, James P. *Elements of Agricultural Trade Policies*. Prospect Heights, IL: Waveland Press, Inc., 1986; reissued, 1992.

Hudson, William J. "The Basic Elements of Agricultural Competitiveness. In Three Parts: Economics and Policy, Geography, History." Misc. Publ. 1510. Washington, DC: Agriculture and Trade Analysis Division, Economic Research Service, USDA. 1993.

Kastens, Terry L., and Barry K. Goodwin. "An Analysis of Farmers' Policy Attitudes and Preferences for Free Trade," *Journal of Agricultural and Applied Economics*, 26(2):497–505 (1994).

Lee, David R. "Western Hemisphere Economic Integration: Implications and Prospects for Agricultural Trade," *American Journal of Agricultural Economics*, 77(5):1274–1282 (1995).

McCalla, Alex F. "Agricultural Trade Liberalization: The Ever-Elusive Grail," *American Journal of Agricultural Economics*, 75(5):1102–1112 (1993).

Ndayisenga, Fidele, and Jean Kinsey. "The Structure of Nontariff Trade Measures on Agricultural Products in High-Income Countries," *Agribusiness*, 10(4):275–292 (1994).

Salvatore, Dominick. *International Economics*, 5th ed. New York: Macmillan Publishing Co., 1995.

Sanders, Larry, Mike Woods, Warren Trock, and Hal Harris. "Local Impacts of International Trade." In *Increasing Understanding of Public Problems and Policies: 1994*, Steve A. Halbrook and Teddee E. Grace (eds.). Oak Brook, IL: Farm Foundation, 1995. Pp. 135–149.

Seitz, Wesley D., Gerald C. Nelson, and Harold G. Halcrow. *Economics of Resources, Agriculture, and Food*. New York: McGraw-Hill, Inc., 1994. Ch. 9.

Sharples, Jerry. "World Events Shaping U.S. Agricultural Trade," *Choices*, 9(2):4–9 (1994).

Slavin, Stephen L. *Economics*, 4th ed. Chicago: Richard D. Irwin, Inc., 1996. Ch. 34.

Tweeten, Luther. *Agricultural Trade: Principles and Policies*. Boulder, CO: Westview Press, Inc., 1992.

30

Economic Development in the Developing Areas of the World

In this chapter we discuss the situation developing countries face as they seek economic growth and development. We also discuss factors responsible for levels of development, assistance provided by the developed countries, and techniques for implementing economic development.

Setting the Stage for Studying Economic Development

Economic development is a complex process, with many aspects different and unique for each country. Thus, there is no prescribed set of steps for a government to follow to achieve economic growth and development. However, there are some common themes to consider when approaching the study of economic development.[1]

First, developing countries can be understood only by examining their place in a global system of social and economic relationships. Categorizing countries based strictly on economic system, such as industrial capitalist, industrial socialist, or less developed, poor, and rural, is inadequate because doing so fails to consider cultural and social factors. Links between cultural and economic aspects are apparent, for example, in sports competitions, economic trade, and tourism, illustrating how interconnected peoples of the world have become.

---

[1] Adapted from Andrew Webster, *Introduction to the Sociology of Development* (London: Macmillan Education, Ltd. 1984), pp. 4–14.

We must also consider cultural and economic features particular to given societies. Because societies of the world are becoming more interconnected does not mean that the societies are becoming more similar. For example, industrialization is a major force affecting global development. However, its impact on the world has been uneven, both culturally and economically. Thus, it is important to study features of a society that will affect its development.

It is also important to study how the relationship between cultural and economic processes influences social development. This involves considering the way people are likely to respond to new cultural and economic influences at work in their communities. For example, they may adapt, rather than abandon, the traditional way of doing things.

Study of history is another important factor in setting the stage for economic development. If we want to know where we are today, we need to know where we have come from and what aspects of our history continue to influence our society. Social and economic change is a dynamic process.

Finally, understanding the political makeup of governments, with their underlying philosophies and objectives, is also important to successful economic development. The implementation of development plans, a complex process, involves political decisions about access to resources, who receives the benefits, who pays, and so on.

Economic and Social Indicators of Developing Countries

Rather than grouping all developing countries together, economists are grouping them geographically—Asian, African, Latin American, and Middle Eastern. Because of cultural similarities, many countries within a particular grouping share similar problems.

The nature of agricultural production is governed mainly by climate. Heat and moisture in the tropical zones support the growth of tropical fruits, nuts, spices, sugar cane, rice, coffee, rubber, oil palm, and other more exotic plants.

Many developing countries have an annual gross domestic product (GDP) of less than $300 per capita, but the average GDP of developing countries in 1993 was $918 per capita (1988 dollars). By region, the GDP per capita ranged from $546 in China and $652 in Africa to over $3,400 in West Asia. In contrast, the GDP per capita in developed economies averaged $18,995 (1988 dollars) in

1993 (Table 30–1). Social indicators reveal that the developing countries are usually higher in at least two categories—namely, birth rates and death rates—but considerably lower in life expectancy and percentage of literacy (Table 30–2).

Table 30–1

Population, GDP, and GDP per Capita for Developed and Developing Countries, 1983 and 1993

	Population (Millions)		Growth Rate of Population (Annual Percentage Change)	GDP (Billions of 1988 Dollars)		GDP per Capita (1988 Dollars)		Growth of Real GDP per Capita (Annual Percentage Change)
	1983	1993	1984–1993	1983	1993	1983	1993	1984–1993
Developed Countries:	765	812	0.6	11,774	15,419	15,393	18,995	2.1
United States	234	258	1.0	4,057	5,291	17,322	20,521	1.7
European Union	337	347	0.3	4,158	5,230	12,346	15,063	2.0
Japan	119	125	0.5	2,333	3,357	19,544	26,864	3.2
Developing Countries (by Region):	3,494	4,295	2.1	2,667	3,943	763	918	1.9
Latin America	378	460	2.0	743	946	1,967	2,058	0.5
Africa	495	666	3.0	359	434	725	652	–1.1
West Asia	101	142	3.5	470	494	4,650	3,469	–2.9
South and East Asia	1,412	1,737	2.1	720	1,281	510	737	3.8
China	1,040	1,205	1.5	253	658	243	546	8.4
Mediterranean	72	85	1.7	122	131	1,694	1,541	–0.9

Source: Department of Economic and Social Information and Policy Analysis, United Nations, *World Economic and Social Survey, 1994: Current Trends and Policies in the World Economy* (New York: United Nations, 1994).

It is estimated that from 1970 to 1991, total world agricultural production increased at an average rate of about 2.1 percent per year. Average per capita utilization of cereals increased from 690 pounds to 765 pounds. However, population in the developing countries increased at a much faster rate than in the developed countries. As a result, food production per capita in some developing countries failed to keep pace with the rate of population increase. For example, in sub-Saharan Africa, average per capita utilization of cereals decreased from 350 pounds to 337 pounds in the early 1990s.

Table 30-2

Comparative Statistics for Selected Countries, 1994[1]

	Birth Rate per 1,000	Death Rate per 1,000	Infant Mortality Rate per 1,000 Live Births	Life Expectancy	Literacy Rate	Labor Force in Agriculture	GDP from Agriculture	GDP per Capita
	No.	*No.*	*No.*	*Years*	*Percent*	*Percent*	*Percent*	*Dollars*
United States	15.3	8.4	7.9	76.0	97	2.9	2	25,850
Australia	14.1	7.4	7.1	77.8	100	6.1	5	20,720
Japan	10.7	7.5	4.3	79.4	99	7.0	2	20,200
Uzbekistan	29.5	6.4	52.0	68.8	97	43.0	N/A	2,400
Pakistan	41.8	12.1	99.5	57.9	35	46.0	24	1,930
Honduras	34.1	6.0	43.4	68.0	73	62.0	28	1,820
Mali	51.9	19.3	104.5	46.4	19	80.0	50	600

[1]Some data are for earlier years, due to reporting lags.
Source: Central Intelligence Agency, *The World Factbook, 1995* (Washington, DC: CIA), 1995.

Among developed countries, the trend of per capita food production has been increasing relatively slowly since the early 1980s. Among the developing regions, South America and North Africa have shown a tendency for slight improvement in per capita food production, but Central America and the rest of Africa have shown a declining trend since 1980. Per capita food production has increased dramatically in South Asia and Southeast Asia. U.S. production has often served as a safety valve for world food supplies during the past half century.

The rate of population growth is a top concern of long-range planners. The world population was about 5.7 billion in 1990. It has been estimated that the world population will increase by 1 billion people every 12 years through about 2023. The world population is projected to be about 8.5 billion in 2030. Accordingly, to feed this growing population, food production must increase. However, food production will need to expand at a faster rate than population growth, because as per capita incomes in developing countries increase, consumers in those countries will demand not only more food but also a more diverse and nutritious diet.

Although some countries frown upon birth control campaigns, family planning programs have been introduced into many developing countries in

Distribution of world population by continent, 1990, 2025, and 2100

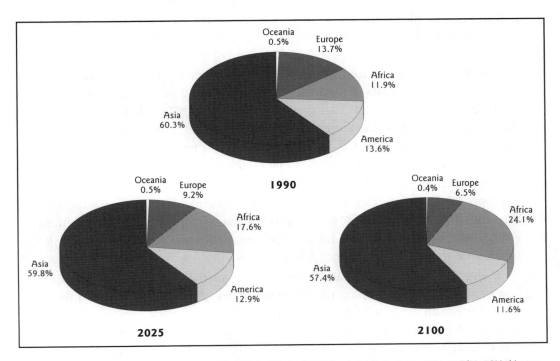

Source: Edward Bos, My T. Vu, Ernest Massiah, and Rodolfo A. Bulatao, *World Population Projections—1994–95 Edition* (Washington, DC: The World Bank, 1994).

Figure 30–1

an effort to provide population management. India, Pakistan, Japan, Korea, Taiwan, Sri Lanka, Singapore, Indonesia, and China are examples in Asia. In Latin America and Africa, family planning is being adopted very slowly because of religious and cultural resistance and the relatively high level of illiteracy in those countries. While for some countries, birth control programs have slowed population growth, for most their overall effect has fallen short of expectations.

Obstacles to Economic Development

The first requirement for economic development in any country is that the government and its political leaders be determined to overcome obstacles, provide encouragement, and instill confidence among agricultural and industrial producers. Without political stability, no government can achieve growth and development, no matter what it tries. In an increasingly interconnected world, there is also the need to promote education and literacy of the population.

Considerable agricultural research has been done by the more advanced countries, but the results are not generally transferable in their entirety to the developing countries, particularly since most of the research has been conducted traditionally in the temperate zones. The developing countries, for the most part, are located in the tropics; therefore, differences in climate, soils, technology, and management skills must be taken into account.

Another obstacle is the lack of development capital. Developing countries hampered by poverty lack the funds with which to modernize production. This difficulty is made even more severe by inadequate or nonexistent infrastructure—for example, roads for transportation, storage and processing facilities, water systems for irrigation, and ports for shipping and receiving goods.

Using development capital, should it be available, to modernize production agriculture in developing countries introduces other problems to address. Such moves can actually work counter to the interests of developing countries with large populations, especially in rural areas. Labor-intensive agriculture gives occupations to the people in these countries, while mechanization results in greater unemployment, unless there are opportunities available in other sectors and adequate training to prepare those leaving agriculture for those opportunities. This problem is aggravated in areas where large families are considered desirable for working on the farm.

With the growth of cities and the growth of commercial activities in metropolitan areas, traditional subsistence farming must make way, at least in part, for market-oriented production. Changing some of the habits, attitudes, and practices to achieve a more progressive agriculture is essential. Education, research, and extension can serve as the foundation for economic development and improved production and distribution.

Another critical element is the time factor. It takes decades to educate people and additional decades to carry on the necessary research. Obviously, the developing countries will be lagging behind indefinitely unless provided with technical assistance from the more developed countries.

Training of teachers, scientists, leaders, technicians, and administrators, along with aid to develop institutional structures, would appear to be the minimal requirement for promoting positive change in developing countries. Success will depend on trained and dedicated personnel who recognize the potentials for improvement.

Credit institutions have been very helpful to all types of enterprises in developed countries. Many of the remarkable success stories in developed agriculture can be traced to the use of capital made available through the farm credit banks. Similar credit institutions are needed in the developing countries. Foreign capital should be regarded as "seed capital" when providing economic assistance. Over the long haul, there should be formation and flow of indigenous capital (from domestic savings) because of increased earnings and savings.

Unfortunately, many poor countries do not have the natural resources required to provide for their own needs, whether those needs be for land, water, minerals, or vegetation. In this case, special skills must be developed to promote processing and services, to utilize imported raw products, and to market skillfully. Technical assistance is particularly important in such situations.

One aspect of foreign aid that often creates problems is that of unilateral aid and the suspicion that the donor country manifests domineering attitudes and imperialistic ambitions. Ideologies of the capitalistic countries invariably clash with socialist or communist ideologies. Each side competes for certain favors or privileges in struggles for political power. Thus, the object of producing for, and aiding the development of, a foreign country is often superseded by the self-serving goals of a donor nation.

Fortunately, the more advanced countries express a common interest in providing food aid and capital assistance. The United Nations serves as the worldwide agency to coordinate and dispense aid to needy countries on a multilateral basis. All governments have an interest in this international institution, and increasing support is needed for its broadly diversified efforts.

Many charitable organizations have assisted the developing countries with health services, food distribution, education, and training. Although these efforts have been conducted on a modest scale, many countries can acclaim this type of assistance.

Supplies needed for production must be made available at the appropriate time; hence, proper planning and coordination of service agencies, including transportation, must be assured. This is also true for the marketing system handling the products when harvested. A lack of adequate infrastructure is often the primary limiting factor, especially in areas away from the main population centers. For example, in the more isolated regions of many African countries, dirt roads are impassable during the rainy season. This often means that basic supplies must be transported by air, which is usually considered unjustifiable because of the expense.

It has been generally recognized that the food problems of developing countries can be largely alleviated by increasing productivity and improving profitability and distribution of crop and animal enterprises. Such inputs as improved crop varieties, fertilizer, water for irrigation, improved cultural techniques, and modernized management are made more readily available through technical assistance programs. Improved diets and medical care also contribute to the welfare and productivity of developing countries.

Public and private investments in marketing and processing are also badly needed. In many countries, the problem lies with inadequate storage and transportation facilities rather than with the ability to produce. Improving aspects of the agribusiness system other than just production agriculture, i.e., transportation, storage, processing, and distribution, should receive focus in development attempts. Good management and strategic planning are needed to increase efficiency in the system.

In the history of food aid provided by the United States, it is possible to note that the food aid program was largely a disposal program for U.S. agricultural surpluses. More recently, the amounts and types of food supplied to needy countries have been reduced in volume but show more variety.

Changes in the nature of food aid by the United States have been prompted by two primary factors. First, the world market for feed grains has absorbed all the surplus stocks, and food aid has shifted from concessional trade to commercial trade. Second, an appraisal of aid accomplishments leads to the conclusion that food aid is not the most effective form of aid to encourage economic development. Food aid will alleviate hunger and is indispensable in times of emergency. It also permits the recipient nation to save foreign currency that is usually in short supply. It has been recognized, however, that prolonged food aid may actually retard agricultural development by depressing the market price for domestic production, thus, removing the needed incentive to local producers. Moreover, food aid to some countries is like welfare to some people; it leads to complacency and lack of initiative. In the final analysis, instead of direct food donations, the assistance should be aimed at providing the inputs and management skills needed to develop local production and encourage the developing countries to provide for themselves. Some governments encourage U.S. food aid to depress domestic food prices and make food cheaper for their increasing nonfarm populations.

The main problems of the developing countries are unemployment, poor income distribution, and an increasing dependency on the developed countries for food supplies. Development specialists agree that policies to promote economic development need to be modified. The suggestions range from the orthodox recommendations of bringing the costs of foreign exchange, capital, and labor to levels closer to their real market values to proposals for increasing public expenditures, improving their allocation, emphasizing agricultural production, and financing all of these with more progressive rates of taxation. Land reform also continues to gain favor as a technique for improving technology, creating employment, distributing income equitably, and reducing migration to the cities.

Many developing countries have encouraged the formation of labor unions in efforts to improve wages in urban centers. As a result, organized labor receives higher wages and yields a somewhat lower volume of output per worker as compared to non-union labor. Since labor costs are relatively high, one would expect capital costs also to be high. Instead, capital costs to governments and large private enterprises are relatively low, owing their access to external development loans and suppliers' credits. In light of these relationships, employers are often induced to substitute capital for labor, aggravating unemployment. Moreover, aid to developing countries has accentuated the capital bias by insisting on design specifications based on the experience of industrial countries and by recommending sophisticated labor-saving techniques. However, labor is the one comparative advantage developing countries have in a world market. Failure to adequately utilize their labor resources could slow development in such countries.

Promoting Economic Development

U.S. Response to World Hunger

The United States is the world's largest donor of foreign food aid, providing over 50 percent of the annual total. One of the most notable aid programs was the Food for Peace program, authorized by Public Law 480 (P.L. 480) when Senate Bill 2475 was passed on July 10, 1954. It attempted to do two things—first, to provide an outlet for U.S. agricultural surpluses and, second, to improve purchasing power of foreign countries by making it possible for them to buy from the United States using their own local currencies to overcome the shortage of trade dollars. At some time during the past 40 years, nearly every country in the world received P.L. 480 food aid under one type of program or another. About $43 billion has been provided for food needs around the world, with an additional $3.5 billion in commodity value provided through other U.S. food aid programs.

P.L. 480 focuses on combating world hunger, promoting sustainable development, expanding U.S. export trade, and fostering the development of private enterprise in developing markets. Title I provides government-to-government financing for sales of U.S. agricultural commodities at below-market interest rates, with repayment terms of up to 30 years. Title II provides commodities that are distributed overseas by the recipient government, private voluntary organizations, and international organizations such as the World Food Program. Title III provides government-to-government food aid grants to support economic growth in developing countries.

The USDA also administers the Section 416(b) program authorized by the Agricultural Act of 1949. Under this authority, surplus Commodity Credit Corporation (CCC) stocks obtained through the CCC's price support programs are donated to needy people overseas through foreign governments, public nonprofit humanitarian agencies, and international agencies such as the World Food Program.

The third program through which the USDA provides humanitarian assistance is the Food for Progress program. Under this program, which was established by the Food Security Act of 1985, the USDA provides commodities on grant or concessional credit terms to needy countries and emerging democracies to reward efforts toward agricultural reform and free enterprise. The United States can enter into agreements with recipient governments, private voluntary organizations, nonprofit agricultural organizations, and coopera-

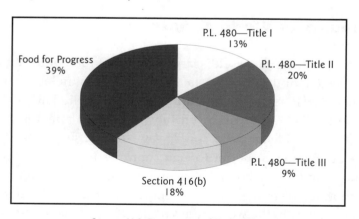

U.S. food aid, FY 1993
Programmed commodity value $2.5 billion

Source: U.S. Department of Agriculture

Figure 30–2

tives. This program was used in 1993 to provide 5.8 million metric tons of agricultural commodities valued at $939 million to countries of the former Soviet Union.[2]

Food aid has been used to provide food for wages in building some farm-to-market roads in developing countries. Certain positive aspects are attached to this form of payment: It enables a government to carry out development projects with a minimum of deficit financing, it creates employment for the rural unemployed and underemployed, and it supplies food to the worker and the worker's family. On the other hand, many criticisms can be leveled at this practice from an economic standpoint: First, it is a regressive step, from a money economy to a barter economy. Second, it disrupts the market for local food. Third, it destroys incentive for local production. Fourth, it leads to deception and dishonesty among workers and merchants who illicitly trade the food aid items for other necessities. Fifth, it perpetuates a subsistence economy instead of developing a commercial one.

Fortunately, food aid administration in the United States has changed significantly since the inception of the P.L. 480 program. One example was in 1966, when Congress insisted that developing countries display positive efforts at self-help in order to qualify for aid. The requirement that food aid be from surplus stocks was also removed. Since food surpluses are no longer of serious concern in the United States, more emphasis is directed to providing technical and financial assistance to foreign countries instead of just providing food aid. Finally, the acceptance of local currencies to pay for imported U.S. goods has gradually shifted to use of dollars, thereby moving the developing countries from "aid" status to "trade" status.

Other Aids

In light of the increasing agricultural competition from the European Community, as well as from other countries, there is a growing reluctance in the United States to advance aid to potential competitors. This is a fact of life not to be ignored, and, philosophically, one must admit that the willingness to provide assistance is strongly influenced by business interests as well as humanitarian concerns. In fact, specific restrictions have sometimes been included in the types of aid authorized if that aid could result in additional competition for U.S. products. Fear of growing competition from developing

---

[2] U.S. Department of Agriculture, *Agriculture Fact Book, 1994*, pp. 71–72.

countries has been largely dispelled, however, by the fact that countries that have made the most rapid economic progress have also made the most rapid increases in imports of U.S. agricultural and other products.

International trade and monetary policies must be geared to encourage balanced economic growth in all countries and permit developing countries to earn foreign exchange through commercial trade rather than through grants and concessional loans.

American universities continue to make valuable contributions, both by training and educating visiting students and by producing researchers, teachers, and consultants to aid in building educational and other institutions in developing countries.

Resource Assistance to Developing Areas

A relatively large amount of capital and technical assistance has been provided for research, consulting, and improvement of infrastructures. In this respect, some economic integration may be necessary among the smaller countries to achieve economies of scale not possible when functioning alone. For example, a nitrogen fertilizer plant, a steel mill, or a hydroelectric plant may not be a feasible undertaking for every small country. Such a plant may be too large and costly to serve a single country, but by economic grouping and by cost and benefit sharing, such a development may become feasible.

Government

Economic growth in any developing country requires, first of all, the determination of its government to overcome obstacles in the way of development. An evaluation of the country's resources, as to quantity and quality, is an essential first step. Then, productivity and marketing potential must be assessed. To accomplish these ends, research studies and educational programs are critically needed in most countries. This type of assistance is gaining increasing support from donor countries and private foundations. Education is receiving impressive financial support both for student training in local schools and for advanced education of visiting students in higher-level institutions of developed countries.

Land

Through the ages, expansion of agricultural production has been achieved by bringing additional land into cultivation. Currently, the more accessible and productive lands are being used. The unoccupied areas are either inaccessible or costly to put into production.

There are an estimated 3.5 billion acres of arable cropland and 8.3 billion acres of permanent pastures in the world. The total irrigated area increased about 43 percent from 1970 to 1990. Many potential cropping areas require irrigation water, others require drainage, and practically all of them need supplemental nutrients and organic matter. Capital is critically needed to supply the water, to drain wet areas, to build transportation and communication facilities, and to obtain necessary operating inputs, such as seeds and fertilizer.

Capital

Developing the infrastructures and additional cropland in developing countries requires a tremendous amount of capital. The Aswan Dam, for example, cost more than $1 billion and required about 15 years of planning and construction to increase the cropland area in Egypt. While major projects such as this have important benefits, there are also costs that must be carefully anticipated and considered. For example, some argue that the Aswan Dam has destroyed traditional cropping systems in regions below it.

The input of foreign capital is generally regarded as "seed capital" necessary to generate economic activity that will not only sustain itself but will also lead to formation of additional capital for further investment and economic development. In this sense there must be continuing opportunities and incentives for savings and investment.

On a global scale, but excluding most of the communist countries, the major sources of agricultural finance have been the International Bank for Reconstruction and Development (IBRD), commonly known as the World Bank, and the International Development Association (IDA).

Although government aid and private capital are very important sources of finance for development, exports by low-income countries are by far the largest source of foreign exchange. Unfortunately, the growth in export earnings has been slow in relation to these countries' need for imports to support reasonable development programs. For example, the increase in demand for products these countries export may not keep pace with the growth of income in high-income countries. In addition, the inflationary rise in prices of imported industrial products puts the economic squeeze on the low-income countries even more.

Research

As countries develop, cities grow and a more complex marketing system materializes. This changes the habits, practices, and attitudes of farm people and necessitates a more complex set of institutions for education, research, and extension. In the developing countries, most of the past research was concentrated in the development of commercial exports—mainly, raw products, such as rubber, cocoa, coffee, oil palm, bananas, and sugar cane—primarily to colonial powers. There was much less comparable investment in research for basic food production for the indigenous population.

Research is one of the most pressing needs in developing countries. Adaptive research holds considerable promise for developing improved varieties, increasing yields, and boosting total production. Technical assistance from donor countries can provide the expertise by making teachers, scientists, and technicians available.

Developing countries have limited resources available for research purposes. To assist in alleviating this problem, 18 agricultural research centers have been organized and financed by private foundations. Also, international agencies have been created to encourage scientists to do practical, adaptive, and basic research. One of the first research institutes was the International Rice Research Institute (IRRI), organized in 1960 in the Philippines. The research institutes are supported by an informal organization of 40 public and private donors called the Consultative Group on International Agricultural Research (CGIAR), formed in 1971. Programs carried out by the CGIAR system fall into six broad categories: productivity research, natural resource management, improvement of the policy environment, institution building, germplasm conservation, and building of linkages with national partners.

Extension

Conveying research information to the farmers and assisting them in applying this new knowledge is the responsibility of extension. New varieties and new agronomic techniques will not benefit the developing countries unless they are utilized by producers.

Farmers everywhere tend to be conservative. If they have always grown a crop in the same manner as their parents, they feel they do not need advice.

This sociological problem is perhaps the greatest obstacle to overcome. Demonstration plots on farmers' fields are frequently used to show potential production increases that can be achieved by using improved seeds and recommended cultural practices. In fact, farmers often are encouraged to perform the necessary cultural activities under the guidance of an agricultural extension agent.

As additional incentives to get farmers to participate in development programs, field day demonstrations are organized, and neighboring farmers are invited to observe applied techniques for land preparation, fertilization, and planting. At harvest time, field days are again utilized to display the results of the demonstration plots. On these occasions, farmers are encouraged to explain the results and the techniques that were applied.

Subsidies in the form of grants or lowered unit costs are frequently used to introduce new seeds and fertilizers. Likewise, farm credit is made available for the purchase of necessary supplies and equipment.

Techniques Used to Assist Developing Countries

Planning

Planning activities are sometimes carried out by individual agencies, but more commonly they are done on a cooperative basis, utilizing representatives from international organizations, such as the United Nations, the International Labour Organization (ILO), and the World Health Organization (WHO), among many others.

In order to tackle the world food problem, the Food and Agriculture Organization (FAO) has been working on an International Agricultural Adjustment Policy. This agency feels that the time has come for all countries to recognize that a "one world" view of agriculture is essential. The world has now become so interdependent that national agricultural policies in one country inevitably have repercussions on events in other countries. Moreover, a world food security plan has been proposed to build up food reserves by having all countries maintain, according to their circumstances, minimum security stocks.

Cooperatives

Cooperatives are an excellent means of carrying out development projects and disseminating benefits to the working classes. Weaknesses often center on the lack of good financial management rather than on the lack of capital. The service and supply cooperatives (marketing, processing, irrigation, credit, fertilizer, and the like) are usually more successful than production cooperatives, where the productive resources are owned and used collectively. In the short run, government aid to cooperatives may be essential, but in the long run, cooperatives should be self-sustaining.

Professional Training

During recent decades, many governments have been concerned about the "brain drain," that is, the flight of professionally trained people (those with at least 16 years of education) from their homelands. Several techniques have been employed to stem this loss, but in the main, the effort has been directed toward providing new opportunities for professional employment and greater recognition to young professionals through responsible positions and better salaries in their homelands.

Not all of the professional training acquired outside the students' homelands has proven beneficial. Many return with the wrong type of experiences or education or find themselves extremely frustrated because of lack of supporting technicians, lack of facilities, or lack of challenging employment. Some return at the same salaries to the same positions they held before receiving the extended training. Organized national institutes, seminar programs, pilot projects, and field days are some additional techniques for keeping professionals interested and involved so as to overcome isolation, boredom, and neglect. During the past two decades a great deal of attention has been directed to problems of education and training. Now it is necessary to direct attention to making the best use of educated and trained people. Human resource planning is increasingly important, and this must be related to national planning. Along with the opportunity for education and training, there is the need for effective coordination and synchronization of human resources, production plans, and opportunities for employment.

Technical Assistance

One example of technical aid is the U.S. Peace Corps. The Peace Corps began with an emphasis on people rather than on programs. The first Peace Corps Act and subsequent legislation specified three principal aims: (1) to perform a needed job requested by a host country, (2) to give foreigners a better idea of what Americans are like, and (3) to teach Americans more about foreign countries through the knowledge transmitted by returned volunteers. Peace Corps efforts have ranged from a token presence in a host country to a significant role in training locals to improve production, increase earnings, improve marketing, and enter into the cash economy of the country.

Recommendations

Sustained improvement in agricultural productivity is a key step in achieving progress in overcoming rural poverty in developing countries. Some of the measures that can be taken to achieve such progress are (1) land reform, specifically private ownership; (2) better access to credit; (3) an assured water supply; (4) expanded research and extension facilities; (5) improved infrastructure, especially transportation, storage, and communications, but also irrigation and drainage, schools, and health centers; and (6) development of competitive markets that encourage low-cost production of goods that consumers desire.

Developing countries frequently try to adopt technology from industrialized countries without sufficient concern for the long-term effect upon their own economic and social structures. To guard against such pitfalls, basic principles have been suggested by actions in certain developing countries. A major thrust is to diffuse technical control of industrial facilities to broad segments of the population and to enlist local participation in technical design and innovation. Industrialization policies should lean toward (1) substitution of labor for capital, (2) dispersal of industry to rural areas, (3) formation of extensive teaching and training facilities utilizing available materials and human resources, and (4) promotion of national research and technology suited to the resource endowments of the country.

Topics for Discussion

1. Where are most of the developing countries located?
2. What are the characteristics ascribed to developing countries?

3. What are the important obstacles to development in the developing countries?

4. What proportion of the world's population is found in the developing regions?

5. Why are yields of grain per hectare so much higher in developed countries than in the developing countries?

6. Describe the P.L. 480 food aid program, and explain its shortcomings in promoting development.

7. Explain the lack of foreign exchange as an obstacle to development.

8. Why should the United States provide aid to countries whose increased production will mean more competition for U.S. products?

9. Explain the role of education, research, and extension in developing areas.

Problem Assignments

1. Compare the rates of population growth among the developed and developing regions of the world.

2. Imagine that you have somehow been transplanted to a developing country, one with a per capita income of only $300. Discuss how this would affect your life style. What luxuries would you be forced to forego?

Recommended Readings

Black, Jan Knippers. *Development in Theory and Practice: Bridging the Gap.* Boulder, CO: Westview Press, 1991.

Boussard, Jean-Marc. *The Impact of Structural Adjustment on Smallholders.* FAO Economic and Social Development Paper 103. Rome: Food and Agriculture Organization of the United Nations, 1992.

Food and Agriculture Organization of the United Nations. *Rural Poverty Alleviation: Policies and Trends.* FAO Economic and Social Development Paper 113. Rome: Food and Agriculture Organization of the United Nations, 1993.

Scott, James C. *The Moral Economy of the Peasant: Rebellion and Subsistence in Southeast Asia.* New Haven, CT: Yale University Press, 1976.

Tisch, Sarah J., and Michael B. Wallace. *Dilemmas of Development Assistance: The What, Why, and Who of Foreign Aid.* Boulder, CO: Westview Press, 1994.

Webster, Andrew. *Introduction to the Sociology of Development.* London: Macmillan Education, Ltd., 1984.

GLOSSARY

Glossary

Absolute advantage The ability of one region or country to produce a certain good more efficiently than another region or country. Having an absolute advantage in the production of a good does not necessarily mean that the region or country will produce that good for trade; the region or country must also have comparative advantage in that good.

Agrarian fundamentalism The notion popularized by Thomas Jefferson that farming is morally superior to other occupational work. Also called "Jeffersonian fundamentalism."

Agribusiness A subfield of economics that includes not only agricultural economics but also segments of labor, industrial, business, and consumer economics pertaining to agriculturally related industries and businesses. Agribusiness includes all the vital activities performed in any of the three broad categories of the food and fiber system—the agricultural input sector, the production sector, and the processing-manufacturing sector.

Agricultural economics A specialized and applied branch of general economics that deals with the allocation and utilization of scarce resources (land, labor, capital, and management) among different choices of crop, livestock, orchard, timber, and other enterprises to produce goods and services that satisfy human wants.

Amortization A process by which the level of debt is reduced when regular, periodic debt payments that cover interest and principal are made until the entire loan is repaid.

Average fixed cost (AFC) Total fixed cost divided by the total units of output at any given output level.

Average physical product (APP) Total physical product divided by the number of units of input at a given level of input use.

Average revenue (AR) Total revenue divided by the total units of output at any given output level; the revenue generated by each unit of output.

Average total cost (ATC) The average cost of producing a unit of output at each level of output, obtained by dividing total cost by the total units of output at any given output level.

Average variable cost (AVC) Total variable cost divided by the total units of output at any given output level.

Balance of payments A measure of a nation's total receipts from, and total payments to, the rest of the world.

Barter The exchange of goods or services for other goods or services, with no money changing hands.

Bilateral trade Trade between any two nations.

Call A contract that gives the call-option purchaser the right, but not the obligation, to buy a futures contract at a specific price during a specific period.

Capital Goods used to produce other goods and services.

Capitalism An economic system in which goods and services are produced for profit using privately owned capital goods and wage labor.

Ceteris paribus Letting specific variables change while holding the remaining variables constant.

Change in demand The movement or shift of a demand schedule.

Change in supply The movement or shift of a supply schedule.

Change in the quantity demanded A movement along a demand schedule; refers to what happens to the quantity purchased when the price of a product varies.

Change in the quantity supplied A movement along a supply schedule; refers to what happens to the quantity offered for sale when the price of a product varies.

Communism A totalitarian system of government in which a single authoritarian party controls state-owned means of production with the professed aim of establishing a stateless society.

Comparative advantage The theory that it is best for a region or country to devote its energies not to all lines of production in which it may have superiority, but to those in which its superiority is relatively greatest.

Competitive enterprise An activity that detracts from the success of another.

Complementary enterprise An activity that contributes to the success of another.

Conservation Reserve Program (CRP) A program first authorized by the Food Security Act of 1985 designed to reduce erosion on 40 to 45 million acres of cropland. Under the program, producers who signed contracts agreed to convert highly erodible cropland to approved conserving uses for no less than 10 years. In exchange, participating producers received annual rental payments and cash or payments-in-kind to share up to 50 percent of the cost of establishing permanent vegetative cover.

Controlling The management function that involves the activities necessary to ensure that the policies of the business are, in fact, being carried out.

Corporation A legal entity, authorized by the state or federal government, that owns its own assets and must meet its own liabilities. The entity is separate from the individuals who own it.

Cross elasticity of supply The change in quantity supplied of one good or service in response to a change in the price of another good or service.

Cross-price elasticity of demand The percentage change in the quantity demanded of one good or service, given a change in the price of another good or service, *ceteris paribus*.

Decoupling The gradual process of separating a farmer's production decisions from income supplements or price support considerations.

Deficiency payment A government payment made to farmers who participate in support programs for the "basic commodities"—e.g., feed grain, wheat, rice, and upland cotton. The deficiency payment rate is based on the difference between the target price and the market price or the loan rate, whichever difference is less. The total payment equals the payment rate multiplied by the eligible production.

Deflation A sharp decrease in consumer spending, a declining price level, a tight money supply, and a reduction in credit.

Demand curve The amount of a good or service a consumer is willing to purchase at different prices in a given period, *ceteris paribus*.

Demand schedule *See* Demand curve.

Depreciation A decline in the value of an asset brought about by wear and tear or by obsolescence.

Devaluation A reduction in the value of a country's currency in relation to another currency.

Diminishing marginal utility A law of economics that states that the additional utility or satisfaction gained from consuming a good or service decreases as more units of that good or service are consumed.

Diversification Engaging in several enterprises or activities to provide protection against the risk of adversity.

Easement A technique for land use control in which a governmental unit purchases the development rights to land, but not the land itself, for a fixed number of years.

Economics The science of allocating scarce resources (land, labor, capital, and management) among different and competing choices and utilizing them to best satisfy human wants.

Effective interest rates Actual or true interest, which depends not only on the nominal rate quoted but also on the frequency of repayment required.

Embargo A government order restricting imports or exports of a good.

Eminent domain A right assumed by public governing bodies to condemn private property, with compensation at fair market prices, for public use.

Equilibrium The point where the demand curve and the supply curve intersect.

Equilibrium price The price in a competitive market established by the intersection of the demand and supply curves, where the quantity demanded equals the quantity supplied.

Export subsidies Payments to exporters by the government designed to increase the competitiveness of excess domestic supplies in the world market. Two types are:

1. *Specific export subsidy*—A fixed amount is paid per unit of the good exported.
2. *Ad valorem export subsidy*—The amount paid is a set percentage of the value of the exported good.

Factor-factor model The use of two variable inputs in the production of one output.

Factor-product model The use of one variable input in the production of one output.

Factors of production Inputs (land, labor, capital, management) used in a production process.

Farm As defined by the U.S. Bureau of the Census, an entity that has or would have had $1,000 or more in gross sales of farm products during a year.

Fascism A centralized, autocratic national organization with intensely nationalistic policies exercising regimentation of industry, commerce, and finance.

Financial performance The results of production and financial decisions over one or more periods.

Financial position The total resources controlled by a business and the total claims against those resources at a single point in time.

Fixed asset theory The theory explaining why producers do not respond quickly to changes in the marketplace. There are often few alternative uses for many resources (land, equipment, and so on) used in production agriculture. Because these resources are immobile, they are kept in production even though the returns they generate may be very low.

Fixed costs Costs that do not change when variable input usage and the corresponding output level are varied; also called "sunk costs." Examples include property taxes, insurance, and land and machinery payments.

Fixed inputs Inputs that are used in a constant quantity over a given period in the production process, regardless of the quantity of output produced.

Flexibility With respect to farm program participation, this term relates to restrictions imposed by the farm bill on what crops farmers can plant.

Form utility The utility gained by changing the form of a product (i.e., through processing) into another product that provides consumers with more utility.

Green payments Sometimes called "stewardship payments," these are potential replacements for current price and income support programs; they will provide financial assistance to producers who adopt approved soil, water, and wildlife management practices.

Gross domestic product (GDP) Gross national product minus net earnings from the rest of the world. It measures only output produced by factors in the United States.

Gross national product (GNP) The total dollar value of final goods and services produced in a country during a given period, usually one year.

Highly erodible land (HEL) Soils with a natural erosion potential at least eight times their tolerance (T) value, the rate of soil erosion above which long-term soil productivity may be depleted.

Horizontal integration A business structure in which several plants operating at the same stage of production are owned by one firm.

Implementing The management function that involves overseeing and directing how plans are executed, including delegating the work appropriately and communicating expectations clearly.

Import quota A limit on the amount of a good that may be imported in a given period.

Income elasticity The percentage change in the quantity demanded of a good or service, given a percentage change in income, *ceteris paribus*.

Independent enterprise An enterprise that neither adds to nor detracts from another enterprise.

Indifference curve A curve connecting combinations of goods that provide the same level of utility or satisfaction.

Inferior good A good whose consumption will decrease as income increases.

Inflation A situation in the economy in which the supply of money expands while the supply of goods and services remains the same or expands at a slower rate, causing the prices of goods and services to increase.

International trade The exchange of goods and services between countries.

Isoquant A line connecting all possible combinations of two variable inputs that produce a given amount of an output.

Jeffersonian fundamentalism *See* Agrarian fundamentalism.

Laissez-faire Noninterference by government, leaving coordination of an individual's wants to be controlled by the market.

Law of demand The quantity of a good or service purchased is inversely related to the price of that good or service, *ceteris paribus*.

Law of diminishing marginal returns As additional units of a variable input are used in combination with one or more fixed inputs, after a point, each additional unit of input produces less and less additional output.

Law of diminishing marginal utility As a consumer uses additional units of a good or service, the additional satisfaction or utility gained from the last units consumed decreases.

Liquidity The ability to meet financial obligations as they come due in the ordinary course of business without disrupting normal business operations.

Loan rate A price per unit of a program commodity set by the government to determine the amount extended in a nonrecourse loan. The loan rate serves as a price floor.

Macroeconomics The area of economics that deals with a nation's economy as a whole; the study of the effects of changes in the production of goods and services and employment and how they interact to influence economic performance.

Marginal costs Additional costs incurred from producing the last unit of output.

Marginal input cost (MIC) The change in total variable cost, given a unit change in the input level.

Marginal physical productivity (MPP) The change in output that results from an incremental change in the variable input.

Marginal revenue (MR) The change in total revenue that results from producing the last unit of output.

Marginal value product (MVP) The amount added to total value product as a result of adding the last unit of the variable input.

Marketing The processes, functions, and services performed in connection with handling food and fiber, from securing production inputs, through production, and up to and including delivery into the hands of the consumer.

Market potential An estimate of the maximum possible sales opportunities present in a particular market segment open to all sellers of a product or service during a specific future period.

Means testing A farm policy concept in which the only eligible participants in federal farm programs would be those who "need" assistance, based on certain income or size criteria.

Microeconomics The area of economics that deals with individual decision units—people, firms, or markets—within the economy.

Money A common medium of exchange whose chief characteristics are acceptability, stability, durability, transportability, and divisibility.

Monopolistic competition The market structure characterized by a large number of firms, each selling a good or service that is differentiated in some manner, real or perceived, from the goods or services produced by its competitors.

Most-favored nation A nation that is charged as low a tariff on its exports as any other country. The most-favored nation principal provides equal tariff treatment for imports from all GATT members.

Multilateralism Cooperation among a number of countries in trade negotiations and other means to serve the interests of the participating countries. The General Agreement on Tariffs and Trade (GATT) is an example.

Neutral good A good whose consumption is not affected by changes in income.

Nominal interest rates Stated annual interest rates compounded periodically; the percentages usually quoted by lenders.

Nonrecourse loans A price-support mechanism of the U.S. Commodity Credit Corporation (CCC). Farmers put up their crops as collateral for a loan from the CCC. If the market price is below the government loan rate at maturity of the loan, farmers may choose to pay back the loan with their crops, thereby transferring crop ownership to the government.

Oligopoly The market structure in which a few firms sell a given good or service and rival firms closely observe each other's prices and pricing policies.

Oligopsony The market structure in which there are few buyers for a given good or service.

Opportunity cost The cost of using a resource in one way compared to the return that could be obtained from using that resource in its best alternative use.

Options A tool for managing price risk that gives the purchaser the right, but not the obligation, to a position in the underlying futures market.

Organizing The management function of combining the people, resources, and business activities necessary to bring the goals and objectives identified in the planning stage into reality.

Own-price elasticity The percentage change in the quantity demanded of a good or service, given a percentage change in the price of that good or service, *ceteris paribus*.

Partnership An unincorporated association of two or more persons to carry on a business for profit as co-owners.

Place utility The utility or satisfaction gained by transporting goods to places desired by consumers.

Planning The management function that considers all the activities that determine the future direction or course of action for the business.

Planning (for land use) Developing a comprehensive plan detailing the optimum utilization of a given land area.

Policy A definite course of action, given current conditions, to determine present and future decisions in the accomplishment of agreed-upon goals and objectives.

Premium (1) The amount for which a given futures contract sells over another futures contract. (2) The additional payment an exchange allows for delivery of a higher-than-required-quality commodity against a futures contract. (3) The price an option buyer pays to an option seller for the right to buy or sell a futures contract at a specific price during the life of the option.

Price elasticity of demand The percentage change in quantity demanded, given a percentage change in price, *ceteris paribus*.

Price elasticity of supply The percentage change in quantity supplied, given a percentage change in price, *ceteris paribus*.

Price index A ratio showing the price of a basket of goods and services in various years in relation to the price of the basket in a base year.

Principle of equimarginal returns The economic principle that states that to optimally allocate scarce resources, they should be allocated among alternative uses in such a way that returns from the last unit employed in each use equal returns from the last unit employed in all other uses.

Principle of specialization A phenomenon in which each producing unit tends to specialize in the activity to which it is best suited and then sells its surplus production to others engaged in different specialties.

Processing Any activity that alters the condition or nature of a product substantially or provides form utility.

Production function The technical relationship that transforms inputs (resources) into outputs (commodities) within a given period.

Product-product model The production of two or more outputs using one or more variable inputs.

Profit The difference between total revenue and total cost (including opportunity costs), assuming total revenue is greater.

Profitability The extent to which a business generates a financial gain from the use of its resources.

Public policy A defined course of action or guiding principle for governments to follow to achieve stated goals or solve problems of public concern.

Pure competition A market organization with many firms in an industry, a homogeneous product, and the freedom of firms to enter or leave the industry. No one firm can control supply or influence price.

Put A contract that gives a put option buyer the right, but not the obligation, to sell a futures contract at a specific price during a specified period. The put option seller is obligated to buy futures from the put option buyer if the put option buyer exercises his or her option.

Rate of inflation The percentage increase in the price level from one year to the next.

Regional trade agreements Trade negotiations among countries within a given region.

Regulatory trade restrictions Government-imposed rules and regulations that limit imports.

Revenue assurance A policy proposal that would replace deficiency payments with an insurance-type program. Farmers would presumably receive payments when their crop revenue (from both yields and prices) fell below some targeted amount.

Sales forecast An estimate of sales, in dollars or physical units, for a future period under a particular marketing program and an assumed set of economic and other factors outside the firm for which the forecast is made.

Sales potential An estimate of the maximum possible sales opportunities present in a particular market segment open to a particular firm selling a product or service during a specific future period.

Slippage The failure of supply management programs, such as the Acreage Reduction Program, to meet their objective of adequately reducing supply. A major reason slippage occurs is that farmers set aside their poorest land to satisfy program requirements and then farm the remaining land more intensively.

Social contract An implicit contract between society and agriculture dating to the Depression Era. Society agreed to protect farm incomes and prices, while agriculture agreed to provide an abundant supply of high-quality, low-cost food and fiber for society.

Socialism A political and economic theory of social organization based on collective or government ownership and administration of the essential elements needed for production and distribution of goods.

Sodbuster A provision, first authorized by the Food Security Act of 1985, designed to discourage bringing highly erodible cropland into intensive agricultural production.

Sole proprietorship An unincorporated type of business firm in which an individual owner supplies the working capital, directs or manages the firm, and receives all profits or bears all losses.

Solvency The amount of borrowed capital (debt), leasing commitments, and other expense obligations a business has relative to the amount of owner equity invested in the business.

Specialization The engagement of an individual, firm, region, or country in the production of one or a few goods or services.

Stages of production Classification of the classical production function into three stages, or regions. Stage I is that area where marginal physical product (MPP) is greater than average physical product (APP) and APP is increasing; Stage II is that area where MPP is less than APP but is still positive; Stage III is that area where MPP is negative. Stage II is the rational region of production for producers with no constraints limiting their production decisions.

Standardization The establishment of recognized standards for such characteristics or features as quality, quantity, size, weight, and color.

Strike price The price at which an option contract may be exercised.

Superior good A good whose consumption increases as income increases.

Supplementary enterprise An activity that utilizes otherwise idle resources.

Supply The quantity of a particular good or service available for purchase at prevailing market prices.

Supply curve The amount of a good or service a producer is willing to offer for sale at different prices in a given period, *ceteris paribus*.

Supply schedule *See* Supply curve.

Swampbuster A provision of the Food Security Act of 1985 that discourages the conversion of natural wetlands to cropland use.

Target price A price set by the government for certain commodities. If market price is below the target price, participating producers receive a deficiency payment to make up the difference.

Tariffs Government-imposed taxes on goods when they cross national boundaries. Three types of import tariffs include:

1. *Specific tariff*—A fixed charge per unit of the good being imported.

2. *Ad valorem tariff*—A percentage of the value of the good being imported.

3. *Countervailing duty*—A tariff designed to offset an export subsidy of a competing country.

Time utility The utility or satisfaction gained (primarily through storage) by delivering a product at the desired time.

Time value of money The concept that a dollar received today is worth more than a dollar received at some point in the future, using an interest rate as the pricing mechanism.

Total cost (TC) Total variable cost plus total fixed cost.

Total physical product (TPP) Output; total quantity of a product that is produced at each level of variable input use, given the levels of the fixed inputs.

Total revenue (TR) Output price multiplied by total physical product at any output level; the total dollar amount a firm receives from the sale of a good or service.

Unilateralism An approach to international trade in which each nation designs its own trade policies without consultation or collaboration in wider trade forums.

Utility The satisfaction an individual gets from consuming goods and services.

Variable costs Costs of purchasing those variable inputs that vary directly with the level of production.

Variable input An input whose quantity used in a production process can be changed, depending on the desired quantity of output to be produced.

Variable levy An import tax imposed when the world price falls below a government-set minimum import price, yielding a domestic price at the predetermined level.

Vertical integration A business structure in which a single firm owns and controls different stages of a production and/or marketing process.

Voluntary export restraint An agreement in which foreign governments are asked to limit their exports of specific commodities to a given quantity.

Wetlands Reserve Program (WRP) A program, authorized by the Food, Agriculture, Conservation, and Trade Act of 1990, in which participating producers agree to implement an approved wetland restoration and protection plan and provide either a permanent easement or one of 30 years or more. In return, participating producers receive payments over a 5- to 20-year period, or they receive one lump sum if they grant a permanent easement.

World Trade Organization (WTO) An intergovernmental organization made up of the nations that ratified the GATT. It is intended to provide for more effective decision making and greater involvement in trade relations.

Zoning A legal method employed by government under its police power to restrict the conduct of certain activities in specified areas.

INDEX

Index

A

Absolute advantage 176
Accounts payable turnover 479
Accounts receivable turnover 479
Acid-test ratio 476
Acreage Reduction Program 412
Advertising 309, 488
Agrarian fundamentalism 393
Agribusiness 6, 405
Agricultural Act of 1949 427
Agricultural Act of 1956 419
Agricultural Adjustment Act of 1933 395
Agricultural and Food Act of 1981 417
Agricultural Credit Act of 1987 140
Agricultural Credit Association 133, 140
Agricultural Marketing Act of 1929 15
Agricultural Marketing Act of 1946 317
Agricultural Marketing Agreements Act 316
Agricultural policy 383, 403
Agriculture, U.S.
    accomplishments of 19
    and the GDP 76
    balance sheet of 129
    history of 9
    structure of 396
Agriculture and Consumer Protection Act 417
Air resource 114
Aldrich-Vreeland Act 435
American Farm Bureau Federation 14, 17
Amortized loans
    decreasing-payment 136
    equal-payment 136
Assemblers of farm products 324
Assembly of farm products 302
Average costs
    fixed 194
    total 196
    variable 194
Average physical product 188
Average revenue 201

B

Balanced budget amendment 61
Balance of payments 513
Balance sheet 469
Bank for Cooperatives 132, 142
Banking Act of 1933 440
Banking Act of 1935 440

Banking holidays 54, 440
Banks (commercial) 54, 133, 139
Barter 51
Bilateral trade 512
Biotechnology 378
Board of Governors 440
Bonus Incentive Commodity Export Program 508
Break-even point analysis 489
Budgeting 462
Business decision-making process 466

C

Capital 127
Capital investment analysis
   net present value 149
   payback period 147
   simple rate of return 148
Capitalism 24
Capper-Volstead Act 272
Cardinal utility 242
Cash flow budget 480
Clayton Act 14, 271
Cobweb theorem 251
   convergent 251
   divergent 252
Commodity Credit Corporation 141, 414
Commodity Futures Trading Commission 307
Common Agricultural Policy 512
Common market 511
Communications, market 309
Communism 9, 28

Comparative advantage 173, 177, 495
Conservation compliance provisions 420, 430
Conservation Reserve Program 419, 430
Consolidated Farm Service Agency 135
Consumer
   expenditures for farm products 359
   group action 380
   protection 376
   trends 379, 402
Consumer Price Index 368
Consumption of agricultural products 357, 362
Cooperative 45, 292, 296, 327, 400, 538
Cooperative marketing 327
Corporation 43, 290, 292
   farm business organization 164
   subchapter S 44
Credit 131
Cross elasticity of supply 227
Cross-price elasticity of demand 240
Current asset turnover 476
Current ratio 476
Customs union 511

D

Debt
   current 138
   other 138
   real estate 132
Debt ratio 477
Debt-to-equity ratio 477
Decoupling 427
Deficiency payment 416

Demand 231
    curve 233
    law of 231
Depreciation 73, 472
    declining balance 473
    straight-line 472
    sum-of-the years' digits 474
Devaluation 63
Diminishing Marginal Returns, Law of 184
Diminishing Marginal Utility, Law of 232
Distribution 315
Diversification 173
Domestic Food Assistance Programs 373

E

Easement 123
Economic development 523
    obstacles to 527
Economic federation 511
Economic order quantity 493
Economics 3
    agricultural 5
    consumer 6
    of food buying 370
Economic system 23
Economic union 511
Effective demand 235
Effective interest rates 143
Elasticity
    income 242
    of demand 239
    of supply 225
Embargo 502

Eminent domain 123
Enterprise relationships
    competitive 172
    complementary 172, 463
    independent 172, 463
    joint 463
    supplementary 171, 463
Environmental Quality Incentive Program 430
Equilibrium 248
    price 248
Equimarginal return, principle of 213
Export Enhancement Program 422, 508
Export Payment-in-Kind 508
Export restitution 512
Exports (agricultural)
    historical perspectives 10, 17, 18
    of food and fiber products 336, 515
Export subsidies 506

F

Factor-factor model 449
Factor-product model 179
Factors of production 23
Farm
    economic classes 158
    income 159
    sizes 157
    tenure 161, 163
Farm Credit Act of 1933 16
Farm Credit Act of 1971 132
Farm Credit System 132, 140
Farmer-Owned Reserve 415
Farmers for the Future 430
Farmers Home Administration 135
Farming areas, types of 155

Farm machinery
    distribution of   292
    full-line companies   292
    short-line companies   293
Farm supplies
    distribution of   288
    stores   295
Farm-to-retail price spread   344, 346
Farm value of retail food   344
Fascism   27
Federal Agricultural Improvement and Reform Act of 1986   419, 427, 429, 517
Federal Food Stamp Program   430
Federal Land Bank Associations   133, 140
Federal Land Credit Associations   133
Federal Open Market Committee   440, 443
Federal Reserve Act   56, 438
Federal Reserve System   19, 55, 435, 438
Federal Seed Act   316
Federal Trade Commission Act   14, 272
Feed, distribution of   288
Fertilizer, distribution of   291
Financial performance   475
Financial position   475
Fiscal policy   59
Fixed asset theory   389
Fixed costs   193
Fixed input   180
Food, Drug, and Cosmetics Act   316
Food for Progress   508
Food Security Act of 1985   419, 421
Food service   334
Food spending   365
    distribution of   347

Freedom to Farm   428
Free-trade area   511

G

General Agreements on Tariffs and Trade   401, 509
Grading farm products   302
Great Depression   16, 388, 394, 409, 497
Green payments   429
Grocery stores   330
Gross domestic product   67, 76
Gross national product   67, 80
Gross profit margin ratio   478

H

Hatch Act of 1887   14
Hedging   312
Homestead Act   13, 387

I

Import quota   500
Imports of food and fiber products   336, 519
Income elasticity   242
Income statement   470
Indifference curve analysis   242
Industrial uses of farm products   338
Infant industry theory   11, 504
Inflation   19, 61, 80
    cost-push   62, 81
    demand-pull   63
    rate of   62
Insurance   151

Integration
    horizontal   304
    vertical   304, 399
International trade   406, 495
Interstate Commerce Act   14, 306
Inventory management   492
Inventory turnover ratio   478
Invisible hand   23
Isocost line   455
Isoquant   451

J

Jefferson, Thomas   11, 394
Jeffersonian agrarianism   393

K

Kinked demand curve   275
Knapp, Seaman A.   15

L

Labor in agriculture   164
Laissez-faire   23
Land
    development   125
    ownership   109
    uses   105
    value of farm land   115
Land-Grant College Act   13
Land Ordinance Act of 1785   11
Lease
    cash   162
    share   162
Leverage ratio   477
Life insurance companies   134

Linear programming   462
Liquidity   145, 470, 475
Loan rate   413
Loss-leader pricing   487

M

Macroeconomic equation   76
Macroeconomics   5
Management functions   465
Marginal cost   196
Marginal input cost   196
Marginal physical product   180, 186
Marginal rate of technical substitution   450
Marginal revenue   202
Marginal value product   202
Market
    conduct   276
    performance   276
    structure   263, 276
Market Access Program   430
Marketing   301
    bill   348
    channels   321
    functions   301
    regulation   315
    research   315
Market potential   481
Marx, Karl   28
Mass selling   488
McNary-Haugen Bill   15
Means testing   428
Microeconomics   4
Military demand for farm products   339
Mill, John Stuart   498

Mineral resources   113
Monetary policy   64
Monetary system, U.S.   51, 435
Money   51
Monopolistic competition   267
Monopoly   266, 268
   regulation of   279
Monopsony   268
Most-favored nation   509
Multilateralism   509

N

National Farmers Organization   17
National Farmers Union   14
National Farm Loan Act   15
National Grange   13
National income   70
National Monetary Commission   435, 438
Natural Resources Conservation Service   123, 430
Necessary product   24
Net profit margin ratio   478
Net present value   149
Net worth statement   469
New Deal program   16
Nominal interest rate   143
Nonrecourse loan   141, 413
Nontariff trade barriers   503
Normal flex acres   417
North American Free Trade Agreement   380
Nutrition labeling   380
Nutrition Labeling and Education Act   318

O

Oligopoly   268
Oligopsony   268
Operating budget   479
Opportunity cost   213
Optional flex acres   417
Options (commodity)   313
Ordinal utility   242

P

Packaging   304
Packers and Stockyards Act   316
Partnership   42, 290
    farm business organization   164
    general   42
    limited   42
Payback period   147
Payment-in-kind   405
Penetration pricing   487
Perishable Agricultural Commodities Act   317
Personal selling   489
Pesticides, distribution of   294
Planning, land use   121
P.L. 480 (Food for Peace)   430, 508, 531
Policy   384
Policy-making process   384
Pollution
   nonpoint sources   112
   point sources   112
Price
   controls   261
   cycles   254
   indexes   258

influence on consumer decision-making   366
parity   259
role of   5
trends   254, 287
Pricing   310
differential   297
flexible markup   484
full-cost   484
gross margin   485
profit margin   486
with unknown product costs   484
Processing   304
Processors of farm products   325
Production control   410
Production Credit Associations   140
Production function   179
Product-product model   458
Profit   201
maximum   205, 212
Profitability   477
Profit-and-loss statement   471
Pro forma financial statements   480
Promotion   275, 488
Publicity   488
Public policy   384
Pure competition   264

Q

Quick ratio   476

R

Reclamation Act of 1902   14, 388
Regional trade agreement   511
Regulatory trade restrictions   503

Repayment capacity   146
Repurchase agreements   445
Retailing   315, 330
Return on investment   478
Return on owner equity   478
Returns   146
Revenue assurance   428
Ricardo, David   498
Ridgelines   453
Risk   147, 312
Robinson-Patman Act   271
Rural living   96
Rural people, characteristics of   87

S

Sales forecast   482
Sales policies   483
Sales potential   482
Segment pricing   487
Sherman Antitrust Act   14, 271
Simple rate of return   148
Skim pricing   487
Slide-down pricing   487
Slippage   412
Smith, Adam   23, 497
Smith-Hughes Act of 1916   15
Smith-Lever Act of 1914   15
Smoot-Hawley Tariff Act of 1930   505
Social contract   395
Socialism   32
Sodbuster   420, 430
Soil Bank Act of 1956   17, 419
Soil conservation   123, 419

Sole proprietorship   40, 290
    farm business organization   164
Solvency   476
Specialization   173, 495
    principal of   174
Stages of production   188
Standard Containers Acts   317
Standardization of farm products   302
Storing farm products   303
Supply   219
    curve   219
Supply management   410
Surplus product   24
Swampbuster   420, 430

T

Target price   415
Tariffs   500
Times interest earned ratio   477
Time value of money   149
Tobacco Inspection Act   318
Total cost   193
Total physical product   180
Total product   24
Total revenue   199
Trade creditors   138
Transcontinental Railroad Act   13
Transportation   305
Treadmill Theory   391
Trust   48
Truth in lending   144

U

Unemployment   80
Unilateralism   513
United Nations   529, 537
Uruguay Round of GATT   509
U.S. Cotton Standards Act   318
U.S. Grain Standards Act   318
U.S. Warehouse Act   317
Utility   23, 232
    form   301, 304
    place   301, 305
    time   301, 303, 305

V

Variable costs   191
Variable input   180
Variable levy   503
Voluntary export restraint   503

W

Water resources   109
    quality   111
    rights   112
Webb-Pomerene Act   272
Wetland Reserve Program   419, 430
Wholesaling   315, 328
Wool Standards Act   318
World Trade Organization   510

Z

Zoning   122